Lecture Notes in Computer Science 6658

Commenced Publication in 1973
Founding and Former Series Editors:
Gerhard Goos, Juris Hartmanis, and Jan van Leeuwen

Xiaoyi Jiang   Miquel Ferrer
Andrea Torsello (Eds.)

Graph-Based Representations in Pattern Recognition

8th IAPR-TC-15 International Workshop, GbRPR 2011
Münster, Germany, May 18-20, 2011
Proceedings

 Springer

Volume Editors

Xiaoyi Jiang
University of Münster
Department of Mathematics and Computer Science
Einsteinstraße 62, 48149 Münster, Germany
E-mail: xjiang@uni-muenster.de

Miquel Ferrer
Universitat Politècnica de Catalunya
Institut de Robòtica i Informàtica Industrial
C. Llorens i Artigas 4-6, 08028 Barcelona, Spain
E-mail: mferrer@iri.upc.edu

Andrea Torsello
Università Ca' Foscari di Venezia
Dipartimento di Scienze Ambientali, Informatica, Statistica
Via Torino 155, 30172 Venezia Mestre, Italy
E-mail: torsello@dsi.unive.it

ISSN 0302-9743   e-ISSN 1611-3349
ISBN 978-3-642-20843-0   ISBN 978-3-642-20844-7 (eBook)
DOI 10.1007/978-3-642-20844-7
Springer Heidelberg Dordrecht London New York

Library of Congress Control Number: 2011926691

CR Subject Classification (1998): I.5, I.3, I.4, I.2.10, G.2.2

LNCS Sublibrary: SL 6 – Image Processing, Computer Vision, Pattern Recognition,
and Graphics

Typesetting: Camera-ready by author, data conversion by Scientific Publishing Services, Chennai, India

Printed on acid-free paper

Springer is part of Springer Science+Business Media (www.springer.com)

Preface

It was an honor and a pleasure to organize the 8th IAPR-TC-15 Workshop on Graph-Based Representations in Pattern Recognition (GbR 2011) in Münster, Germany. GbR has been held biennially for 14 years: Lyon, France (1997); Haindorf, Austria (1999); Ischia, Italy (2001); York, UK (2003); Poitiers, France (2005); Alicante, Spain (2007); and Venice, Italy (2009).

The Technical Committee TC15 Graph-Based Representations (`http://www.greyc.ensicaen.fr/iapr-tc15/`) of the International Association for Pattern Recognition (IAPR) was founded in 1996 to encourage research work at the intersection between pattern recognition and graph theory. Since then TC15 has been very active in organizing the biennial GbR workshops, sponsoring related special sessions at conferences, and promoting special issues in journals.

All the papers presented in these proceedings were reviewed by two, and in most cases, three reviewers. They cover a wide range of topics related to graphs, ranging from novel theoretic and algorithmic development to graph-based applications. It is our goal to encourage the use of graph-theoretic concepts and algorithms for solving real problems. The challenges derived from the practice also pose new fundamental graph-theoretic problems to be tackled. It is this interaction of theory and applications which makes GbR a unique forum for presenting and discussing the most recent progress in graph-based representations from a pattern recognition perspective at a high-quality level. Our efforts in reflecting both the theoretic and application aspects in the workshop program were further substantiated by two invited talks: "Partial Difference Equations (PdE) on Graphs for Image and Data Processing" (Olivier Lezoray) and "Graph Algorithmic Techniques for Biomedical Image Segmentation" (Milan Sonka).

The success of GbR 2011 would not have been possible without the support of many institutions and people. First of all, we would like to thank all authors of the submitted papers and the invited speakers for their contributions. All Program Committee members and additional reviewers listed here deserve great thanks for their timely and competent reviews. We are grateful to our sponsors for their support as well. Also, the cooperation with Münster City Marketing was very pleasant and helpful. Many thanks go to the members of the Local Organizing Committee. Finally, we would like to thank Springer for giving us the opportunity to continue publishing the GbR proceedings in the LNCS series.

Founded in 793, Münster belongs to the historical cities of Germany. It is most famous as the site of signing of the Treaty of Westphalia ending the Thirty Years' War in 1648. Today, it is acknowledged as a city of science and learning (and the capital city of bicycles, Germany's Climate Protection Capital, and more).

It was our great pleasure to offer the participants the platform in this multi-faceted city for a lively scientific exchange and many other relaxed hours. Finally, to the readers of this proceedings book: Enjoy it!

May 2011   Xiaoyi Jiang
                                                                    Miquel Ferrer
                                                                  Andrea Torsello

Organization

Co-chairs

Xiaoyi Jiang   University of Münster, Germany
Miquel Ferrer   Universitat Politècnica de Catalunya, Spain
Andrea Torsello   Università Ca' Foscari Venezia, Italy

Program Committee

Luc Brun (France)
Horst Bunke (Switzerland)
Roberto M. Cesar (Brazil)
Diane Cook (USA)
Robert P.W. Duin (The Netherlands)
Francisco Escolano (Spain)
Edwin R. Hancock (UK)
Jean-Michel Jolion (France)
Walter G. Kropatsch (Austria)
Tetsuji Kuboyama (Japan)
Josep Lladós (Spain)

Marcello Pelillo (Italy)
Antonio Robles-Kelly (Australia)
Alberto Sanfeliu (Spain)
Christian Schellewald (Norway)
Francesc Serratosa (Spain)
Ali Shokoufandeh (USA)
Salvatore Tabbone (France)
Koji Tsuda (Japan)
Ernest Valveny (Spain)
Mario Vento (Italy)
Changshui Zhang (China)

Additional Reviewers

Donatello Conte
Rogério Feris
Ana B.V. Graciano
Yll Haxhimusa
Nina Hirata

Adrian Ion
Takashi Kaburagi
David Martins
Tetsuhiro Miyahara
Alexandre Noma

Gennaro Percannella
Nicola Rebagliati
Kaspar Riesen
Samuel Rota Bulò
Hiroto Saigo

Local Organizing Committee

Hildegard Brunstering
Klaus Broelemann
Michael Fieseler

Lucas Franek
Fabian Gigengack

Michael Schmeing
Daniel Tenbrinck

Sponsoring Institutions

University of Münster, Germany
International Association for Pattern Recognition

Table of Contents

Graph Representation and Characterization

Graph Matching, Classification, and Querying

Graph-Based Learning

Graph-Based Segmentation

Applications

A Global Method for Reducing Multidimensional Size Graphs[*]

Andrea Cerri[1], Patrizio Frosini[2], Walter G. Kropatsch[1], and Claudia Landi[3]

[1] Vienna University of Technology, Faculty of Informatics,
Pattern Recognition and Image Processing Group, Austria
{acerri,krw}@prip.tuwien.ac.at
[2] Università di Bologna,
ARCES and Dipartimento di Matematica, Italy
frosini@dm.unibo.it
[3] Università di Modena e Reggio Emilia,
Dipartimento di Scienze e Metodi dell'Ingegneria, Italy
clandi@unimore.it

Abstract. This paper introduces the concept of discrete multidimensional size function, a mathematical tool studying the so-called size graphs. These graphs constitutes an ingredient of Size Theory, a geometrical/topological approach to shape analysis and comparison. A global method for reducing size graphs is presented, together with a theorem stating that size graphs reduced in such a way preserve all the information in terms of multidimensional size functions. This approach can lead to simplify the effective computation of discrete multidimensional size functions, as shown by examples.

1   Introduction

In the last twenty years, Size Theory has revealed to be a suitable geometrical/topological approach to shape analysis and comparison, which are probably two of the main issues in the fields of Computer Vision, Computer Graphics and Pattern Recognition. In this context, the main tool proposed by Size Theory is the concept of *size function*, a shape descriptor able to capture the qualitative aspects of a shape, and describing them quantitatively. More precisely, the basic idea is to model a shape by means of a topological space $\mathcal{M}$ and a continuous function $\varphi : \mathcal{M} \to \mathbb{R}$, called *measuring function*. The role of the measuring functions is to describe those properties that are considered relevant for the shape comparison or the shape analysis problem at hand. In this setting, the size function $\ell_{(\mathcal{M},\varphi)}$ encodes part of the topological changes occurring in the lower level sets induced on $\mathcal{M}$ by φ. In this way, the starting problem of comparing shapes modeled by pairs $(topological\ space, measuring\ function)$ can then be recast into the one of comparing the associated size functions. For details and more references about Size Theory the reader is referred to [3].

---

[*] Partially supported by the Austrian Science Fund (FWF) grant no. P20134-N13.

X. Jiang, M. Ferrer, and A. Torsello (Eds.): GbRPR 2011, LNCS 6658, pp. 1–11, 2011.

More recently, similar ideas have been re-proposed by Persistent Homology from the homological point of view [13]. More precisely, the notion of size function coincide with the one of 0th persistent homology group.

Since their introduction, size functions have been extensively used especially in the fields of Computer Vision [8,12,23], where the objects under study are images, and Computer Graphics, comparing, e.g, 3D-models [4]. The success of size functions in such applicative contexts is due to the fact that they admit a very simple and compact representation [15], and they are stable with respect to a suitable distance [10]. Moreover, size functions show resistance to noise and modularity [14]: In particular, they inherit their invariance properties directly from the considered measuring functions. For example, in [8] an effective system for content-based retrieval of figurative images, based on size functions, is presented. Three different classes of image descriptors are introduced and integrated, for a total amount of 25 measuring functions. The evaluation of this fully automatic retrieval system has been performed on a benchmark database of more than 10,000 real trademark images, supplied by the United Kingdom Patent Office. Comparative results have been performed, showing that the proposed method actually outperforms other existing whole-image matching techniques.

As the previous considerations suggest, a common scenario in applications is when two or more properties characterize the objects under study. Moreover, sometimes it could be desirable to consider properties of shapes that are intrinsically multidimensional, such as the coordinates of a point into the 3-dimensional space or the representation of color in the RGB model. These motivations recently drove the attention to extending Size Theory to a multidimensional context [2] (see [6,7] for multidimensional generalizations of Persistent Homology). Here the term multidimensional (or, equivalently, k-dimensional) refers to the fact that the measuring functions take values in $\mathbb{R}^k$ and has no reference with the dimensionality of the objects under study. Therefore, such an enlarged setting leads to model a shape as a pair $(\mathcal{M}, \varphi)$, with $\varphi : \mathcal{M} \to \mathbb{R}^k$, and consequently to consider the so called *multidimensional size functions*.

Even in this multidimensional setting Size Theory gave encouraging results when applied to shape analysis and comparison problems, see, e.g., [2,5]. In those papers it has been shown that, besides enabling the study of multidimensional properties of the objects under study, the advantage of working with multidimensional measuring functions is that shapes can be simultaneously investigated by k different 1-dimensional measuring functions. In other words, k different functions cooperate to produce a single shape descriptor. The higher discriminatory power of multidimensional size functions in comparison to 1-dimensional ones has been formally proved in [2].

Motivations and contributions of the paper. Obviously, dealing with applications involves the development of a discrete counterpart of the theory. In this perspective, a shape can be discretized by a graph $G = (V(G), E(G))$ endowed with a function $\varphi : V(G) \to \mathbb{R}^k$, being $V(G)$ the set of vertices of G. This leads to consider pairs of the type (G, φ), called *size graphs*. In this mathematical setting, *discrete k-dimensional size functions* count the number of connected components in $G\langle \varphi \preceq \boldsymbol{y} \rangle$ containing at least one vertex of $G\langle \varphi \preceq \boldsymbol{x} \rangle$

where, for $t \in \mathbb{R}^k$, $G\langle \varphi \preceq t \rangle$ is defined as the subgraph of G obtained by erasing all vertices of G at which φ_i takes a value strictly greater than t_i, for at least one index $i \in \{1, \ldots, k\}$, and all the edges connecting those vertices to others.

Therefore, in computing discrete k-dimensional size functions, we have to count the connected components of particular subgraphs of a size graph. It is reasonable to argue that, the greater the dimension k, the higher the discriminatory power of k-dimensional size functions. On the other hand, the smaller the graph, the faster the computation. Moreover, in applications we often have to deal with big graphs, implying high computational costs. According to these considerations, it follows that the problem of reducing a size graph without changing the associated discrete k-dimensional size function is a desirable target.

In previous works ([11,16]), it has been proved that, in the case $k = 1$, a size graph can be reduced by means of a global method (its application requires the knowledge of all the size graph) and a local method (it requires only a local knowledge of a size graph), obtaining a very simple structure.

In this paper, we present a first attempt for a reduction procedure for size graphs in the case $k > 1$. More precisely, we provide a global reduction method for size graphs, together with a theorem stating that reduced size graphs preserve all the information in terms of k-dimensional size functions. Based on an idea presented in [9], the present contribution differentiates from that one in two main respects: (i) It focuses on the formal proof of Theorem 1, that has never been published or presented before; (ii) Experiments have been performed to support the theoretical results.

Related works. The ideas underlying the concept of size functions are partly shared by the so-called *maximally stable extremal regions (MSERs)* [19]. MSERs are image elements useful in wide-baseline matching. Given a gray-level image I, the basic intuition is to study the evolution of the thresholded image I_t, varying the parameter t. The formal definition of MSERs is then derived by considering the set of all connected components of all thresholded images I_t. These image elements are characterized by a number of nice properties, such as the invariance to affine transformation of image intensities and stability.

Besides being related to [11] and [16], the present work fits in the current research and interest in strategies for reducing data structures preserving some topological/homological information, motivated by Pattern Recognition and data analysis problems. For example, in [22] the authors propose a method for computing homology groups and their generators of a 2D pixel image, by using a hierarchical structure called irregular graph pyramid. Their method is based on two operations, preserving the homology information contained in each region of an image, but progressively simplifying the starting graph representing the image, and constituting the base level of the pyramid. The desired homological information is then computed at the top of the pyramid. This approach finds its roots in a more general framework first introduced in [18]. Motivated by problems coming from discrete dynamics, in [17] the authors propose an algorithm for computing homology of a finitely generated chain complex. Such an algorithm is based on reducing the size of the complex preserving homology information in each step of the reduction. Computing the homology of the chain complex is then postponed until the complex is acceptably small. The same philosophy leads the authors of [21] to provide a reduction algorithm for simplifying the computation of homology information for cubical sets and polytopes.

The remainder of the paper is organized as follows. In Section 2 we introduce the standard facts and some basic definitions about discrete multidimensional size functions. Section 3 is devoted to present our main result, together with some experiments. Some discussions in Section 4 conclude the paper.

2   Basic Definitions

The following relations are introduced in $\mathbb{R}^k$: for every $\boldsymbol{x} = (x_1, \ldots, x_k)$ and $\boldsymbol{y} = (y_1, \ldots, y_k)$, we shall say $\boldsymbol{x} \preceq \boldsymbol{y}$ (resp. $\boldsymbol{x} \prec \boldsymbol{y}$, $\boldsymbol{x} \succeq \boldsymbol{y}$) if and only if $x_i \leq y_i$ (resp. $x_i < y_i$, $x_i \geq y_i$) for every index $i = 1, \ldots, k$. Moreover, we shall write $\boldsymbol{x} \not\preceq \boldsymbol{y}$ (resp. $\boldsymbol{x} \not\succeq \boldsymbol{y}$) when the relation between $\boldsymbol{x}$ and $\boldsymbol{y}$ expressed by the operator $\preceq$ (resp. $\succeq$) is not verified. Finally, we recall that Δ^+ is defined as the open set $\{(\boldsymbol{x}, \boldsymbol{y}) \in \mathbb{R}^k \times \mathbb{R}^k : \boldsymbol{x} \prec \boldsymbol{y}\}$.

Definition 1 (Size Graph). *Let* $G = (V(G), E(G))$ *be a finite, ordered simple graph with* $V(G)$ *set of vertices and* $E(G)$ *set of edges. Assume that a function* $\boldsymbol{\varphi} = (\varphi_1, \ldots, \varphi_k) : V(G) \to \mathbb{R}^k$ *is given. The pair* $(G, \boldsymbol{\varphi})$ *is called a* size graph.

Definition 2. *For every* $\boldsymbol{y} = (y_1, \ldots, y_k) \in \mathbb{R}^k$, *we denote by* $G\langle \boldsymbol{\varphi} \preceq \boldsymbol{y} \rangle$ *the subgraph of* G *obtained by erasing all vertices* $v \in V(G)$ *such that* $\boldsymbol{\varphi}(v) \not\preceq \boldsymbol{y}$, *and all the edges connecting those vertices to others. If* $v_a, v_b \in V(G)$ *belong to the same connected component of* $G\langle \boldsymbol{\varphi} \preceq \boldsymbol{y} \rangle$, *we shall write* $v_a \cong_{G\langle \boldsymbol{\varphi} \preceq \boldsymbol{y} \rangle} v_b$.

We are now ready to introduce discrete k-dimensional size functions.

Definition 3 (discrete k-dimensional size function). *We shall call* discrete k-dimensional size function *of the size graph* $(G, \boldsymbol{\varphi})$ *the function* $\ell_{(G, \boldsymbol{\varphi})} : \Delta^+ \to \mathbb{N}$ *defined by setting* $\ell_{(G, \boldsymbol{\varphi})}(\boldsymbol{x}, \boldsymbol{y})$ *equal to the number of connected components in* $G\langle \boldsymbol{\varphi} \preceq \boldsymbol{y} \rangle$ *containing at least one vertex of* $G\langle \boldsymbol{\varphi} \preceq \boldsymbol{x} \rangle$.

Example 1. Figure 1 shows an example of size graph, together with the related discrete 1-dimensional size function. We remark that in the case $k = 1$ the symbols $\boldsymbol{\varphi}, \boldsymbol{x}, \boldsymbol{y}$ are replaced by φ, x, y respectively. As can be seen, in the 1-dimensional case the domain Δ^+ of $\ell_{(G, \varphi)}$ is a subset of the real plane. In each region of Δ^+, the value of $\ell_{(G, \varphi)}$ in that region is displayed.

For example, to compute the value of $\ell_{(G, \varphi)}$ at the point (a, b), it is sufficient to count how many of the three connected components in the subgraph $G\langle \varphi \leq b \rangle$ contain at least one vertex of $G\langle \varphi \leq a \rangle$: It can be easily checked that $\ell_{(G, \varphi)}(a, b) = 2$.

In what follows, we will assume that the set of vertices $V(G)$ of the graph G is a subset of a Euclidean space.

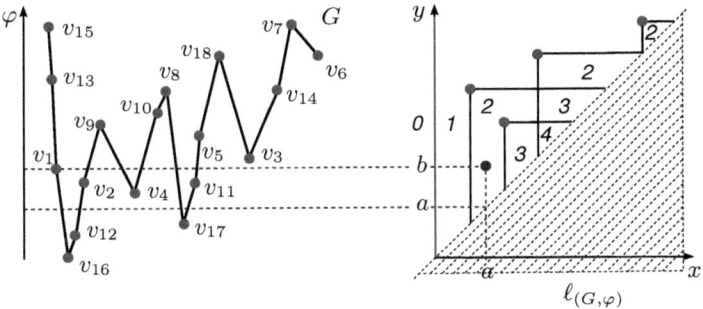

Fig. 1. A size graph and the associated discrete size function

3   A Global Method for Reducing (G, φ): the $\mathcal{L}$-Reduction

As stressed before, our goal is to reduce a size graph (G, φ) without changing the related discrete k-dimensional size function: This can be done by erasing all those vertices of G that do not contain, in terms of discrete k-dimensional size functions, "meaningful information". Indeed, in order to compute the discrete k-dimensional size function of (G, φ), we are only interested in capturing the "birth" of new connected components and the "death", i.e. the merging, of the existing ones: As will be shown, these events are strongly related to particular vertices of G, that can be seen, in some sense, as "critical points" of the function φ with respect to the relation $\preceq$. The proposed reduction method allows us to detect these particular vertices and to introduce the concept of $\mathcal{L}$-*reduction of* (G, φ), a new size graph $(G_{\mathcal{L}}, \varphi_{\mathcal{L}})$ that is obtained by considering *only* such vertices instead of the entire set $V(G)$. The importance of the $\mathcal{L}$-reduction is shown by our main result, stated in Theorem 1, which will be formally proved at the end of this section and can be rephrased as follows:

Theorem 1 (rephrased). *The k-dimensional size functions of $(G_{\mathcal{L}}, \varphi_{\mathcal{L}})$ and (G, φ) coincide.*

From now on, we assume that a size graph (G, φ) is given. Moreover, for every $v_i \in V(G)$ we define A_i as the set of the "lower adjacent vertices" for v_i, i.e.
$A_i = \{v_j : (v_i, v_j) \in E(G), \varphi(v_j) \preceq \varphi(v_i)\} \cup \{v_i\}$.

Definition 4 (Single step descent flow operator). *Let $L : V(G) \to V(G)$ be the function defined as follows: For every $v_i \in V(G)$ let $B_i \subseteq A_i$ be the set whose elements are the vertices $w \in A_i$ for which the Euclidean norm $\|\varphi(w) - \varphi(v_i)\|$ takes the largest value. Finally, we choose the vertex $v_h \in B_i$ for which the index h is minimum. Then, we set $L(v_i) = v_h$. We shall call L the* single step descent flow operator.

From the definition of L and the finiteness of $V(G)$, if follows that for every $v \in V(G)$ there must exist a minimum index $m(v) \leq |V(G)|$ such that $L^{m(v)}(v) = L^{m(v)+1}(v)$ (if $L(v) = v$ we will set $m(v) = 0$).

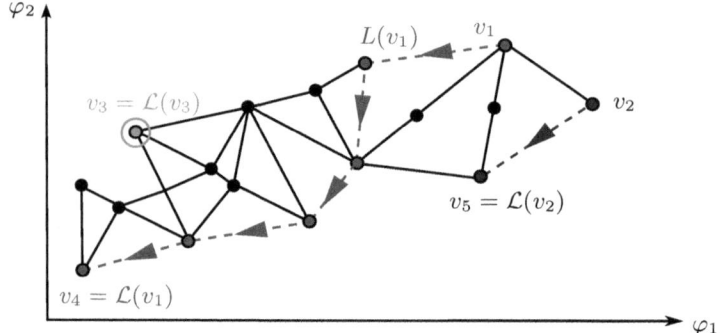

Fig. 2. The operators L and $\mathcal{L}$ in action: Some examples when $\varphi = (\varphi_1, \varphi_2)$

Definition 5 (Descent flow operator). *For every $v \in V(G)$ we set $\mathcal{L}(v) = L^{m(v)}(v)$. We shall call the function $\mathcal{L} : V(G) \rightarrow V(G)$ the descent flow operator.*

In other words, the descent flow operator takes each vertex $v_i \in V(G)$ to a sort of "local minimum" $v_j = \mathcal{L}(v_i)$ of the function φ, with respect to the relation $\preceq$. This implies that, starting from v_j we are not able to reach a vertex w adjacent to it with $\varphi(w) \preceq \varphi(v_j)$, strictly decreasing the value of at least one component of φ. During the descent, indexes are used to univocally decide the path in case the set B_i contains more than one vertex.

Example 2. Figure 2 shows some possible cases arising from the action of the operators L and $\mathcal{L}$ when $\varphi = (\varphi_1, \varphi_2)$. As can be seen, the vertex v_1 is taken by the operator L to $v_4 = L^5(v_1)$. Since it is not possible to reach another vertex from v_4 decreasing the values of both φ_1 and φ_2, we shall set $v_4 = \mathcal{L}(v_1)$. Analogously, we have $v_5 = \mathcal{L}(v_2)$. The last considered case is represented by the vertex v_3: it can be seen as a fixed point with respect to the operator L, i.e. it holds that $L(v_3) = v_3$, so we shall set $\mathcal{L}(v_3) = v_3$.

Definition 6 (Minimum vertex). *Each vertex v for which $\mathcal{L}(v) = v$ is called a minimum vertex of (G, φ). Call M the set of minimum vertices of (G, φ).*

We point out that M is the set of all those vertices representing the "birth" of new connected components in (G, φ): By increasing the values of $\varphi_1, \ldots, \varphi_k$, such an event occurs only when the values labeling a minimum vertex are reached.

The following two definitions characterize the "death-points" of existing connected components of (G, φ).

Definition 7 (Ridge pair). *Let $v_{j_1}, v_{j_2} \in V(G)$ be two distinct minimum vertices of (G, φ). Suppose v_{i_1}, v_{i_2} are two adjacent vertices of G, such that $\{\mathcal{L}(v_{i_1}), \mathcal{L}(v_{i_2})\} = \{v_{j_1}, v_{j_2}\}$; we shall call $\{v_{i_1}, v_{i_2}\}$ a ridge pair adjacent to the minimum vertices v_{j_1} and v_{j_2}.*

Definition 8 (Main saddle). *Let $v_{j_1}, v_{j_2} \in V(G)$ be two distinct minimum vertices of (G, φ). Let also $\{v_{i_1}, v_{i_2}\}$ be a ridge pair adjacent to the minimum vertices v_{j_1}, v_{j_2} such that the following statements hold:*

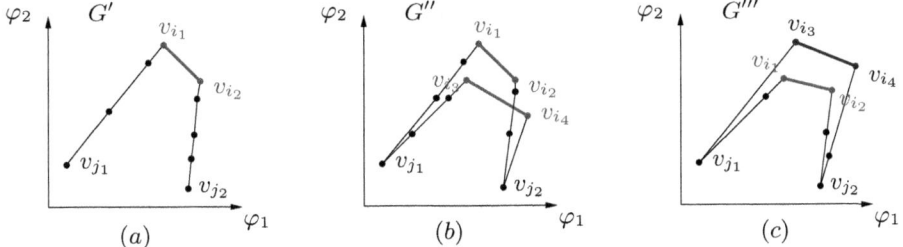

Fig. 3. Some examples of ridge pairs and main saddles

1. *there does not exist another ridge pair $\{v_{i_3}, v_{i_4}\}$ adjacent to the minimum vertices v_{j_1}, v_{j_2} with*

$$(1) \begin{cases} \max\{\varphi_h(v_{i_3}), \varphi_h(v_{i_4})\} \leq \max\{\varphi_h(v_{i_1}), \varphi_h(v_{i_2})\}, \ h = 1, \dots, k; \\ \exists \bar{h} : \max\{\varphi_{\bar{h}}(v_{i_3}), \varphi_{\bar{h}}(v_{i_4})\} < \max\{\varphi_{\bar{h}}(v_{i_1}), \varphi_{\bar{h}}(v_{i_2})\}, \ \bar{h} \in \{1, \dots, k\}; \end{cases}$$

2. *if $\{v_{i_3}, v_{i_4}\}$ is another ridge pair adjacent to the minimum vertices v_{j_1}, v_{j_2} with*

$$(2) \quad \max\{\varphi_h(v_{i_3}), \varphi_h(v_{i_4})\} = \max\{\varphi_h(v_{i_1}), \varphi_h(v_{i_2})\}, \ h = 1, \dots, k,$$

then (i_1, i_2) precedes (i_3, i_4) in the lexicographic order. We shall call the set $\{v_{i_1}, v_{i_2}\}$ the main saddle adjacent to the minimum vertices v_{j_1}, v_{j_2} and S the set of main saddles of (G, φ).

Roughly, the set of ridge pairs of (G, φ) can be partially ordered by means of the relation $\preceq$. In this sense, the main saddles will be the lowest ridge pairs.

Example 3. Figure 3(a), 3(b), 3(c) shows some examples of ridge pairs and main saddles, when function φ takes values in $\mathbb{R}^2$.

To clarify the role of main saddles, let us study the changing in the number of connected components of the subgraphs $G'\langle\varphi \preceq y\rangle$, $G''\langle\varphi \preceq y\rangle$ and $G'''\langle\varphi \preceq y\rangle$, with $y \in \mathbb{R}^2$, just for $y \succeq (\max\{\varphi_1(v_{j_1}), \varphi_1(v_{j_2})\}, \max\{\varphi_2(v_{j_1}), \varphi_2(v_{j_2})\})$: Indeed, we want to capture the merging of the connected components arising from the minimum vertices v_{j_1} and v_{j_2} in the three instances. According to this consideration, by means of the chosen assumption on y we ensure that both v_{j_1} and v_{j_2} belong to the subgraphs $G'\langle\varphi \preceq y\rangle$, $G''\langle\varphi \preceq y\rangle$ and $G'''\langle\varphi \preceq y\rangle$.

In Figure 3(a) a main saddle adjacent to the minimum vertices v_{j_1} and v_{j_2} is displayed. In this setting, by varying the values taken by y under the assumption $y \succeq (\max\{\varphi_1(v_{j_1}), \varphi_1(v_{j_2})\}, \max\{\varphi_2(v_{j_1}), \varphi_2(v_{j_2})\})$, it holds that for $y \not\succeq (\max\{\varphi_1(v_{i_1}), \varphi_1(v_{i_2})\}, \max\{\varphi_2(v_{i_1}), \varphi_2(v_{i_2})\})$ the subgraph $G'\langle\varphi \preceq y\rangle$ consists of the two connected components arising from v_{j_1} and v_{j_2}, reducing to a unique one when $y \succeq (\max\{\varphi_1(v_{i_1}), \varphi_1(v_{i_2})\}, \max\{\varphi_2(v_{i_1}), \varphi_2(v_{i_2})\})$.

Figure 3(b) shows an example of two ridge pairs that can be considered "uncomparable", due to the fact $\max\{\varphi_1(v_{i_1}), \varphi_1(v_{i_2})\} < \max\{\varphi_1(v_{i_3}), \varphi_1(v_{i_4})\}$, while $\max\{\varphi_2(v_{i_1}), \varphi_2(v_{i_2})\} > \max\{\varphi_2(v_{i_3}), \varphi_2(v_{i_4})\}$. Thus, both $\{v_{i_1}, v_{i_2}\}$ and $\{v_{i_3}, v_{i_4}\}$ will be main saddles. In this case, when y varies under the assumption $y \succeq (\max\{\varphi_1(v_{j_1}), \varphi_1(v_{j_2})\}, \max\{\varphi_2(v_{j_1}), \varphi_2(v_{j_2})\})$, the number of the connected components in the subgraph $G''\langle\varphi \preceq y\rangle$ decreases (from 2 to 1) when the relation $\boldsymbol{y} \succeq (\max\{\varphi_1(v_{i_1}), \varphi_1(v_{i_2})\}, \max\{\varphi_2(v_{i_1}), \varphi_2(v_{i_2})\})$ (or, alternatively, the relation $\boldsymbol{y} \succeq (\max\{\varphi_1(v_{i_3}), \varphi_1(v_{i_4})\}, \max\{\varphi_2(v_{i_3}), \varphi_2(v_{i_4})\})$) becomes true.

Finally, Figure 3(c) shows two comparable ridge pairs, hence the "lower" one, that is $\{v_{i_1}, v_{i_2}\}$, will be a main saddle, while the other will be not. Consider $G'''\langle \boldsymbol{\varphi} \preceq \boldsymbol{y} \rangle$, assuming that $\boldsymbol{y}$ varies according to the restriction $\boldsymbol{y} \succeq (\max\{\varphi_1(v_{j_1}), \varphi_1(v_{j_2})\}, \max\{\varphi_2(v_{j_1}), \varphi_2(v_{j_2})\})$: It consists of two connected components arising from v_{j_1} and v_{j_2}, merging into a unique one as soon as the relation $\boldsymbol{y} \succeq (\max\{\varphi_1(v_{i_1}), \varphi_1(v_{i_2})\}, \max\{\varphi_2(v_{i_1}), \varphi_2(v_{i_2})\})$ becomes true.

As Example 3 suggests, S is the set of all those couples of vertices representing the "death", i.e. the merging, of existing connected components in the given size graph $(G, \boldsymbol{\varphi})$.

We are now ready to introduce the concept of $\mathcal{L}$-reduced size graph:

Definition 9 ($\mathcal{L}$-reduced size graph). *Let $G_{\mathcal{L}} = (V(G_{\mathcal{L}}), E(G_{\mathcal{L}}))$ be the graph with $V(G_{\mathcal{L}}) = M \cup S$ and $E(G_{\mathcal{L}})$ defined as follows: $(u, v) \in E(G_{\mathcal{L}})$ (and hence u and v are adjacent) if and only if either u or v is a minimum vertex and the other is a main saddle adjacent to it (in the sense of Definition 8). Let also $\boldsymbol{\varphi}_{\mathcal{L}} : V(G_{\mathcal{L}}) \to \mathbb{R}^k$ be a function defined in this way: $\boldsymbol{\varphi}_{\mathcal{L}}(v) = \boldsymbol{\varphi}(v)$ if $v \in M$ and $\boldsymbol{\varphi}_{\mathcal{L}}(u) = (\max\{\varphi_1(v_{i_1}), \varphi_1(v_{i_2})\}, \ldots, \max\{\varphi_k(v_{i_1}), \varphi_k(v_{i_2})\})$ if $u = \{v_{i_1}, v_{i_2}\} \in S$. The size graph $(G_{\mathcal{L}}, \boldsymbol{\varphi}_{\mathcal{L}})$ is called the $\mathcal{L}$-reduction of $(G, \boldsymbol{\varphi})$.*

Remark 1. We stress that each main saddle $\{v, w\}$ of a size graph $(G, \boldsymbol{\varphi})$ will be represented, in the $\mathcal{L}$-reduced size graph, by a *unique* vertex labeled by the k-tuple $(\max\{\varphi_1(v), \varphi_1(w)\}, \ldots, \max\{\varphi_k(v), \varphi_k(w)\})$.

Remark 2. The global reduction method we have just defined is strongly related to the concept of Pareto-Optimality, a well-known topic in Economy, especially in the field of Multi-Objective Optimization. For a detailed treatment about Pareto-Optimality, the reader is referred to [20].

The importance of the $\mathcal{L}$-reduction is shown by our main result, stating that discrete k-dimensional size functions are invariant with respect to this global reduction method.

Theorem 1. *For every $(\boldsymbol{x}, \boldsymbol{y}) \in \Delta^+$, it holds that $\ell_{(G,\boldsymbol{\varphi})}(\boldsymbol{x}, \boldsymbol{y}) = \ell_{(G_{\mathcal{L}},\boldsymbol{\varphi}_{\mathcal{L}})}(\boldsymbol{x}, \boldsymbol{y})$.*

In order to prove Theorem 1, we need the following lemma.

Lemma 1. *Let v_1, v_2 be two minimum vertices of $(G, \boldsymbol{\varphi})$. Then, for every $\boldsymbol{y} \in \mathbb{R}^k$, it holds that $v_1 \cong_{G\langle\boldsymbol{\varphi}\preceq\boldsymbol{y}\rangle} v_2$ if and only if $v_1 \cong_{G_{\mathcal{L}}\langle\boldsymbol{\varphi}_{\mathcal{L}}\preceq\boldsymbol{y}\rangle} v_2$.*

Proof. Suppose that $v_1 \cong_{G\langle\boldsymbol{\varphi}\preceq\boldsymbol{y}\rangle} v_2$. Then, by definition there exists a sequence $(v_1 = v_{j_1}, v_{j_2}, \ldots, v_{j_{m-1}}, v_{j_m} = v_2)$ such that $(v_{j_n}, v_{j_{n+1}}) \in E(G)$ for every $n = 1, \ldots, m-1$, and $v_{j_n} \in G\langle\boldsymbol{\varphi} \preceq \boldsymbol{y}\rangle$ for every $n = 1, \ldots, m$. Consider the sequence $(\mathcal{L}(v_1) = v_1, \mathcal{L}(v_{j_2}), \ldots, \mathcal{L}(v_{j_{m-1}}), \mathcal{L}(v_2) = v_2)$ of minimum vertices. Substituting each subsequence of equal consecutive vertices by a unique vertex representing such a subsequence, we obtain a new sequence $(v_1 = w_1, w_2, \ldots, w_{s-1}, w_s = v_2)$ (in other words, the sequence $(u_1, u_1, \ldots, u_1, u_2, u_2 \ldots, u_2, \ldots, u_n, u_n, \ldots, u_n)$ is substituted with $(u_1, u_2, \ldots, u_n)$). It is easy to prove that, for every index $j < s$, there exists at least one main saddle σ_j adjacent to w_j and w_{j+1}, such that $\sigma_j \in G\langle \varphi \preceq y \rangle$. Then, consider the sequence $(w_1, \sigma_1, v_2, \sigma_2, \ldots, w_{s-1}, \sigma_{s-1}, w_s)$: such a sequence proves that $v_1 \cong_{G_{\mathcal{L}}\langle \varphi_{\mathcal{L}} \preceq y \rangle} v_2$.

Conversely, suppose that $v_1 \cong_{G_{\mathcal{L}}\langle \varphi_{\mathcal{L}} \preceq y \rangle} v_2$. By definition there exists a sequence $(v_1 = w_1, \sigma_1, w_2, \sigma_2, \ldots, w_{s-1}, \sigma_{s-1}, w_s = v_2)$ of vertices of $G_{\mathcal{L}}\langle \varphi_{\mathcal{L}} \preceq y \rangle$ such that every vertex w_j is a minimum vertex and every σ_j is a main saddle adjacent to w_j and w_{j+1}. Therefore, we can modify such a sequence in order to obtain the following one: for every index $j < s$, between w_j and $\sigma_j = \{v_{i_j}, v_{n_j}\}$ insert the sequence $(L^{m(v_{i_j})-1}(v_{i_j}), L^{m(v_{i_j})-2}(v_{i_j}), \ldots, L^2(v_{i_j}), L(v_{i_j}))$, while between σ_j e w_{j+1} insert the sequence $(L(v_{n_j}), L^2(v_{n_j}), \ldots, L^{m(v_{n_j})-2}(v_{n_j}), L^{m(v_{n_j})-1}(v_{n_j}))$ (we are assuming $w_j = \mathcal{L}(v_{i_j})$ and $w_{j+1} = \mathcal{L}(v_{n_j})$). Finally, by substituting the vertices v_{i_j} e v_{n_j} (taken in this order) for every main saddle σ_j, we obtain a new sequence proving that $v_1 \cong_{G\langle \varphi \preceq y \rangle} v_2$.

Now we are ready to prove Theorem 1.

Proof. Let $(x, y) \in \Delta^+$. We have to prove that there exists a bijection $F : G\langle \varphi \preceq x \rangle / \cong_{G\langle \varphi \preceq y \rangle} \to G_{\mathcal{L}}\langle \varphi_{\mathcal{L}} \preceq x \rangle / \cong_{G_{\mathcal{L}}\langle \varphi_{\mathcal{L}} \preceq y \rangle}$. For every equivalence class $C \in G\langle \varphi \preceq x \rangle / \cong_{G\langle \varphi \preceq y \rangle}$ we choose a minimum vertex $v_C \in C$. Obviously, v_C is also a vertex of $G_{\mathcal{L}}\langle \varphi_{\mathcal{L}} \preceq x \rangle$. Therefore in $G_{\mathcal{L}}\langle \varphi_{\mathcal{L}} \preceq x \rangle / \cong_{G_{\mathcal{L}}\langle \varphi_{\mathcal{L}} \preceq y \rangle}$ there exists an equivalence class D containing v_C. We shall set $F(C) = D$. From Lemma 1 it follows that F is injective. The surjectivity of F is trivial, since each equivalence class in $G_{\mathcal{L}}\langle \varphi_{\mathcal{L}} \preceq x \rangle / \cong_{G_{\mathcal{L}}\langle \varphi_{\mathcal{L}} \preceq y \rangle}$ contains at least one minimum vertex of $G\langle \varphi \preceq x \rangle$.

Remark 3. The $\mathcal{L}$-reduction of a size graph (G, φ) is not unique: Changing the ordering of the set $V(G)$ can produce different, non-isomorphic $\mathcal{L}$-reduced size graphs. On the other hand, Theorem 1 shows that we will always obtain $\mathcal{L}$-reductions of (G, φ) endowed with the same discrete k-dimensional size function.

Therefore, Theorem 1 allows us to evaluate the discrete k-dimensional size function of a size graph (G, φ) directly dealing with one of its $\mathcal{L}$-reductions.

3.1 Experimental Results

Table 1 shows how our global reduction method can facilitate the computation of $\ell_{(G,\varphi)}$, simplifying the structure of (G, φ) but preserving the same information in terms of discrete k-dimensional size functions. We considered four graphs obtained from as many triangle meshes (available at [1]) by taking the 0-dimensional simplexes as vertices and the 1-dimensional simplexes as edges.

For each graph, we considered the 2-dimensional measuring function $\varphi = (\varphi_1, \varphi_2)$ taking each vertex v of coordinates (x, y, z) to the pair $\varphi(v) = (|x|, |y|)$.

Table 1, from row 1 to 4, shows respectively the number of vertices $|V(G)|$ and edges $|E(G)|$ for each considered size graph (G, φ) and for the associated $\mathcal{L}$-reduction $(G_{\mathcal{L}}, \varphi_{\mathcal{L}})$ (i.e. $|V(G_{\mathcal{L}})|$ and $|E(G_{\mathcal{L}})|$). In particular, if M and S are respectively the set of minimum vertices and of main saddles for a size graph (G, φ), it easily follows from Definition 9 that $|V(G_{\mathcal{L}})| = |M \cup S|$ and $|E(G_{\mathcal{L}})| = 2|S|$. For each considered (G, φ), the last row of Table 1 shows respectively the

Table 1. Some experimental results

	tie	space_shuttle	x_wing	space_station
$|V(G)|$	2014	2376	3099	5749
$|E(G)|$	5944	6330	9190	15949
$|V(G_\mathcal{L})|$	588	262	571	1935
$|E(G_\mathcal{L})|$	826	328	838	2778
$V\% - E\%$	29.2% - 13.9%	11% - 5.2%	18.4% - 9.2%	33.66% - 17.42%

ratios $V\% = |V(G_\mathcal{L})|/|V(G)|$ and $E\% = |E(G_\mathcal{L})|/|E(G)|$, expressing them in percentage points. In other words, the lower those ratios, the higher the reduction rate. As can be seen, these experiments gave encouraging results, enabling the reduction of a size graph up to the 11% of the starting number of vertices and the 5.2% of the starting number of edges (space_shuttle case).

To conclude, we report the most salient part of the algorithm we implemented to obtain our experimental results, i.e. the computation of the descent flow operator introduced in Definition 5. All the rest can be easily derived from the theoretical setting discussed in the previous sections, combined with what follows. The symbol $SSDFO(v)$ denotes the single step descent flow operator computed at a vertex v.

Algorithm 1. Computation of the descent flow operator $\mathcal{L}(v)$ for $v \in V(G)$

> $L(v) \leftarrow SSDFO(v)$
> **while** $L(v) \neq v$ **do**
> $\quad v \leftarrow L(v)$;
> $\quad L(v) \leftarrow SSDFO(v)$;
> **end while**
> $\mathcal{L}(v) \leftarrow L(v)$

4   Conclusion

In this paper we presented a global method for reducing size graphs together with a theorem, stating that discrete multidimensional size functions are invariant with respect to this reduction method. This result can lead us to easily and fast compute discrete multidimensional size functions for applications, as highlighted by some experiments showing the feasibility of the proposed reduction scheme. This work can be seen as a contribution in finding reduction methods for data structure encoding multidimensional information of shapes, in a way that the topological/homological information carried with them is preserved. For the next future, it could be interesting to study the existence of a local reduction method for k-dimensional size graphs preserving the information in terms of multidimensional size functions.

References

1. http://gts.sourceforge.net/samples.html
2. Biasotti, S., Cerri, A., Frosini, P., Giorgi, D., Landi, C.: Multidimensional size functions for shape comparison. J. Math. Imaging Vision 32(2), 161–179 (2008)
3. Biasotti, S., De Floriani, L., Falcidieno, B., Frosini, P., Giorgi, D., Landi, C., Papaleo, L., Spagnuolo, M.: Describing shapes by geometrical-topological properties of real functions. ACM Comput. Surv. 40(4), 1–87 (2008)
4. Biasotti, S., Giorgi, D., Spagnuolo, M., Falcidieno, B.: Size functions for comparing 3d models. Pattern Recogn. 41(9), 2855–2873 (2008)
5. Biasotti, S., Cerri, A., Giorgi, D.: k-dimensional size functions for shape description and comparison. In: ICIAP, pp. 795–800. IEEE Computer Society, Los Alamitos (2007)
6. Cagliari, F., Di Fabio, B., Ferri, M.: One-dimensional reduction of multidimensional persistent homology. Proc. Amer. Math. Soc. 138(8), 3003–3017 (2010)
7. Carlsson, G., Zomorodian, A.: The theory of multidimensional persistence. Discrete & Computational Geometry 42(1), 71–93 (2009)
8. Cerri, A., Ferri, M., Giorgi, D.: Retrieval of trademark images by means of size functions. Graph. Models 68(5), 451–471 (2006)
9. Cerri, A., Frosini, P., Landi, C.: A global reduction method for multidimensional size graphs. Electronic Notes in Discrete Mathematics 26, 21–28 (2006)
10. d'Amico, M., Frosini, P., Landi, C.: Natural pseudo-distance and optimal matching between reduced size functions. Acta. Appl. Math. 109, 527–554 (2010)
11. d'Amico, M.: A New Optimal Algorithm for Computing Size Functions of Shapes. In: CVPRIP Algorithms III, pp. 107–110. Atlantic City (2000)
12. Dibos, F., Frosini, P., Pasquignon, D.: The use of size functions for comparison of shapes through differential invariants. J. Math. Imaging Vis. 21(2), 107–118 (2004)
13. Edelsbrunner, H., Letscher, D., Zomorodian, A.: Topological persistence and simplification. Discrete & Computational Geometry 28(4), 511–533 (2002)
14. Frosini, P., Landi, C.: Size theory as a topological tool for computer vision. Pattern Recognition and Image Analysis 9, 596–603 (1999)
15. Frosini, P., Landi, C.: Size functions and formal series. Appl. Algebra Engrg. Comm. Comput. 12(4), 327–349 (2001)
16. Frosini, P., Pittore, M.: New methods for reducing size graphs. Intern. J. Computer Math (70), 505–517 (1999)
17. Kaczynski, T., Mrozek, M., Slusarek, M.: Homology computation by reduction of chain complexes. Computers & Mathematics with Applications 35(4), 59–70 (1998)
18. Kropatsch, W.: Building irregular pyramids by dual graph contraction. In: IEE-Proc. Vision, Image and Signal Processing, pp. 366–374 (1995)
19. Matas, J., Chum, O., Martin, U., Pajdla, T.: Robust wide baseline stereo from maximally stable extremal regions. In: Proceedings of British Machine Vision Conference, London, vol. 1, pp. 384–393 (2002)
20. Miettinen, K.: Nonlinear multiobjective optimization. Kluwer Academic Publishers, Boston (1999)
21. Mrozek, M., Batko, B.: Coreduction homology algorithm. Discrete Comput. Geom. 41, 96–118 (2009)
22. Peltier, S., Ion, A., Haxhimusa, Y., Kropatsch, W., Damiand, G.: Computing Homology Group Generators of Images Using Irregular Graph Pyramids. In: Escolano, F., Vento, M. (eds.) GbRPR 2007. LNCS, vol. 4538, pp. 283–294. Springer, Heidelberg (2007)
23. Uras, C., Verri, A.: Computing size functions from edge maps. International Journal of Computer Vision 23(2), 169–183 (1997)

Graph Descriptors from B-Matrix Representation

Wojciech Czech

AGH University of Science and Technology, Kraków, Poland
czech@agh.edu.pl

Abstract. In this paper we propose graph descriptors derived from B-matrices, which are built on the basis of distances between graph vertices. The B-matrices, being invariant under graph isomorphism, are a rich source of information about graph structure. We explore this representation and propose several new graph characteristics that can be used for efficient graph comparison. Experiments on clusterization and classification with synthetic and real-world data revealed, that new descriptors allow for distinguishing graphs with non trivial structural differences. Moreover, they appear to outperform descriptors based on heat kernel matrix, being at the same time more effective computationally.

1 Introduction

In the last decade we have been observing extensive use of graph structures in various research fields ranging from physics to sociology. Such an increase of interest on structured data is epitomized by recent advances in theory of complex networks that provided deep insight into topology and dynamics of real-world networks and a common viewpoint for their analysis [1]. New perspectives of graph-based representations are also emerging in such fields as image vision, image processing or sensor networks [2].

Describing a system as a set of binary relations or interactions among its elements is convenient for humans. Such a fine-grained bottom-up approach reduces complexity and makes the problem easier to comprehend. Graphs allow for integration of large amounts of data into one high-level structure capturing system as a whole. This is particularly useful in biomedical or chemical applications where high-throughput experiments produce high-volume data which cannot be easily tackled without previous synthesis. Constant growth of computing power allow for increasing size of structured datasets subject to analysis. Nevertheless, the development of efficient and robust algorithms is still a challenging task.

Graph comparison is a crucial research method finding its applications in such tasks involving structured data as confronting models and simulation results with real-world data, building structured databases, classification and clusterization. The result of graph comparison is dissimilarity or similarity measure that can be used in pattern recognition methods. Due to non-vectorial nature of graphs, their comparison poses some intrinsic problems that cannot be simply neglected during the development of new graph matching algorithms. The direct comparison

X. Jiang, M. Ferrer, and A. Torsello (Eds.): GbRPR 2011, LNCS 6658, pp. 12–21, 2011.

of two graphs requires enumeration of all sub-substructures and tackling with elements order. The exponential cost of such procedure makes construction of efficient graph metric infeasible. Graph comparison algorithms should also give results invariant under isomorphism, what becomes cumbersome when typical graph representations in a form of adjacency matrices or neighborhood lists are considered.

The practical approach to graph matching uses graph invariants to construct feature vector and embed graph into metric space. The question how to quantitatively capture relevant structural properties of graphs provoked the development of graph measures such as *clustering coefficient, efficiency* or *betweenness centrality* [3]. Today the abundance of scalar graph descriptors makes the selection of relevant features a difficult task that can be tackled e.g. with a help of information-theoretic tools [4]. Elegant methods of graph features generation use invariants computed on the basis of graph matrices with a help of permutation invariant functions. Spectral decomposition of graph Laplace matrix was used to construct high-dimensional pattern vector, effective in clustering graphs representing images [5]. More recently, Xiao and Hancock proposed robust descriptors based on heat kernel matrix, obtained by exponentiating the Laplacian eigensystem [6]. The values of these characteristics can be tuned by time parameter, that allows for navigation between local and global features. Embedding vertices of graph into vector space using heat kernel matrix was used to create generative model for graph matching that allows for finding correnspondences between vertices and capturing structural differences over sets of graphs [7]. Different approach of collecting several scalar features such as degree distribution measures or motif profile measures is present in metabolic networks comparison, where biological relevance of features is desirable [8]. When vertex or edge labels are known *a priori* the construction of vector representations is easier as permutation invariant functions are no longer necessary. In this case graph feature vector can be obtained by enumeration of vertex or edge descriptors. Such approach was used for clustering organisms based on metabolic networks [9].

In this paper we explore feature vectors derived from graph B-matrices, which are non-complete graph invariants constructed on the basis of the shortest path lengths [10]. We show that two types of graph B-matrices constitute rich source of information about graph structure and can be used to account for graph comparison. To this end we present feature vectors constructed using aggregated statistics of B-matrices rows and test them on artificial and real-world data. The experiments on clusterization and classification of selected datasets are reported and the results are compared to the ones obtained for graph characteristics derived from heat kernel. In the end we draw conclusions with some ideas for further development of this concept.

2   Graph B-Matrices

Typical graph matrix representations such as adjacency matrix or Laplace matrix depend on vertex ordering, therefore they are not identical for the same graphs with permuted labels. In [10], Bagrow et al. proposed approach that circumvent this inconvenience. The authors introduced B-matrix that serve as a "portrait" of network and can be used for graph visualization and comparison. Let us define l-shell of vertex v_i as subset of graph vertices at distance l from v_i. Degree of order l is a size of respective l-shell. The nearest neighbors of vertex v_i form its 1-shell and the size of this 1-shell is degree of vertex v_i. The vertex B-matrix of a graph with n vertices is defined as follows.

$$B_{l,k}^V = \text{number of nodes that have } k \text{ members in their } l\text{-shells,} \quad (1)$$

where $l \leq n$ and $k \leq n$. The B-matrix stores information about distribution of l-shells sizes (degrees of order l) and in fact it is a sequence of histograms. The number of non-zero rows is given by the graph diameter. The example of vertex B-matrix of random graph with 100 vertices and 322 edges is depicted in Fig. 1b. To compute B-matrix we need to enumerate shell members for each vertex of the graph. Using Breadth-First Search this can be achieved with $\mathcal{O}(n^2)$ steps for dense and $\mathcal{O}(n)$ steps for sparse graphs. Therefore, construction of B-matrix for sparse graph has $\mathcal{O}(n^2)$ time-complexity.

Maximum value of k for which $B_{l,k} \neq 0$ in each row, reflects network branching. As shown in [10] vertex B-matrix can be used for visual comparison and classification of various networks. It is capable of capturing such high-level properties of networks as *assortativity/disassortativity, small-worldliness, regularity* and many more, far beyond the degree distribution. The higher rows of B-matrix encode more specific information about structure of the graph. This information can be extracted using row aggregated statistics or by capturing differences between rows. The B-matrices can be used for pairwise comparison of graphs [10] however in this work we present different approach of using B-matrix for graph feature vector generation. In case of graphs of different size B-matrices have to be scaled accordingly by padding zero-rows or columns.

Despite its robustness in encoding information about graph, vertex B-matrix is not a complete graph invariant. This means that there exist pairs of non-isomorphic graphs possessing the same vertex B-matrix representation. The example of such a pair is dodecahedral and Desaurges graphs. In [10], the authors conjectured that more general edge B-matrix together with vertex B-matrix can identify graph univocally. In order to introduce edge B-matrix we define distance from a vertex v_i to an edge (v_j, v_k) as the mean of distances $d(v_i, v_j)$ and $d(v_i, v_k)$. The l-edge-shell of vertex v_i is a subset of graph edges at distance l from v_i. In this case l can have half-integer values. Equation 2 defines edge B-matrix of a graph.

$$B_{i,k}^E = \text{number of nodes that have } k \text{ edges in their } (\tfrac{1}{2}i)\text{-edge-shells.} \quad (2)$$

The edge B-matrix of sample random graph is depicted in Fig. 1c. The maximum size of edge B-matrix is $2n \times m$, where n denotes number of vertices and m number of edges in the graph. For dense graphs this becomes computationally cumbersome, as a number of columns reaches the order of n^2. Nevertheless, owing to narrow row distributions, after removing zero columns, effective size can be decreased significantly. Even rows of edge B-matrix account for edges

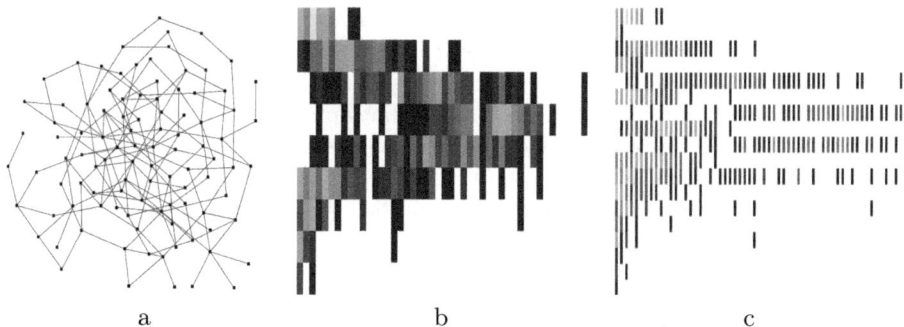

Fig. 1. a. Connected random graph G_1 with 100 vertices, 322 edges and diameter 9, b. Vertex B-matrix of G_1, c. Edge B-matrix of G_1 (dark blue: 1, red: 32).

whose endpoints are equidistant from a certain vertex, therefore graphs without odd cycles (equivalently bipartite graphs) such as trees or square graphs possess empty rows for i divisible by 2.

3   Pattern Vectors from B-Matrices

As consecutive rows of B-matrices describe distribution of respective l-shells, we use aggregated statistics to generate per-row characteristics and combine them into one feature vector. Let $\star$ stand for V or E symbol, so that $B^\star$ denote B^V or B^E matrix.

$$D^\star_{rstd}(l) = \frac{\sigma^\star(l)}{\mu^\star(l)} \tag{3}$$

$D^\star_{rstd}(l)$ is relative deviation of row l, whereby $\mu^\star(l)$ denotes average and $\sigma^\star(l)$ standard deviation of l-shell size (only nonzero entries of $B^\star$ are taken into account). High values are achieved for left-skewed broad l-shell size distributions while low values indicate right-skewed narrow distributions. Therefore $D^\star_{rstd}(l)$ can be used as a measure of l-shell regularity.

The Shannon entropy $D^\star_{ent}(l)$ measures unpredictability of l-shell size. Low values of this descriptor indicate that l-shells of certain size dominate, whereas higher values are obtained for broad, uniform l-order degree distributions.

$$p(l, k) = \frac{B^\star_{l,k}}{\sum_k B^\star_{l,k}} \tag{4}$$

$$D^\star_{ent}(l) = -\sum_k p(l, k) \log(p(l, k)) \tag{5}$$

Assessing B-matrices inter-row diversity yields relevant graph features. The difference of average $(l-1)$-shell size and average l-shell size reflect average offset between consecutive distributions.

$$D^\star_{avgd}(l) = \mu^\star(l - 1) - \mu^\star(l) \tag{6}$$

Negative values of $D^{\star}_{avgd}(l)$ indicate that l-shells have on average more members than $(l-1)$-shells. Large absolute values are obtained for dense graphs while small ones for sparse graphs. The change from negative to positive value of $D^{\star}_{avgd}(l)$ reflects "turning point" (see Fig. 1b, row 4), that appears in B-matrices of certain graph types due to finite size effects (after l passes average path length, l-shells start to posses less and less members).

B-matrices are graph invariants, therefore their rows or columns can be packed to form a long pattern vector.

$$D^{\star}_{long}(l_{min}, l_{max}, k_{min}, k_{max}) = [B^{\star}_{l,k}]$$
$$l_{min} \leq l \leq l_{max}, \; k_{min} \leq k \leq k_{max} \tag{7}$$

Such an approach allows for retaining all the information stored in a B-matrix, what is impossible for descriptors defined earlier. Using parameter l_{min}, l_{max}, k_{min} and k_{max} one can extract the B-submatrix that combines selected subset of low and high-level features. For instance by choosing $1 \leq l \leq 2$ and $1 \leq k \leq n$, where n is number of graph vertices, we extract local information about nearest and second-nearest-neighbors. In turn, after setting $1 \leq l \leq diameter$ and $1 \leq k \leq 3$, $D^{\star}_{long}$ stores information about l-shells with small number of elements i.e. counts vertices that are relatively distinct from "typical" ones either because of their location in the center of the graph or due to low local branching factor.

4 Experiments

In this section we present experiments on graph clusterization and classification performed to test descriptors derived from B-matrices. Two datasets were prepared: artifical one created on the basis of four seed graphs using random edit operations and a set of real-world graphs representing satellite photos obtained from *Google Earth*. After extracting corners [11], the images were transformed to graphs with the use of Delaunay triangulation.

We compare proposed graph descriptors with two measures generated on the basis of heat kernel [6], that is heat kernel content invariant:

$$D_{hkc}(t) = \sum_{u \in V} \sum_{v \in V} \sum_{k=1}^{n} \exp(-\lambda_k t)\phi_k(u)\phi_k(v), \tag{8}$$

and heat kernel content coefficients invariant:

$$D_{hkcc}(m) = \sum_{k=1}^{n} \left\{ \left(\sum_{u \in V} \phi_k(u) \right)^2 \right\} \frac{(-\lambda_k)^2}{m!}, \tag{9}$$

where V denotes vertices set, $n = |V|$ and (λ_k, ϕ_k) is k-th eigenpair of graph normalized Laplace matrix. We also use control feature vector D_{con} that contains seven well-known scalar graph descriptors such as *efficiency* and average values and standard deviations of *degree*, *clustering coefficient* and *betweenness*. The results of graph embeddings are evaluated using clustering validation indices: C

index, *Davies-Bouldin index* and *Rand index* for which we use 5-nearest neighbor algorithm to establish agreements in cluster assignments [12]. The computation of graph descriptors was performed using *Graph Investigator* application [13].

4.1   Artificial Graphs

In order to generate the first dataset we selected four seed graphs of size 100: *bb-seed* is 3-regular graph with 150 edges, *me-seed* is irregular 2D mesh with 200 edges, *rm-seed* is 2D lattice with 180 edges and *rr-seed* is random connected Erdős-Rényi graph with 143 edges. For each seed graph we performed 100 series of random edit operations that can be described by three parameters (*ne*, *spread*, *type*). Parameter *ne* denotes a number of edit operations performed, *spread* is a maximal random number added to *ne* in order to increase group variance and *type* describes type of edit operations performed (edge addition, deletions or both). Clusters *bb*, *me* and *rr* were generated using parameters (40, 20, *both*) while *rm* with (20, 30, *deletion*). In this manner we obtained groups that cannot be easily separated by graph density. The sample graphs from the dataset are depicted in Fig. 2. Except for *rr* sample, the planar embeddings appear to be structurally similar. The differences can be captured more accurately with the use of vertex B-matrices.

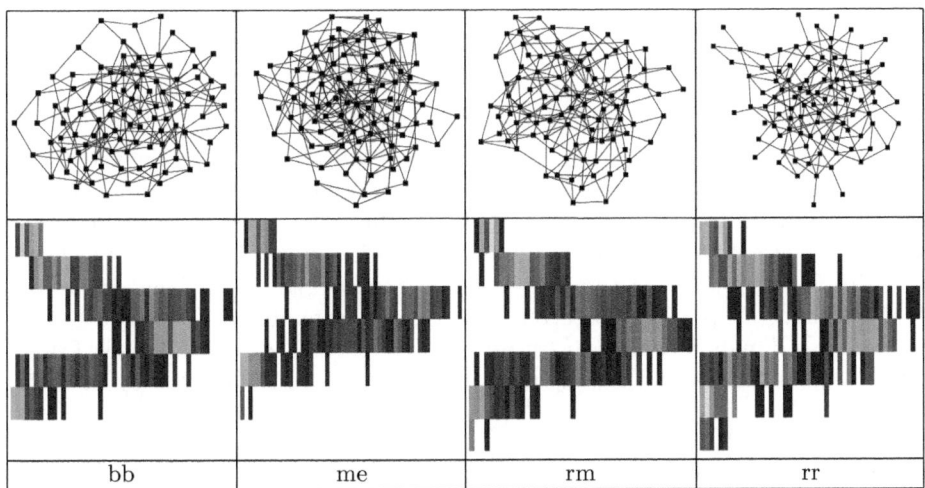

Fig. 2. Visualization of sample graphs from artificial dataset using planar embeddings and vertex B-matrices

The descriptors based on B-matrices were computed and mapped onto 2D and 3D space using two unsupervised dimensionality reduction algorithms PCA (Principal Component Analysis) and LPMIP (Locality-Preserved Maximum Information Projection) [14]. The latter one controls between-locality and within-locality simultaneously. The tradeoff between local and global structure of the data is adjusted using parameter α. In this manner we can mix properties of

Table 1. Low-dimensional embeddings of descriptors derived from graph B-matrices evaluated using indices: *C index, Davies-Bouldin index* and *Rand index* (the order of indices preserved in triples presented in the table)

	Vector	Dim	Dim Red Method	2D	3D
1	$D^E_{long}(1,4,1,25)$	100	PCA	0.07, 1.00, 0.93	0.05, 1.13, 0.96
2	$D^E_{long}(1,4,1,25)$	100	LPMIP(20, 20, 0.1)	**0.04, 0.69, 0.98**	**0.03, 0.74, 0.99**
3	$D^V_{long}(1,4,1,20)$	80	PCA	0.13, 1.62, 0.89	0.14, 1.85, 0.94
4	$D^V_{long}(1,4,1,20)$	80	LPMIP(15, 20, 0.1)	0.08, 2.11, 0.93	0.08, 2.13, 0.93
5	$D^V_{ent}, 1 \le l \le 10$	10	PCA	0.14, 1.28, 0.79	0.15, 1.45, 0.80
6	$D^E_{avgd}, 1 \le l \le 20$	20	LPMIP(20, 20, 0.1)	0.12, 1.67, 0.81	0.15, 1.96, 0.81
7	$\mu^E, \mu^V, \sigma^E, \sigma^V,$ $1 \le l \le 4$	16	PCA	**0.07, 1.16, 0.97**	**0.11, 1.50, 0.97**
8	$D^V_{rstd}, 1 \le l \le 10$	10	LPMIP(20, 20, 0.1)	0.09, 1.2, 0.86	0.15, 1.69, 0.87
9	$D^E_{rstd}, 1 \le l \le 20$	20	LPMIP(20, 20, 0.1)	0.09, 1.37, 0.81	0.16, 1.92, 0.84
10	vectors from row 2 and 7 together	116	LPMIP(20, 20, 0.1)	**0.01, 0.47, 1.0**	**0.04, 0.54, 1.0**
11	$D_{hkc}, 1 \le t \le 10$	10	PCA	0.05, 0.73, 0.89	0.05, 0.73, 0.98
12	$D_{hkc}, 1 \le t \le 20$	20	PCA	0.05, 0.86, 0.86	0.05, 0.86, 0.97
13	$D_{hkcc}, 1 \le m \le 10$	10	PCA	0.07, 1.46, 0.79	0.07, 1.47, 0.78
14	$D_{hkcc}, 1 \le m \le 20$	20	PCA	0.09, 1.16, 0.80	0.09, 1.17, 0.80
15	D_{con}	7	PCA	0.11, 2.01, 0.79	0.11, 2.14, 0.86

manifold learning methods such as LLE (Locally Linear Embedding) or LPP (Locality Preserving Projection) with a general-variance methods as PCA. The discrimination ability of introduced descriptors is evaluated with the use of clustering validation indices computed in low-dimensional spaces. The most interesting results are reported in Table 1. Low values of *C index* (range $[0;1]$) and *Davies-Bouldin index* (range $[0;\infty]$) indicate good clustering, whereby *Davies-Bouldin index* prefers clusters of spherical shape. *Rand index* ranges from 0 to 1, where 1 means perfect clustering. The three parameters of LPMIP algorithm were adjusted manually to ($nearest\_neighbors = 20, \sigma = 20, \alpha = 0.1$), see Table 1. This means that we set the weight 0.1 to between locality and 0.9 to within locality.

The best results (both in 2D and 3D space) were obtained for long feature vectors derived from edge B-matrix (see Table 1, rows 2 and 10). Fig. 3 shows the embedding of graphs in 3D space using $D^E_{long}(1,4,1,25) + \mu^E, \mu^V, \sigma^E, \sigma^V$ for $1 \le l \le 4$ and its associated distance matrix. With the use of edge B-matrices we can generate robust graph characteristics, that can perform better than ones derived from a heat kernel. Moreover, the cost of computing such descriptors is lower ($\mathcal{O}(n^2)$ vs. $\mathcal{O}(n^3)$). The B^E stores more discriminative information than B^V. Long vectors perform better than aggregated statistics (e.g. D^E_{rstd}), nevertheless the latter are still comparable with descriptor D_{hkcc}, that requires $\mathcal{O}(n^3)$ steps (see Table 1, rows 7 and 8).

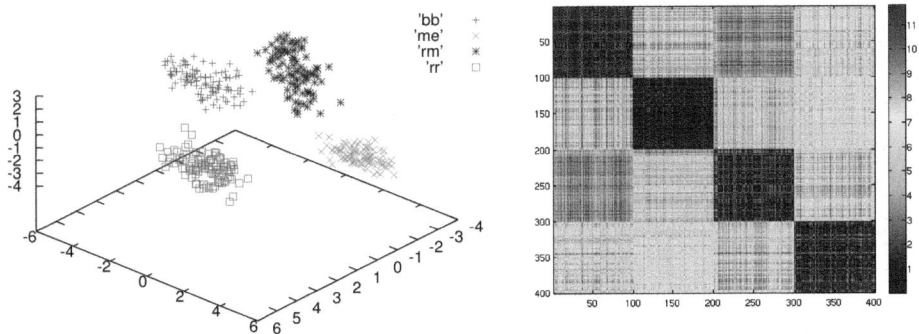

Fig. 3. 3D embedding of combined feature vector build from graph B-matrices (see Table 1, row 10) and its associated distance matrix

4.2 Satellite Photos

In this section we describe experiments on classification of graphs generated from satellite photos. With the help of *Google Earth* application we obtained three groups of photos (size 1412×940) representing fragments of cities. Each group contains 90 samples from approximately the same area taken from altitudes 980m to 1.08km. In order to simulate occlusions and clutter, rotations and translations were performed before exporting a photo. Three examples from this dataset are depicted in Fig. 4. The photos were transformed to graphs using Harris corner detector [11] and Delaunay triangulation. To discard graph size as a main discrimination factor we decided to select 100 most distinct corners. As all photos contained at least 100 corners, 270 graphs of size 100 were generated.

We used this dataset to investigate the performance of the nearest centroid classifier for feature vectors based on graph B-matrices. The obtained results were compared with ones received for graph characteristics derived from heat kernel matrix [6]. We randomly selected 15 graphs of each class to construct the test set and use the remaining graphs as the training set (75 training samples, 15 testing samples). The operation was performed 100 times for each type of graph feature vector. The average and maximum classification accuracy is reported. Additionally, in order to get a meaningful low-dimensional representation of patterns and increase recognition rate we applied dimensionality reduction methods

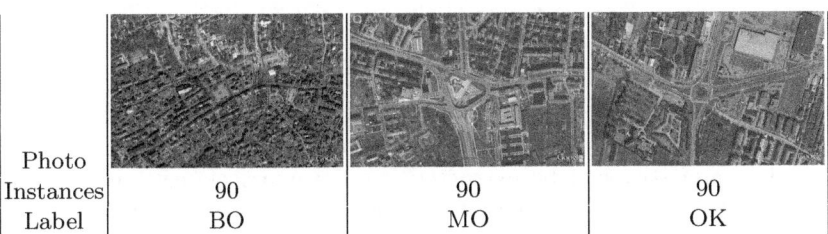

Photo			
Instances	90	90	90
Label	BO	MO	OK

Fig. 4. Three samples from satelite photo dataset obtained using *Google Earth* application

Table 2. Comparison of average and maximal recognition rates for graph descriptors derived from graph B-matrices and heat kernel matrix

Vector	Dim	Dim Red	Target Dim	Avg Accuracy	Max Accuracy	Rand Idx
$D_{avgd}^E, 1 \le l \le 15$	30	PCA	10	0.73	0.86	0.60
$D_{rstd}^E, 1 \le l \le 15$		PCA	15	0.75	0.89	0.63
		LDA	2	**0.77**	**0.91**	0.8
		MMC	2	0.61	0.73	0.63
		MMC	5	0.65	0.80	0.74
$D_{long}^V(1,8,1,30)$	240	MMC	2	**0.76**	**0.91**	0.93
$D_{long}^E(1,20,1,100)$	2000	MMC	100	**0.78**	**0.89**	0.93
D_{hkc} and	30	PCA	10	0.65	0.80	0.61
$D_{hkcc}, 1 \le t \le 15$		PCA	15	0.66	0.80	0.61
		LDA	2	0.64	0.80	0.55
		MMC	2	0.67	0.84	0.56
D_{con}	7	PCA	5	0.60	0.71	0.57
		LDA	2	0.69	0.84	0.63
		MMC	2	0.68	0.80	0.54

PCA (Principal Component Analysis), LDA (Linear Discriminant Analysis) and MMC (Maximum Margin Criterion) [15]. The projection vectors were computed using training data and then applied to obtain lower-dimensional representations of vectors from the testing set, like in Fisherfaces or Eigenfaces technique [16]. The results of classification are depicted in Table 2. The target dimensionalities were selected experimentally to obtain highest accuracies. In case of MMC only the directions associated with eigenvalues greater than 0.5 were considered.

The results obtained for descriptors derived from B-matrices are considerably better than classification rates for D_{hkc}, D_{hkcc} and D_{con}. Even unsupervised, global-variance-oriented dimensionality reduction with PCA gives better accuracy than D_{hkc} with LDA or MMC. The highest average accuracies were achieved for $D_{long}^E(1, 20, 1, 100)$ and for aggregated D_{avgd}^E and D_{rstd}^E descriptors. Dimensionality of $D_{long}^{\star}$ descriptors exceeds the number of samples, therefore, in this case we applied MMC, which does not suffer from small sample size problem and, additionally, is less vulnerable to overfittig (for high input dimensionality LDA has poor generalization capability).

5 Conclusions

In this paper we demonstrated that B-matrices which contain information about l-shell size distributions can be successfully employed to distinguish graphs. Such a representation is convenient to use owing to its invariance under isomorphism. Graph characteristics can be extracted from B-matrices using row aggregated statistics or by selecting submatrix and forming long feature vector. For sparse graphs B-matrices can be constructed with $\mathcal{O}(n^2)$ steps what makes this approach more computationally feasible than the eigendecomposition-based algorithms. In the future we aim to test the presented graph comparison method on subsequent real-world datasets, including metabolic networks and vascular networks. Furthermore, a deeper analysis of B-matrices structure and correlations between their rows can lead to more robust descriptors.

Acknowledgments. This work is financed by the Polish Ministry of Science and Higher Education, Project No. N N519 579338.

References

1. Estrada, E., Fox, M., Higham, D.J., Oppo, G.L.: Network Science: Complexity in Nature and Technology. Springer Publishing Company, Heidelberg (2010)
2. Marfil, R., Escolano, F., Bandera, A.: Graph-Based Representations in Pattern Recognition and Computational Intelligence. Bio-Inspired Systems: Computational and Ambient Intelligence 5517, 399–406 (2009)
3. Costa, L.F., Rodrigues, F.A., Travieso, G., Boas, P.R.V.: Characterization of complex networks: A survey of measurements. Advances in Physics 56(1) (2007)
4. Bonev, B., Escolano, F., Giorgi, D., Biasotti, S.: High-Dimensional Spectral Feature Selection for 3D Object Recognition Based on Reeb Graphs. In: Hancock, E.R., Wilson, R.C., Windeatt, T., Ulusoy, I., Escolano, F. (eds.) SSPR&SPR 2010. LNCS, vol. 6218, pp. 119–128. Springer, Heidelberg (2010)
5. Wilson, R.C., Hancock, E.R., Luo, B.: Pattern vectors from algebraic graph theory. IEEE Transactions on Pattern Analysis and Machine Intelligence 27(7), 1112–1124 (2005)
6. Xiao, B., Hancock, E.R., Wilson, R.C.: Graph characteristics from the heat kernel trace. Pattern Recognition 42(11), 2589–2606 (2009)
7. Xiao, B., Hancock, E.R., Wilson, R.C.: A generative model for graph matching and embedding. Computer Vision and Image Understanding 113(7) (2009)
8. Zhu, D., Qin, Z.S.: Structural comparison of metabolic networks in selected single cell organisms. BMC Bioinformatics 6(1) (2005)
9. Arodz, T.: Clustering Organisms Using Metabolic Networks. Computational Science–ICCS 2008, 527–534 (2008)
10. Bagrow, J.P., Bollt, E.M., Skufca, J.D.: Portraits of complex networks. Europhysics Letters 81, 68004 (2008)
11. Harris, C., Stephens, M.: A combined corner and edge detector. In: Alvey Vision Conference, vol. 15 (1988)
12. Günter, S., Bunke, H.: Validation indices for graph clustering. Pattern Recognition Letters 24(8), 1107–1113 (2003)
13. Czech, W., Goryczka, S., Arodz, T., Dzwinel, W., Dudek, A.: Exploring complex networks with Graph Investigator research application. Computing and Informatics 30 (2011) (in press)
14. Wang, H., Chen, S., Hu, Z., Zheng, W.: Locality-preserved maximum information projection. IEEE Transactions on Neural Networks 19(4), 571–585 (2008)
15. Li, H., Jiang, T., Zhang, K.: Efficient and robust feature extraction by maximum margin criterion. IEEE Transactions on Neural Networks 17(1) (2006)
16. Belhumeur, P.N., Hespanha, J.P., Kriegman, D.J.: Eigenfaces vs. fisherfaces: Recognition using class specific linear projection. IEEE Transactions on Pattern Analysis and Machine Intelligence 19(7) (2002)

Dimensionality Reduction for Graph of Words Embedding

Jaume Gibert[1], Ernest Valveny[1], and Horst Bunke[2]

[1] Computer Vision Center, Universitat Autònoma de Barcelona
Edifici O Campus UAB, 08193 Bellaterra, Spain
{jgibert,ernest}@cvc.uab.es
[2] Institute for Computer Science and Applied Mathematics, University of Bern,
Neubrückstrasse 10, CH-3012 Bern, Switzerland
bunke@iam.unibe.ch

Abstract. The Graph of Words Embedding consists in mapping every graph of a given dataset to a feature vector by counting unary and binary relations between node attributes of the graph. While it shows good properties in classification problems, it suffers from high dimensionality and sparsity. These two issues are addressed in this article. Two well-known techniques for dimensionality reduction, kernel principal component analysis (kPCA) and independent component analysis (ICA), are applied to the embedded graphs. We discuss their performance compared to the classification of the original vectors on three different public databases of graphs.

1 Introduction

A graph based representation is a powerful tool to represent patterns which lately has been gaining popularity among the pattern recognition community. The major advantage over statistical feature vectors arises from the fact that graphs are able to describe structural relations between parts of the patterns to be represented. However, on the other hand, the complexity of the related algorithms makes the treatment and processing of graphs a hard problem; while there are many pattern analysis algorithms for feature vector based problems available, in the case of graphs we still lack well established and efficient procedures to process them in the context of learning and pattern classification.

The classical techniques for graph matching [1] do not allow the transition between graph representations and the domain of statistical pattern recognition, which is quite wealthy in terms of algorithms. However, there exist recent approaches that allow such a transition, and much effort is being put on investigating how to take as much profit as possible of both worlds. One important case are graph embeddings. Embedding a graph into a vector space permits one to use every algorithm from machine learning originally developed for feature vectors. Examples of graph embeddings can be widely found in the literature. For instance, in [2], the authors approach the problems of graph clustering and graph visualization by extracting different features from an eigen-decomposition

X. Jiang, M. Ferrer, and A. Torsello (Eds.): GbRPR 2011, LNCS 6658, pp. 22–31, 2011.
© Springer-Verlag Berlin Heidelberg 2011

of the adjacency matrices of the graphs. In [3], the nodes of the graph are embedded into a metric space and then the edges are interpreted as geodesics between points on a Riemannian manifold. The problem of matching nodes to nodes is viewed as the alignment of the resulting point sets. An interesting approach is the one in [4], where graphs are mapped to feature vectors such that each component is the edit distance to a graph prototype. Finally, in [5], to solve the problem of molecules classification, the authors associate a feature vector to every molecule by counting unary and binary statistics in the molecule; these statistics indicate how many times every atomic element appears in the molecule, and how often there is a bond between two specific atoms.

The idea in [5] exhibits interesting properties regarding efficiency and representation. However, the original proposal is only for discrete node attributes and it suffers from dimensionality and sparsity problems. In this work, we generalize this idea for continuous attributed graphs and we specifically deal with the dimensionality problem the embedding exhibits. In the next section, we recall formally the embedding procedure and clarify why the dimensionality has to be reduced. Then, in Section 3, we expose the two techniques that have been used to reduce the number of dimensions. In Section 4, the experimental setup is explained and the results are shown and discussed. Finally, Section 5, draws conclusions from this work.

2   Graph of Words Embedding

Although the embedding of graphs into vectors spaces provides a way to be able to apply statistical pattern analysis techniques to the domain of graphs, the existing methods still suffer from the main drawback that the classical graph matching techniques also did, this is, their computational cost. The Graph of Words Embedding tries to avoid these problems by just visiting nodes and edges instead of, for instance, travelling along paths in the graphs, computing edit distances or performing the eigen-decomposition of the adjacency matrix. In this section we first briefly explain the motivation of this approach and then formally define the procedure.

2.1   Motivation

In image classification, a well-known image representation technique is the so-called bag of visual features, or just bag of words. It first selects a set of feature representatives, called words, from the whole set of training images and then characterizes each image by a histogram of appearing words extracted from the set of salient points in the image [6].

The graph of words embedding proceeds in an analogous way. The salient points in the images correspond to the nodes of the graphs and the visual descriptors are the node attributes. Then, one also selects representatives of the node attributes (words) and counts how many times each representative appears in the graph. This leads to a histogram representation for every graph. Then, to take profit of the edges in the original graphs, one also counts the frequency of the relation between every pair of words. The resulting information is combined with the representative's histogram in a final vector.

2.2   Embedding Procedure

A graph is defined by the 4-tuple $g = (V, E, \mu, \nu)$, where V is the set of nodes, $E \subseteq V \times V$ is the set of edges, μ is the nodes labelling function, assigning a label to every node, and ν is the edges labelling function, assigning labels to every edge in the graph. In this work we just consider graphs whose nodes attributes are real vectors, this is $\mu : V \rightarrow \mathbb{R}^d$ and whose edges remain unattributed, this is, $\nu(e) = \varepsilon$ for all $e \in E$ (where ε is the null label).

Let $\mathcal{P}$ be the set of all nodes attributes in a given dataset of graphs $\mathcal{G} = \{g_1, \ldots, g_M\}$. From all points in $\mathcal{P}$ we derive n representatives, which we shall call *words*, in analogy to the bag of words procedure. Let this set of words be $\mathcal{V} = \{w_1, \ldots, w_n\}$ and be called *vocabulary*. Let also λ be the node-to-word assignment function $\lambda(v) = \arg\min_{w_i \in \mathcal{V}} d(v, w_i)$, this is, the function that assigns a node to its closest word. Then, before assigning a vector to each graph, we first construct an intermediate graph that will allow us an easier embedding. This intermediate graph, called *graph of words* $g' = (V', E', \mu', \nu')$ of $g = (V, E, \mu, \nu) \in \mathcal{G}$ with respect to $\mathcal{V}$, is defined as:

- $V' = \mathcal{V}$
- E' is defined by: $(w, w') \in E' \Leftrightarrow$ there exists $(u, v) \in E$ such that $\lambda(u) = w$ and $\lambda(v) = w'$
- $\mu'(w) = |\{v \in V \mid w = \lambda(v)\}|$
- $\nu'(w, w') = |\{(u, v) \in E \mid \lambda(u) = w, \lambda(v) = w'\}|$.

Once the graph of words is constructed, we easily convert the original graph into a vector by combining the node and edge information of the graph of words, by keeping both the information of the appearing words and the relation between these words. We consider the histogram

$$\phi_a^{\mathcal{V}}(g) = (\mu'(w_1), \ldots, \mu'(w_n)). \tag{1}$$

and a flattened version of the adjacency matrix of the graph of words $A = (a_{ij})$, with $a_{ij} = \nu'(w_i, w_j)$:

$$\phi_b^{\mathcal{V}}(g) = (a_{11}, \ldots, a_{ij}, \ldots, a_{nn}), \quad \forall i \leq j \tag{2}$$

The final graph of words embedding is the concatenation of both pieces of information,

$$\varphi^{\mathcal{V}}(g) = (\phi_a^{\mathcal{V}}(g), \phi_b^{\mathcal{V}}(g)). \tag{3}$$

In Figure 1, there is an example of the graph of words procedure for a simple vocabulary of size equal to 4.

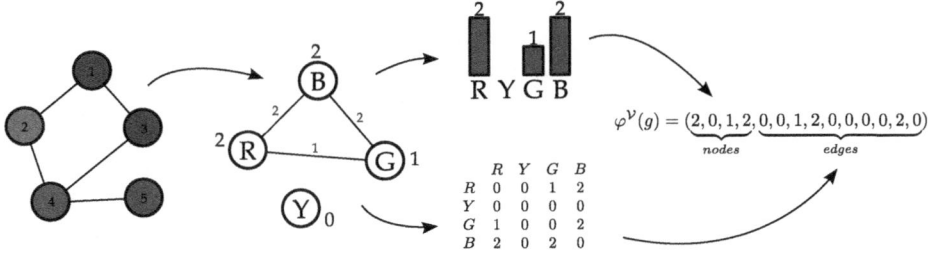

Fig. 1. Example of the graph of words embedding. The graph on the left is assigned to the vector on the right by considering the vocabulary $\mathcal{V} = \{R, Y, G, B\}$. Nodes 1 and 5 are assigned to the R word, 2 and 3 to the B word and 4 to the G word. Note that none is assigned to the Y word. The histogram of words is considered as well as the adjacency matrix. The resulting vector is the concatenation of both types of information.

2.3 Vocabulary Construction

The final configuration of the vectors after the embedding clearly depends on the set of words that has been chosen. In this paper we decided to use the kMeans algorithm in order to build the vocabulary. The kMeans algorithm iteratively builds a set of k cluster centers by first initializing the k centers with some points, assigning each point in the set to the closest center, and then recomputing the centers as the mean of the set of points assigned to the same cluster. The process finishes when there are no center changes between one iteration and the next one.

The initialization points of the clustering under kMeans are usually taken randomly. In our case, in order to avoid uncertainty in the results, we chose to start by selecting uniformly distributed points over the range of nodes attributes. The way we proceed is by first selecting the median vector of $\mathcal{P}$ and then adding at each iteration the point which is furthest away from the already selected ones. This is done until a predefined number of words is obtained.

2.4 Dimensionality and Sparsity

An important issue of the graph of words configuration is the quadratic increase of its dimension with respect to the vocabulary size. In fact, given a vocabulary $\mathcal{V}$ of size n, the dimensionality of $\varphi^{\mathcal{V}}$ is

$$n + \frac{n^2 - n}{2} + n = \frac{n^2 + 3n}{2}. \tag{4}$$

In general, when having small vocabularies, this does not cause any problem. However, the rapid increase of the dimensionality of the vectors constitutes a problem due to the fact that graphs do not really have such large amount of nodes. When nodes are assigned to words, a large vocabulary creates a really sparse vector that needs to be processed. A small number of nodes makes the adjacency matrix of the graph of words to be full of zero entries, since only a few word relations will be present in such graph.

This situation makes necessary to perform feature selection on the vectors obtained from the graph of words, in order to reduce the number of dimensions of the vectors and, therefore, ease the manipulation and the further processing of these vectors in complex and time consuming machine learning algorithms. But also, it seems important to reduce their dimensionality because this may automatically discover the really important words in the vocabulary, or the important relations between words. It seems reasonable to avoid having a feature in the vectors which is expressing a relation between two rarely related words.

3   Dimensionality Reduction

After making clear the need of reducing the number of dimensions of the graph of words vectors, in this section we give a brief description of the two techniques that have been used to perform such reduction.

3.1   Kernel Principal Component Analysis

Given a set of N feature vectors $x_1, \ldots, x_N \in \mathbb{R}^n$, principal component analysis (PCA) finds a linear transformation of the data $y_i = Ax_i \in \mathbb{R}^m$ so that linear correlation among the new features is reduced and these new $m \leq n$ features capture most of the variance. Such transformation is obtained by an orthogonal mapping where each column of the matrix A is an eigenvector of the covariance matrix of the centered original data. These eigenvectors $v_1, \ldots, v_n$ are called principal components and are ordered from greater to smaller variance. By taking $m \leq n$ principal components, the dimensions are reduced and most of the variance is being kept.

Kernel principal component analysis (kPCA) is the natural non-linear generalization of PCA by means of the kernel trick [7]. A kernel function k can always be thought of as a scalar product in a hidden feature space by means of an implicit mapping ϕ from the input space (no matter the nature of it) to a Hilbert one. Hence, given a kernel function $k : \mathcal{X} \times \mathcal{X} \to \mathbb{R}$, there always exists a mapping $\phi : \mathcal{X} \to \mathcal{H}$ such that $k(x_i, x_j) = \langle \phi(x_i), \phi(x_j) \rangle$, where $\mathcal{X}$ is the space of the input patterns, $\mathcal{H}$ is a Hilbert space implicitly defined and $\langle \cdot, \cdot \rangle$ stands for the dot product.

Kernel principal component analysis makes use of this fact in order to find linear behaviors of the data in the hidden space, which in general correspond to non-linear properties of the input patterns. The projection of $\phi(x)$ onto a (non-linear) principal component u_p of the input feature space is given by

$$u_p \cdot \phi(x) = \sum_{j=1}^{N} \beta_j^p k(x_j, x) \qquad (5)$$

where $\beta^p = (\beta_1^p, \ldots, \beta_N^p) \in \mathbb{R}^N$ is the p-th leading eigenvector of the kernel matrix $K = (k(x_s, x_t))_{1 \leq s,t \leq N}$. The final transformation is given by $y_i = (u_1 \cdot \phi(x_i), \ldots, u_n \cdot \phi(x_i))$. Exactly as in PCA, by keeping $m \leq n$ principal components one captures most of the variance in $\mathcal{H}$. The experiments that have been carried out in this article use the Gaussian kernel for kPCA

$$k(x_i, x_j) = \exp(-\gamma \cdot ||x_i - x_j||^2), \quad \gamma > 0. \tag{6}$$

3.2   Independent Component Analysis

The second technique that has been used in this article is independent component analysis (ICA). Originally developed in the field of signal processing, ICA is very-well known and used for pattern recognition problems. Just as PCA, it aims at finding linearly uncorrelated components. However, ICA adds the requirement for these components to be statistically independent. This is done by optimizing the parameters of the linear transformation $y_i = Ax_i$.

Asking for independence in the set of components leads to the assumption for the latent variables to have non-Gaussian distributions. Therefore, the process of estimating the independent components is translated into having as an objective function a measure of non-Gaussianity. Originally from the information theory field, the negentropy of a random variable $\mathcal{Y}$ is used as a non-Gaussianity measure and is defined by $J(\mathcal{Y}) = H(\mathcal{Y}_{gauss}) - H(\mathcal{Y})$, where $\mathcal{Y}_{gauss}$ is a Gaussian variable of the same mean and covariance as $\mathcal{Y}$ and H is the entropy of the variable $(H(\mathcal{Y}) = -\int f(\mathcal{Y}) \log f(\mathcal{Y}) \, d\mathcal{Y}$, with $f(\mathcal{Y})$ being its density function).

Gaussian variables have maximum entropy among all variables of the same variance. Thus, independent components will be found in the directions of maximum negentropy. In [8], the authors propose FastICA, an efficient algorithm that finds independent components by maximizing with respect to an approximated negentropy function, after finding uncorrelated components of unit variance (whitening of data).

4   Experimental Results

We will evaluate the above described dimensionality reduction techniques on different graph datasets by classification rates of a kNN classifier on the reduced vectors. We will compare the results with the results of the classification of the original vectors before being reduced.

4.1   Databases

We have chosen three different datasets from the IAM Graph Database Repository [9], describing both synthetic and real data. The *Letter Database* represents distorted letter drawings. Starting from a manually constructed prototype of every of the 15 Roman alphabet letters that consist only of straight lines, different degrees of distortion have been applied. Each ending point of the lines is attributed with its (x, y) coordinates. The second graph dataset is the *GREC Database*, which represents architectural and electronic drawings under different levels of noise. In this database, intersections and corners constitute the set of nodes. These nodes are attributed with their position on the 2-dimensional plane.

Finally, the *Fingerprint Database* consists of graphs that are obtained from a set of fingerprint images by means of particular image processing operations. Ending point and bifurcations of the skeleton of the processed images constitute the (x, y) attributed nodes of the graphs.

4.2   Experimental Setup

All three described datasets are divided into a training set, a validation set and a test set. All the experiments are done by tuning the involved parameters on the validation set and then testing with the best configuration on the test set. The considered parameters are:

- The size of the vocabulary in the graph of words construction.
- The specific parameters of the dimensionality reduction methods.
- The classification parameters (number of neighbors in the kNN classifier).

Starting from 3, the number of words in the vocabulary has been increased until 100 (in steps of 3). This way, a large range of different configurations is obtained in terms of the behavior of the dimensionality reduction techniques. For every one of these chosen sizes of the vocabulary we perform the embedding and then reduce the dimensions. This forces us to adopt a criterion by which we keep a certain number of dimensions.

Independently of the validation set, in kPCA, we use an energy parameter. Once the vectors are projected into the new set of features, they are arranged from the most to the least important components (in the leading eigenvectors order). The energy of each component is the ratio between the cumulative amount of weight from the first component (the leading one) to itself, and the total amount of weight in the training vectors. Then, by setting a certain amount of energy, we keep those components that are enough to retain such energy. The energy parameter has been set to 0.5, 0.75, 0.9, 0.95, 0.97 and 0.99. In Figure 2, we show how the dimensions using kPCA are drastically reduced compared to the original ones (recall the original dimensions are a quadratic function of the vocabulary size, see equation (4)). Using a 99% degree of energy we obviously keep more information (*i.e.* more components) than using lower ratios. However, the number of dimensions of the resulting vectors with the energy parameter set to 99% is still much less than the dimensionality of the original vectors. For this reason, we use this degree of energy for testing.

In the case of ICA, the criterion is somewhat similar to the one used in kPCA. In this case, however, instead of a cumulative energy, we just keep those components whose importance exceed a certain level. In FastICA, a threshold ξ is set for the eigenvalues of the covariance matrix in the whitening process. After some experimentation, we use $\xi = 10^{-7}$; certainly a low value that generates some more components than kPCA.

In kPCA, we also have to optimize the γ parameter for the kernel function (equation (6)). We took a logarithmic scale for this parameter from 10^{-2} until 10^{2}. In Figure 3 we show the effect of this parameter for the three datasets on the validation set (in terms of accuracy rates), as a function of the number of words

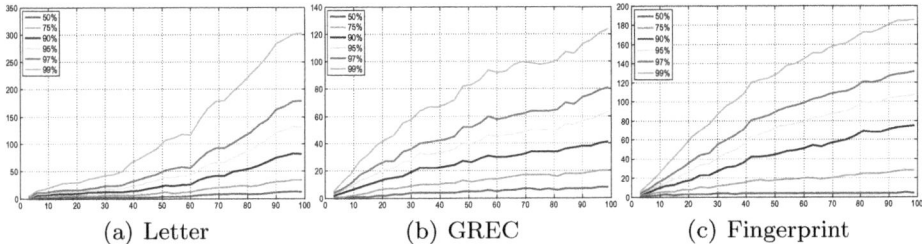

Fig. 2. Kept dimensions (vertical axis) for every vocabulary size (horizontal axis) for different amounts of energy. Results on the validation set for a Gaussian kernel of $\gamma = 1$ in the kPCA reduction method.

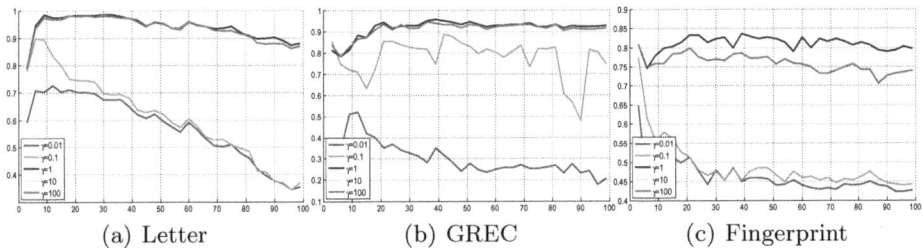

Fig. 3. Accuracy rate on the validation set (vertical axis) for every vocabulary size (horizontal axis). Each line represents a different value for the γ parameter in the Gaussian kernel. The energy rate is set to be 99%. The yellow curve ($\gamma = 10$) is always under the magenta curve ($\gamma = 100$).

in the vocabulary. It seems clear that low values such as 10^{-2} and 10^{-1} always give poor results. High values like 10 or 100 perform always the same (there is no significant difference at these levels). These high values of the parameter are better than the low ones but still not as good as the value $\gamma = 1$, which is the one obtaining the better results while being the most stable. This is why, from here on, the parameter of the kernel is set to be equal to 1.

Finally, in the case of the kNN classifier, different numbers of neighbors have been tried in conjunction with the Euclidean metric.

4.3   Results

The results on the validation set as a function of the number of words in the vocabulary are shown in Figure 4. We show the results of the classification of the original vectors using a kNN classifier, as well as both the reduced vectors by kPCA and ICA.

kPCA performs fairly well in all cases. For the Letter dataset, the tendency is pretty similar to the original vectors, losing accuracy after about 40 words. In the GREC case, between both curves (the original vectors and kPCA) there is never any significant difference, which suggests the kPCA vectors to be used

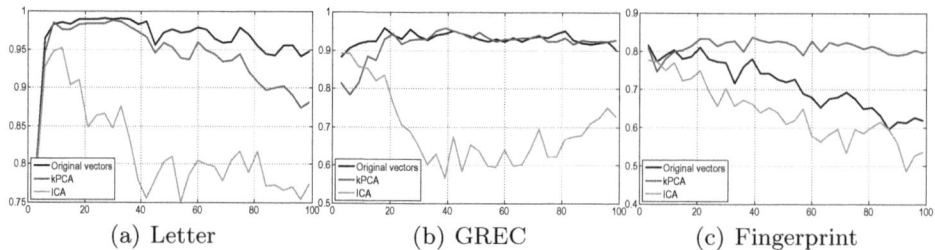

(a) Letter (b) GREC (c) Fingerprint

Fig. 4. Accuracy rate on the validation set (vertical axis) for every vocabulary size (horizontal axis). These figures depict the comparison between the classification of the original vectors (without reduction) with the two dimensionality reduction techniques, kPCA and ICA.

Table 1. Results on the test set for the original vectors and both the kPCA and ICA reduced vectors. Accuracy rates (AR) are shown in %. The dimensionality of the vectors of every system is also shown (Dims).

	Letter		GREC		Fingerprint	
	AR	Dims	AR	Dims	AR	Dims
Original vectors	**98.8**	702	**97.5**	3402	77.7	189
kPCA	97.6	43	97.1	67	**80.6**	108
ICA	82.8	251	58.9	218	63.3	184

since the dimensions are much less, reducing, this way, memory needs and computational complexity in further processing or treatment of the vectors. In the Fingerprint database case, the situation is even better. The kPCA reduced vectors drastically outperform the original features, removing any redundancy or noise in the original vectors and gaining accuracy.

In the case of ICA, the situation is not the desired one and the results advise not to go on this direction. For both the Letter and GREC databases, the performance of ICA rapidly decreases, making its situation not comparable to the other cases. In the case of the Fingerprint graphs, ICA follows a similar behavior as the original vectors, but it still keeps losing accuracy when compared to them.

As already explained in Section 3.2, ICA tries to find the linear transformation from the original vectors to the reduced ones by optimizing the parameters of the transformation in such a way that the new features are uncorrelated and statistically independent. In the case of the graph of words vectors, this seems to be a hard task. Components of the original graph of words vectors are strongly correlated and they hardly depend one on another. The higher the appearance of a specific word, the more relation between this word and others will exist in the final vector representation. Such situation makes ICA not to properly discover important and distinctive components and, therefore, the accuracy is being negatively affected.

Finally, the results on the test set are given in Table 1. We take the best parameters configuration on the validation set and classify the test set under this setup. As expected, ICA performs poorly with respect to the other systems. kPCA, however, takes similar results as the original vectors while reducing drastically the dimensionality of the vectors. In one out of the three databases, kPCA outperforms the original vectors classification.

5 Conclusions

In this article we have addressed the dimensionality and sparsity problem for the graph of words embedding. Kernel principal component analysis (kPCA) and independent component analysis (ICA) have been applied to three different databases of graphs. The experimental results show that kPCA is a proper way to reduce the dimensionality of the vectors since it is able to remain in the same classification accuracy levels while drastically reducing the number of features in the vectors. ICA, on the other hand, does not fit for graph of words embedded vectors since their components are strongly dependent and correlated. Future work is in the direction of other kernel functions for kPCA to check the dependency of the kernel for every database.

References

1. Conte, D., Foggia, P., Sansone, C., Vento, M.: Thirty years of graph matching in pattern recognition. International Journal of Pattern Recognition and Artificial Intelligence 18(3), 265–298 (2004)
2. Luo, B., Wilson, R.C., Hancock, E.R.: Spectral embedding of graphs. Pattern Recognition 36(10), 2213–2230 (2003)
3. Robles-Kelly, A., Hancock, E.R.: A Riemannian approach to graph embedding. Pattern Recognition 40(3), 1042–1056 (2007)
4. Riesen, K., Bunke, H.: Graph Classification and Clustering Based on Vector Space Embedding. World Scientific, Singapore (2010)
5. Gibert, J., Valveny, E., Bunke, H.: Graph of Words Embedding for Molecular Structure-Activity Relationship Analysis. In: Bloch, I., Cesar Jr., R.M. (eds.) CIARP 2010. LNCS, vol. 6419, pp. 30–37. Springer, Heidelberg (2010)
6. Dance, C., Willamowski, J., Fan, L., Bray, C., Csurka, G.: Visual categorization with bags of keypoints. In: ECCV International Workshop on Statistical Learning in Computer Vision, pp. 1–22 (2004)
7. Schölkopf, B., Smola, A., Müller, K.R.: Nonlinear Component Analysis as a Kernel Eigenvalue Problem. Neural Computation 10, 1299–1319 (1998)
8. Hyvärinen, A., Oja, E.: Independent Component Analysis: algorithms and applications. Neural Networks 13, 411–430 (2000)
9. Riesen, K., Bunke, H.: IAM Graph Database Repository for Graph Based Pattern Recognition and Machine Learning. In: da Vitoria Lobo, N., Kasparis, T., Roli, F., Kwok, J.T., Georgiopoulos, M., Anagnostopoulos, G.C., Loog, M. (eds.) S+SSPR 2008. LNCS, vol. 5342, pp. 287–297. Springer, Heidelberg (2008)

Entropy versus Heterogeneity for Graphs

Lin Han, Edwin R. Hancock, and Richard C. Wilson

Department of Computer Science, University of York

Abstract. In this paper we explore and compare two contrasting graph characterizations. The first of these is Estrada's heterogeneity index, which measures the heterogeneity of the node degree across a graph. Our second measure is the the von Neumann entropy associated with the Laplacian eigenspectrum of graphs. Here we show how to approximate the von Neumann entropy by replacing the Shannon entropy by its quadratic counterpart. This quadratic entropy can be expressed in terms of a series of permutation invariant traces, which can be computed from the node degrees in quadratic time. We compare experimentally the effectiveness of the approximate expression for the entropy with the heterogeneity index.

1 Introduction

One of the key problems that arises in the analysis on non-vectorial pattern data such as strings, trees and graphs is how to succinctly characterize such data for the purposes of clustering and classification. Unlike pattern vectors, when the analysis of tree or graph data is attempted then there is frequently no labelling or ordering of the nodes of the structure to hand.

Broadly speaking, there are three ways by which to overcome this problem. The first is to extract characteristics from the graph or tree data to-hand, and then to cluster graphs on the basis of vectors of structural characteristics [12]. The second method is to use a measure of pairwise distance between structures and resort to pairwise clustering methods [16]. The third method involves constructing a class prototype through the union or intersection of different structures[25] [19][20]. These latter two methods can prove very time consuming and even fragile since they require reliable node correspondences to hand [23][24], and this invariably requires inexact graph matching over the dataset to hand.

It is for this reason that the use of graph characteristics has proved to be an attractive one. Although there are a number of simple alternatives that can be used, such as node or edge frequency, edge density, diameter and perimeter, these have proved to be ineffective as a means of characterizing variations in intrinsic structure. Instead, it has proved necessary to resort to more complex representations. One of the most successful of these has been to use graph-spectral methods [21][22]. Here the distribution of the eigenvalues and eigenvectors can be used to construct permutation invariants that do not require node correspondences. Examples here include Laplacian spectra and characteristic polynomials. This study has recently been taken one step further by Xiao, Wilson and Hancock

X. Jiang, M. Ferrer, and A. Torsello (Eds.): GbRPR 2011, LNCS 6658, pp. 32–41, 2011.

[12] who have performed an analysis of the heat kernel for graphs, and have shown that the Riemann zeta function can be used to generate a number of powerful invariants from the normalized Laplacian spectrum. Another route to a unary characterization of graph structure is to define measures of intrinsic complexity. The characterization of graph complexity is a long standing problem, but recently measures based on the heat kernel have proved effective, and these include the use of Birkoff polytopes [18] and heat-flow complexity [17].

Unfortunately, both graph-spectral and heat flow complexity methods can prove computationally burdensome. The reason for this is that the computation of the graph-spectrum is cubic in the number of nodes. A much simpler alternative is the heterogeneity index recently developed by Estrada[1], who defines the heterogeneity based on simple statistics for the distribution of node degree over all pairs of linked nodes. This heterogeneity index can be expressed as a quadratic form of the Laplacian matrix of the graphs, which allows a spectral representation of graph heterogeneity.

Our aim in this paper is to explore whether more efficient complexity characterizations similar in spirit to the heterogeneity plot can be used to characterize differences in graph structure. We commence from the von Neumann entropy of a graph. This is simply the Shannon entropy associated with the spectrum of the normalized Laplacian matrix. We explore how to simplify and approximate the calculation of von Neumann entropy. Our first step is to replace the Shannon entropy by its quadratic counterpart. An analysis of the quadratic entropy reveals that it can be computed from a number of permutation invariant matrix trace expressions. This leads to a simple expression for the approximate entropy in terms of the node-degree. The expression is quadratic in the number of nodes in a graph. We compare our approximate entropy measure with Estrada's heterogeneity index. In the experiment part, we investigate whether the proposed entropy expression and the heterogeneity index are effective on graph clustering and classification tasks. We also compare how the heterogeneity H plots characterize three different kinds of graphs.

2   Graph Representation and the von Neumann Entropy

To commence, we denote the graph under study by $G = (V, E)$ where V is the set of nodes and $E \subseteq V \times V$ is the set of edges. Further, we represent the structure of the graph using a $|V| \times |V|$ adjacency matrix whose elements are

$$A(u, v) = \begin{cases} 1 & \text{if } (u, v) \in E , \\ 0 & \text{otherwise .} \end{cases} \tag{1}$$

The degree matrix of graph G is a diagonal matrix D whose elements are given by $D(u, u) = d_u = \sum_{v \in V} A(u, v)$. From the degree matrix and the adjacency matrix we can construct the Laplacian matrix $L = D - A$, i.e. the degree matrix minus the adjacency matrix. The elements of the Laplacian matrix are

$$L(u, v) = \begin{cases} d_v & \text{if } u = v , \\ -1 & \text{if } (u, v) \in E , \\ 0 & \text{otherwise .} \end{cases} \tag{2}$$

The normalized Laplacian matrix is given by $\hat{L} = D^{-1/2}LD^{-1/2}$ and has elements

$$\hat{L}(u,v) = \begin{cases} 1 & \text{if } u = v \text{ and } d_v \neq 0 \text{ ,} \\ -\frac{1}{\sqrt{d_u d_v}} & \text{if } (u,v) \in E \text{ ,} \\ 0 & \text{otherwise .} \end{cases} \tag{3}$$

The spectral decomposition of the normalized Laplacian matrix is $\hat{L} = \Phi\Lambda\Phi^T$ where $\Lambda = diag(\lambda_1, \lambda_2, ..., \lambda_{|V|})$ is a diagonal matrix with the ordered eigenvalues as elements $(0 = \lambda_1 < \lambda_2 < ... < \lambda_{|V|})$ and $\Phi = (\phi_1|\phi_2|...|\phi_{|V|})$ is a matrix with the corresponding ordered orthonormal eigenvectors as columns. The normalized Laplacian matrix is positive semi-definite and so has all eigenvalues non-negative. The number of zero eigenvalues is the number of connected components in the graph. For a connected graph, there is only one eigenvalue which is equal to zero. The normalization factor means that the largest eigenvalue is less than or equal to 2, with equality only when G is bipartite. Hence all the eigenvalues of the normalize Laplacian matrix are in the range $0 \leq \lambda \leq 2$. The normalized Laplacian matrix is commonly used as a graph representation and the eigenvector ϕ_2 associated with the smallest non-zero eigenvalues λ_2 referred to as the Fiedler-vector [9] is often used in graph cuts [10][11].

The von Neumann entropy of the graph associated with the Laplacian eigenspectrum is defined as [14]

$$S = -\sum_{v=1}^{|V|} \frac{\lambda_v}{2} \ln \frac{\lambda_v}{2} . \tag{4}$$

We approximate the entropy $-\frac{\lambda_v}{2} \ln \frac{\lambda_v}{2}$ by the quadratic entropy $\frac{\lambda_v}{2}(1 - \frac{\lambda_v}{2})$, to obtain

$$S = -\sum_v \frac{\lambda_v}{2} \ln \frac{\lambda_v}{2} \simeq \sum_v \frac{\lambda_v}{2}(1 - \frac{\lambda_v}{2}) = \frac{\sum_v \lambda_v}{2} - \frac{\sum_v \lambda_v^2}{4} . \tag{5}$$

Using the fact that $Tr[\hat{L}^n] = \sum_v \lambda_v^n$, the quadratic entropy can be rewritten as

$$S = \frac{Tr[\hat{L}]}{2} - \frac{Tr[\hat{L}^2]}{4} . \tag{6}$$

Since the normalized Laplacian matrix $\hat{L}$ is symmetric and it has unit diagonal elements, then according to equation (3) for the trace of the normalized Laplacian matrix, we have

$$Tr[\hat{L}] = |V| . \tag{7}$$

Similarly, for the trace of the square of the normalized Laplacian, we have

$$Tr[\hat{L}^2] = \sum_{u \in V} \sum_{v \in V} \hat{L}_{uv} \hat{L}_{vu} = \sum_{u \in V} \sum_{v \in V} (\hat{L}_{uv})^2$$

$$= \sum_{\substack{u,v \in V \\ u=v}} (\hat{L}_{uv})^2 + \sum_{\substack{u,v \in V \\ u \neq v}} (\hat{L}_{uv})^2$$

$$= |V| + \sum_{(u,v) \in E} \frac{1}{d_u d_v} \ . \tag{8}$$

Substituting Equation (7) and (8) into Equation (6), the entropy becomes

$$S = \frac{|V|}{2} - \frac{|V|}{4} - \sum_{(u,v) \in E} \frac{1}{4 \, d_u d_v} = \frac{|V|}{4} - \sum_{(u,v) \in E} \frac{1}{4 \, d_u d_v} \ . \tag{9}$$

As a result, we can approximate the von Neumann entropy using two measures of graph structure. The first is the number of nodes of the graph, while the second is the degree of the nodes of the graph. The approximation bypasses calculating the Laplacian eigenvalues of a graph to estimate its von Neumann entropy.

3 Graph Heterogeneity Index and H Plot

We now turn our attention to network heterogeneity index recently developed by Estrada[1]. To develop the heterogeneity index, Estrada commences by defining a local index which measures the irregularity of an edge in the graph $(u,v) \in E$ as

$$I_{uv} = [f(d_u) - f(d_v)]^2 \ , \tag{10}$$

where $f(d_u)$ is a function of the node degree. Selecting $f(d_u) = d_u^{-1/2}$, the heterogeneity index proposed is defined to be the sum of the irregularity of all edges in the graph,

$$\rho'(G) = \sum_{(u,v) \in E} (d_u^{-1/2} - d_v^{-1/2})^2 \ . \tag{11}$$

The main advantage of defining the index as the sum of square differences of a function of node degree is that the index can be expressed in terms of a quadratic form of the Laplacian matrix of the graph. That is, let $|\mathbf{d}^{-1/2}\rangle = (d_1^{-1/2}, d_2^{-1/2}, ..., d_{|V|}^{-1/2})$ represent a column vector where d_u is the degree of the node u, the index can be written as

$$\rho'(G) = \sum_{(u,v) \in E} (d_u^{-1/2} - d_v^{-1/2})^2 = \frac{1}{2} \langle \mathbf{d}^{-1/2} | L | \mathbf{d}^{-1/2} \rangle \ . \tag{12}$$

The index above can also be stated in terms of the *Randić index* $^1R_{-1/2}$[5] of the graph,

$$\rho'(G) = \sum_{(u,v)\in E} (d_u^{-1/2} - d_v^{-1/2})^2 = |V| - 2\sum_{(u,v)\in E} (d_u d_v)^{-1/2} = |V| - 2\,^1R_{-1/2} \ .$$

(13)

Li and Shi [2] show that for connected graphs the *Randić index* is bounded as follows

$$\sqrt{|V|-1} \leq\, ^1R_{-1/2} \leq \frac{|V|}{2} \ ,$$

(14)

where the lower bound is attained for star graphs and the upper bound is attained for regular graphs with $|V|$ nodes. Then the normalized heterogeneity index is defined as

$$\rho(G) = \frac{|V| - 2\,^1R_{-1/2}}{|V| - 2\sqrt{|V|-1}} = \frac{\displaystyle\sum_{(u,v)\in E} (d_u^{-1/2} - d_v^{-1/2})^2}{|V| - 2\sqrt{|V|-1}}$$

$$= \frac{1}{|V| - 2\sqrt{|V|-1}} \sum_{(u,v)\in E} (\frac{1}{d_u} + \frac{1}{d_v} - \frac{2}{\sqrt{d_u d_v}}) \ .$$

(15)

This is zero for regular graphs and one for star graphs, i.e., $0 \leq \rho(G) \leq 1$. Then heterogeneous starlike graphs are expected to have values of $\rho(G)$ close to one. On the other hand, more regular graphs are expected to have values close to zero.

Finally it is interesting to note that Von Luxburg [8] has shown that $1/d_u + 1/d_v$ is proportional to the commute time (or resistance distance) between nodes for graphs of large degree. Recall that commute time is the average of the outward hitting time and return hitting time, over all paths connecting a pair of nodes. It hence provides a non-local index of connectivity between pairs of nodes, which is non-zero even if there is no connecting edge. Apart from commute time term and constants related to the size of the graph, whereas the entropy depends on $-\sum_{(u,v)\in E} 1/d_u d_v = -\sum_{(u,v)\in E} \hat{L}_{u,v}^2$, the heterogeneity index depends on $-\sum_{(u,v)\in E} 2/\sqrt{d_u d_v} = 2\sum_{(u,v)\in E} \hat{L}_{u,v}$. Hence, the heterogeneity contains measures of both global path length distribution via commute time, and local edge structure via the elements of the normalised Laplacian. The entropy on the other hand is based only on the latter.

Using the Euler theorem [3] the *Randić index* can be expressed as follows

$$^1R_{-1/2} = \frac{1}{2}[\ |V| - \frac{1}{^0R_{-1}} \sum_{v=2}^{|V|} \lambda_v \cos^2 \theta_v] \ ,$$

(16)

we can also establish a link between the normalized heterogeneity index and the spectral representation of graphs

$$\rho(G) = \frac{^0R_{-1}}{|V| - 2\sqrt{|V|-1}} \sum_{v=1}^{|V|} x_v^2 \ ,$$

(17)

where $x_v = \sqrt{\lambda_v}\ \cos\ \theta_v$. λ_v is the vth eigenvalue of the Laplacian matrix of the graph and θ_v is the angle between the orthonormal eigenvector ϕ_v associated with the eigenvalue λ_v and the vector $\mathbf{d}^{-1/2}$ previously defined. Then the $\rho(G)$ can be interpreted as the sum of the squares of the projection of $\sqrt{\lambda_v}\phi_v$ onto the vector $\mathbf{d}^{-1/2}$. Define $y_v = \sqrt{\lambda_v}\ \sin\ \theta_v$, we can represent a graph by plotting x_v vs y_v for all values of v, where the heterogeneity is given by the sum of the squares of the projections of all these points on the abscissa. All the projections on y axis are positive but those on x axis can have positive and negative values. These plots are referred to as heterogeneity plots or H plots.

4   Experiments

In this section, we provide some comparative experimental evaluation of the approximate von-Neumann entropy and the heterogeneity index on both real-word dataset and synthetic dataset. The real-world dataset used is the COIL dataset[15] which consists of images of different views of several objects, with 72 views of each object from equally spaced directions over $360°$. We extract corner features from each image and use the detected feature points as nodes to construct sample graphs by Delaunay triangulation. The synthetic dataset consists of Erdös-Rényi random graphs[6] generated by connecting pairs of nodes in a graph with an equal probability p ($0 \leq p \leq 1$) and scale-free graphs whose degree distribution follows the power-law distribution, the scale-free graphs here are generated with the preferential attachment algorithm of Barabási and Albert[7].

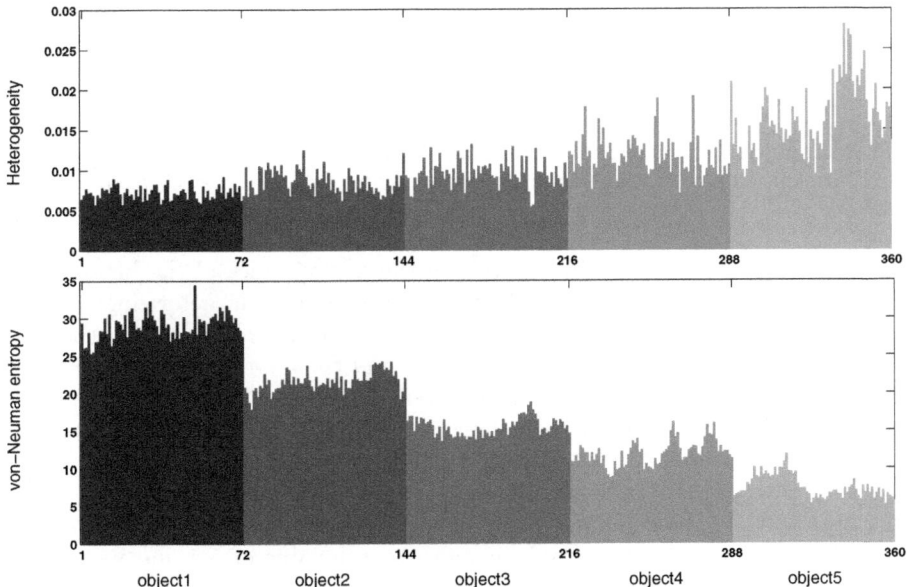

Fig. 1. (top row) The values of the heterogeneity of the Delaunay graphs. (bottom row) The values of the approximate von-Neumann entropy of the Delaunay graphs.

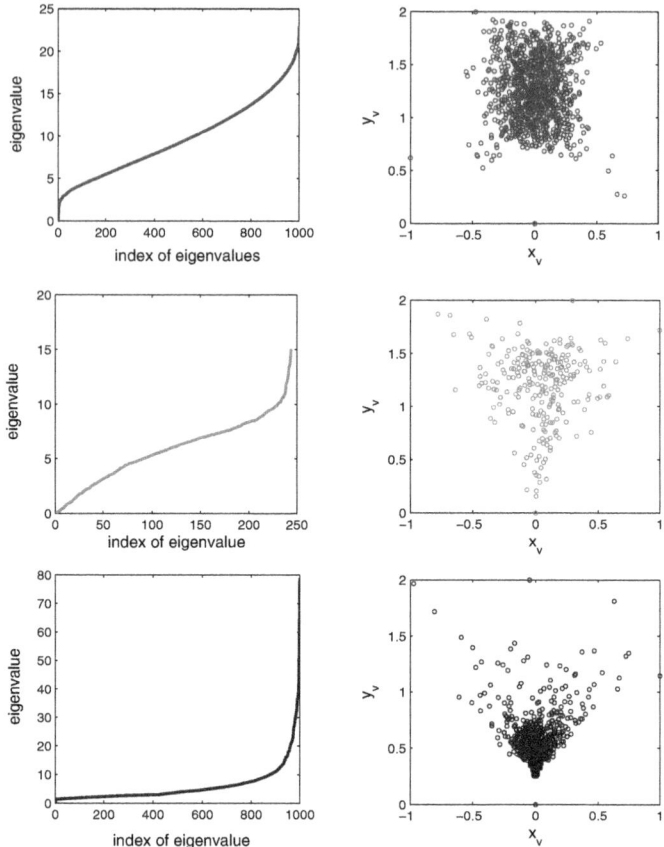

Fig. 2. Laplacian eigenvalue distributions(left column) and H plots(right column) of the ER graph (top row), the Delaunay graph(middle row) and the BA scale-free graph (bottom row)

We commerce our study by comparing the performance of the heterogeneity index and the approximate von-Neumann entropy on characterizing Delaunay graphs from the COIL dataset. To do this, we select 5 objects from the COIL dataset and plot the values of heterogeneity and the approximate von Neumann entropy for all Delaunay graphs from the 5 objects in Figure 1. Figure 1(top row) shows the values of the heterogeneity of the graphs where we use different colors to represents different objects. The values of the heterogeneity of the graphs are very small, indicating the structure of Delaunay graphs are close in structure to that of a regular graph. This can be explained by the fact that the node degree of a Delaunay graph has an average value of 5.5 and its variation is no more than 3. However, the values of the heterogeneity heavily overlap between graphs from different objects. Thus the heterogeneity index can not be used to distinguish Delaunay graphs from different objects. On the other hand, in Figure 1(bottom row), the values of the von-Neumann entropy exhibit good

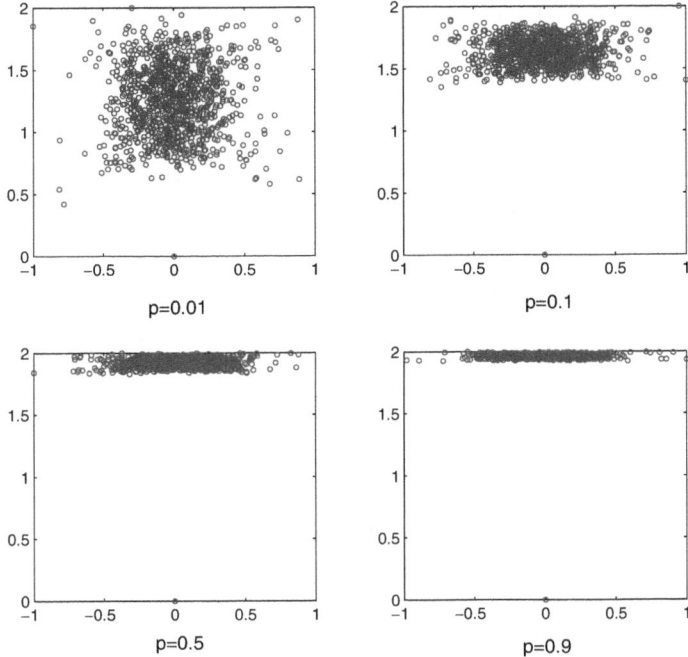

Fig. 3. H plot changes as the p value increases

separation for different objects. When we apply a 3-nearest neighbor classifier to the heterogeneity of the Delaunay graphs, its average classification rate computed using 10-fold cross-validation is 58%, much lower than the classification rate of the approximate von-Neumann entropy which is 92%.

We now turn our attention to comparing the spectral heterogeneity H plot for three kinds of graphs, namely, ER graphs, Delaunay graphs and scale-free graphs. In Figure 2 we illustrate the spectral H plots for an ER graph, a scale-free graph and a Delaunay graph as well as their Laplacian eigenvalue distributions. The ER graph here has a node-pair connecting probability $p = 0.01$, the node number of the ER graph and the scale-free graph is 1000 and that of the Delaunay graph is 250.

The left column in Figure 2 shows the Laplacian eigenvalue distributions of the three graphs. Zhang et al.[4] have observed that for the BA scale-free networks and ER random-graph networks, the Laplacian eigenvalue curves are similar to their node-degree curves. Our result is consistent with their observation by showing that the eigenvalue of the ER graph (top row) has a Poisson-like distribution, while that of the BA scale-free graph (bottom row) has a power-law distribution. Besides, we observe that eigenvalue of the Delaunay graph exhibits a similar distribution to that of the ER graph. The right column in Figure 2 shows the H plots x_v vs y_v for the three graphs where we normalize the values on the x axis between -1 and 1, and those of the y axis between 0 and 2 to have similar length in both scales. We observe that the H plot for the ER graph with $p = 0.01$ has a regular distribution of the points with an almost squared shape. In the case of the Delaunay graph, the H plot is characterized by an inverted triangle shape. The H plot for the BA scale-free graph exhibits a similar shape to that of the Delaunay graph, whereas most of the points in the plot are distributed around (0,0.2).

From the Figure 2 it is clear that the H plot for ER graph with $p=0.01$ has a square shape, we now investigate whether it holds for all ER graphs. To this end, we increase the value of node-pair connecting probability p from 0.01 to 1 and generate ER graphs with different p values. Figure 3 shows the H plot for four ER graphs whose p values are respectively 0.01, 0.1, 0.5 and 0.9. It is clear from Figure 3 that the shape of the H plot for ER graphs has a square shape when the value of p is very small. As p increases, the distribution of the value of y_v becomes condensed and the shape of the H plot becomes rectangle and finally closes to a line.

5   Conclusion

In this paper we show how to use the von Neumann entropy computed from the Laplacian eigenspectrum to characterize graphs. We approximate the Shannon term in the definition of the von Neumann entropy in a quadratic manner. This approximation leads to an expression for the von Neumann entropy in terms of the number of nodes and node degrees. In our experiments, we compare the von Neumann entropy measure with the heterogeneity index on an object classification task and show its effectiveness. Experimental results also reveal that the spectral heterogeneity H plot for the Delaunay graphs exhibits an inverted triangle shape and that of the ER graph tends to a rectangle shape as we increase the node-pair connecting probability p.

References

1. Estrada, E.: Quantifying network heterogeneity. E. Estrada. Quantifying network heterogeneity. Physical Review E 82, 66102 (2010)
2. Li, X., Shi, Y.: A Survey on the Randic Index. MATCH: Communication in Mathematical and in Computer Chemistry 59, 127–156 (2008)
3. Hohn, F.E.: Elementary matrix algebra. Dover, New York (1973)
4. Zhan, C., Chen, G., Yeung, L.F.: On the distributions of Laplacian eigenvalues versus node degrees in complex networks. Physica A 389, 1779–1788 (2010)
5. Randić, M.: Characterization of molecular branching. Journal of the American Chemical Society 97, 6609–6615 (1975)
6. Erdös, P., Rényi, A.: On Random Graphs. Publicationes Mathematicae 6, 290–297 (1959)
7. Barabási, A.L., Albert, R.: Emergence of scaling in random networks. Science 286, 509–512 (1999)
8. Maier, M., von Luxburg, U., Hein, M.: Influence of Graph Construction on Graph-Based Clustering Measures. In: NIPS, pp. 1–9 (2010)

9. Chung, F.R.K.: Spectral Graph Theory. American Mathematical Society, Providence (1997)
10. Robles-Kelly, A., Hancock, E.R.: A Riemannian Approach to Graph Embedding. Pattern Recognition, 1042–1056 (2007)
11. Shi, J., Malik, J.: Normalized Cuts and Image Segmentation. In: CVPR, pp. 731–737 (1997)
12. Xiao, B., Hancock, E.R., Wilson, R.C.: Graph Characteristic from the Heat Kernel Trace. Pattern Recognition 42, 2589–2606 (2009)
13. McKay, B.D.: Spanning Trees in regular Graphs. European Journal of Combinatorics 4, 149–160 (1983)
14. Passerini, F., Severini, S.: The von Neumann entropy of networks. arXiv:0812.2597 (2008)
15. Nene, S.A., Nayar, S.K., Murase,H.: Columbiaobjectimagelibrary(coil100). Technical Report,Department of Computer Science, Columbia University (1996)
16. Torsello, A., Robles-Kelly, A., Hancock, E.R.: Discovering Shape Classes using Tree Edit-Distance and Pairwise Clustering. IJCV 72(3), 259–285 (2007)
17. Escolano, F., Lozano, M.A., Hancock, E.R., Giorgi, D.: What is the complexity of a network? The heat flow-thermodynamic depth approach. In: Hancock, E.R., Wilson, R.C., Windeatt, T., Ulusoy, I., Escolano, F. (eds.) SSPR&SPR 2010. LNCS, vol. 6218, pp. 286–295. Springer, Heidelberg (2010)
18. Escolano, F., Hancock, E.R., Lozano, M.A.: Polytopal graph complexity, matrix permanents, and embedding. In: da Vitoria Lobo, N., Kasparis, T., Roli, F., Kwok, J.T., Georgiopoulos, M., Anagnostopoulos, G.C., Loog, M. (eds.) S+SSPR 2008. LNCS, vol. 5342, pp. 237–246. Springer, Heidelberg (2008)
19. Torsello, A., Hancock, E.R.: Graph Embedding Using Tree Edit-union. Pattern Recognition 40(5), 1393–1405 (2007)
20. Suau, P., Escolano, F.: Bayesian Optimization of the Scale Saliency Filter. Image and Vision Computing 26(9), 1207–1218 (2008)
21. Luo, B., Wilson, R.C., Hancock, E.R.: A Spectral Approach to Learning Structural Variations in Graphs. Pattern Recognition 39(6), 1188–1198 (2006)
22. Wilson, R.C., Hancock, E.R., Luo, B.: Pattern Vectors from Algebraic Graph Theory. IEEE PAMI 27(7), 1112–1124 (2005)
23. Ferrer, M., Valveny, E., Serratosa, F., Bunke, H.: Exact median graph computation via graph embedding. In: da Vitoria Lobo, N., Kasparis, T., Roli, F., Kwok, J.T., Georgiopoulos, M., Anagnostopoulos, G.C., Loog, M. (eds.) S+SSPR 2008. LNCS, vol. 5342, pp. 15–24. Springer, Heidelberg (2008)
24. Ferrer, M., Serratosa, F., Valveny, E.: On the relation between the median and the maximum common subgraph of a set of graphs. In: Escolano, F., Vento, M. (eds.) GbRPR. LNCS, vol. 4538, pp. 351–360. Springer, Heidelberg (2007)
25. Riesen, K., Neuhaus, M., Bunke, H.: Graph embedding in vector spaces by means of prototype selection. In: Escolano, F., Vento, M. (eds.) GbRPR. LNCS, vol. 4538, pp. 383–393. Springer, Heidelberg (2007)

Learning Generative Graph Prototypes Using Simplified von Neumann Entropy

Lin Han, Edwin R. Hancock, and Richard C. Wilson

Department of Computer Science, University of York

Abstract. We present a method for constructing a generative model for sets of graphs by adopting a minimum description length approach. The method is posed in terms of learning a generative supergraph model from which the new samples can be obtained by an appropriate sampling mechanism. We commence by constructing a probability distribution for the occurrence of nodes and edges over the supergraph. We encode the complexity of the supergraph using the von-Neumann entropy. A variant of EM algorithm is developed to minimize the description length criterion in which the node correspondences between the sample graphs and the supergraph are treated as missing data.The maximization step involves updating both the node correspondence information and the structure of supergraph using graduated assignment. Empirical evaluations on real data reveal the practical utility of our proposed algorithm and show that our generative model gives good graph classification results.

1 Introduction

The main obstacle to learning in the graph domain is solving the problem of how to capture variability in graph or tree structure. The main reason for the lack of progress in this area is the difficulty in developing representations that can capture variations in graph-structure. This variability can be attributed to a) variations in either node or edge attributes, b) variations in node or edge composition and c) variations in edge-connectivity. This trichotomy provides a natural framework for analyzing the state-of-the-art in the literature. Most of the work on Bayes nets in the graphical models literature can be viewed as modeling variations in node or edge attributes [1]. Examples also include the work of Christmas et al.[2] and Bagdanov et al. [3] who both use Gaussian models to capture variations in edge attributes. The problems of modeling variations in node and edge composition are more challenging since they focus on modeling the structure of the graph rather than its attributes.

The problem of learning edge structure is probably the most challenging of those listed above. Broadly speaking there are two approaches to characterizing variations in edge structure for graphs. The first of these is graph spectral, while the second is probabilistic. In the case of graph spectra, many of the ideas developed in the generative modeling of shape using principal components analysis can be translated relatively directly to graphs using simple vectorization procedures

X. Jiang, M. Ferrer, and A. Torsello (Eds.): GbRPR 2011, LNCS 6658, pp. 42–51, 2011.

based on the correspondences conveyed by the ordering of Laplacian eigenvectors [4]. Although these methods are simple and effective, they are limited by the stability of the Laplacian spectrum under perturbations in graph-structure. The probabilistic approach is potentially more robust, but requires accurate correspondence information to be inferred from the available graph structure. If this is to hand, then a representation of edge structure can be learned.

In this paper, we focus on the third problem and aim to learn a generative model that can be used to describe the distribution of structural variations present in a set of sample graphs, and in particular to characterize the variations of the edge structure present in the set. We follow Torsello and Hancock [5] and pose the problem as that of learning a generative supergraph representation from which we can sample. However, their work is based on trees, and since the trees are rooted the learning process can be effected by performing tree merging operations in polynomial time. This greedy strategy does not translate tractably to graphs where the complexity becomes exponential, and we require different strategies for learning and sampling. Torsello and Hancock realize both of these using edit operations, here on the other hand we use a soft-assign method for optimization.

In prior work Han, Wilson and Hancock propose a method of learning a supergraph model in [13] which overlooks the complexity of the supergraph model. Here, we take an information theoretic approach to estimating the supergraph structure by using a minimum description length criterion. By taking into account the overall code-length in the model, MDL allows us to select a supergraph representation that trades-off goodness-of-fit with the observed sample graphs against the complexity of the model. We show how to gauge the complexity of the supergraph using von-Neumann entropy[8] (i.e. the entropy associated with the Normalized Laplacian eigenvalues), and how to efficiently approximate this entropy without the need to compute the Laplacian spectrum. We use a variant of EM algorithm to minimize the total code-length criterion, in which the correspondences between the nodes of the sample graphs and those of the supergraph are treated as missing data. In the maximization step, we update both the node correspondence information and the structure of supergraph using graduated assignment. This novel technique is applied to a large database of object views, and used to learn class prototypes that can be used for the purposes of object recognition.

2 Probabilistic Framework

We are concerned with learning a structural model represented in terms of a so-called supergraph that can capture the variations present in a sample of graphs. In Torsello and Hancock's work [5] this structure is found by merging the set of sample trees, and so each sample tree can be obtained from it by edit operations. Here, on the other hand, we aim to estimate an adjacency matrix that captures the frequently occurring edges in the training set. To commence our development we require the *a posteriori* probabilities of the sample graphs given the structure of the supergraph and the node correspondences between each sample graph and the supergraph. To compute these probabilities we use the method outlined in [6].

Let the set of sample of graphs be $\mathcal{G} = \{G_1, ...G_i, ...G_N\}$, where the graph indexed i is $G_i = (V_i, E_i)$ with V_i the node-set and E_i the edge-set. Similarly, the supergraph which we aim to learn from this data is denoted by $\Gamma = (V_\Gamma, E_\Gamma)$, with node-set V_Γ and edge-set E_Γ. Further, we represent the structure of the two graphs using a $|V_i| \times |V_i|$ adjacency matrix D^i for the sample graph G_i and a $|V_\Gamma| \times |V_\Gamma|$ adjacency matrix M for the supergraph model Γ. The elements of the adjacency matrix for the sample graph and those for the supergraph are respectively defined to be

$$D^i_{ab} = \begin{cases} 1 & \text{if } (a,b) \in E_i \\ 0 & \text{otherwise} \end{cases}, \quad M_{\alpha\beta} = \begin{cases} 1 & \text{if } (\alpha,\beta) \in E_\Gamma \\ 0 & \text{otherwise} \end{cases}. \quad (1)$$

We represent the correspondence matches between the nodes of the sample graph and the nodes of the supergraph using a $|V_i| \times |V_\Gamma|$ assignment matrix S^i which has elements

$$s^i_{a\alpha} = \begin{cases} 1 & \text{if } a \to \alpha \\ 0 & \text{otherwise} \end{cases}. \quad (2)$$

where $a \to \alpha$ implies that node $a \in V_i$ is matched to node $\alpha \in V_\Gamma$.

With these ingredients, according to Luo and Hancock [6] the *a posteriori* probability of the graphs G_i given the supergraph Γ and the correspondence indicators is

$$P(G_i|\Gamma, S^i) = \prod_{a \in V_i} \sum_{\alpha \in V_\Gamma} K^i_a \exp[\mu \sum_{b \in V_i} \sum_{\beta \in V_\Gamma} D^i_{ab} M_{\alpha\beta} s^i_{b\beta}]. \quad (3)$$

where $\mu = \ln \frac{1-P_e}{P_e}$ and $K^i_a = P_e^{|V_i| \times |V_\Gamma|} B^i_a$. In the above, P_e is the error rate for node correspondence and B^i_a is the probability of observing node a in graph G_i, the value of which depends only on the identity of the node a. $|V_i|$ and $|V_\Gamma|$ are the number of the nodes in graph G_i and supergraph Γ.

3   Model Coding Using MDL

Underpinning minimum description length is the principle that learning, or finding a hypothesis that explains some observed data and makes predictions about data yet unseen, can be viewed as finding a shorter code for the observed data [9,7]. To formalize this idea, we encode and transmit the observed data and the hypothesis, which in our case are respectively the sample graphs $\mathcal{G}$ and the supergraph structure Γ. This leads to a two-part message whose total length is given by $\mathcal{L}(\mathcal{G}, \Gamma) = LL(\mathcal{G}|\Gamma) + LL(\Gamma)$.

Encoding sample graphs: We first compute the code-length of the graph data. For the sample graph-set $\mathcal{G} = \{G_1, ...G_i, ...G_N\}$ and the supergraph Γ, the set of assignment matrices is $\mathcal{S} = \{S^1, ...S^i, ...S^N\}$ and these represent the correspondences between the nodes of the sample graphs and those of the supergraph. Under the assumption that the graphs in $\mathcal{G}$ are independent samples from the distribution, using the *a posteriori* probabilities from Section 2 the likelihood of the set of sample graphs is

$$P(\mathcal{G}|\Gamma, \mathcal{S}) = \prod_{G_i \in \mathcal{G}} P(G_i|\Gamma, S^i) = \prod_{G_i \in \mathcal{G}} \prod_{a \in V_i} \sum_{\alpha \in V_\Gamma} K_a^i \exp[\mu \sum_{b \in V_i} \sum_{\beta \in V_\Gamma} D_{ab}^i M_{\alpha\beta} s_{b\beta}^i] \ . \tag{4}$$

Instead of using the *Shannon-Fano code*, which is equivalent to the negative logarithm of the above likelihood function, we measure the code-length of the graph data using its average. Our reason is that if we adopt the former measure, then there is a bias to learning a complete supergraph that is fully connected. The reason will become clear later-on when we outline the maximization algorithm in Section 4, and we defer our justification until later. Thus, the graph code-length is $LL(\mathcal{G}|\Gamma) = -\frac{1}{|\mathcal{G}|} \sum_{G_i \in \mathcal{G}} \log P(G_i|\Gamma, S^i)$ which is the average over the set of sample graphs $\mathcal{G}$.

Encoding the supergraph model: Next, we must compute a code-length to measure the complexity of the supergraph. For two-part codes the MDL principle does not give any guideline as to how to encode the hypotheses. Hence every code for encoding the supergraph structure is allowed, so long as it does not change with the sample size N. Here the code-length for describing supergraph complexity is chosen to be measured using the von-Neumann entropy [8] $H = -\sum_k \frac{\lambda_k}{2} \ln \frac{\lambda_k}{2}$ where λ_k are the eigenvalues of the normalized Laplacian matrix of the supergraph $\hat{L}$ whose elements are

$$\hat{L}_{\alpha\beta} = \begin{cases} 1 & \text{if } \alpha = \beta \\ -\frac{1}{\sqrt{T_\alpha T_\beta}} & \text{if } (\alpha, \beta) \in E_\Gamma \\ 0 & \text{otherwise} \end{cases} \ . \tag{5}$$

where $T_\alpha = \sum_{\xi \in V_\Gamma} M_{\alpha\xi}$ and $T_\beta = \sum_{\xi \in V_\Gamma} M_{\beta\xi}$. The normalized Laplacian matrix is commonly used as a graph representation and graph cuts, and its eigenvalues are in the range $0 \leq \lambda_k \leq 2$ [11]. Divided by 2, the value of $\frac{\lambda_k}{2}$ is constrained between 0 and 1, and the von-Neumann entropy derived thereby is an intrinsic property of graphs that reflects the complexity of their structures better than other measures. We approximate the entropy $-\frac{\lambda_k}{2} \ln \frac{\lambda_k}{2}$ by the quadratic entropy $\frac{\lambda_k}{2}(1 - \frac{\lambda_k}{2})$, to obtain

$$H = -\sum_k \frac{\lambda_k}{2} \ln \frac{\lambda_k}{2} \simeq \sum_k \frac{\lambda_k}{2}(1 - \frac{\lambda_k}{2}) = \frac{\sum_k \lambda_k}{2} - \frac{\sum_k \lambda_k^2}{4} \ . \tag{6}$$

Using the fact that $Tr[\hat{L}^n] = \sum_k \lambda_k^n$, the quadratic entropy can be rewritten as $H = \frac{Tr[\hat{L}]}{2} - \frac{Tr[\hat{L}^2]}{4}$. Since the normalized Laplacian matrix $\hat{L}$ is symmetric and it has unit diagonal elements, then according to equation(5) for the trace of the normalized Laplacian matrix we have $Tr[\hat{L}] = |V_\Gamma|$. Similarly, for the trace of the square of the normalized Laplacian, we have

$$Tr[\hat{L}^2] = \sum_{\alpha \in V_\Gamma} \sum_{\beta \in V_\Gamma} \hat{L}_{\alpha\beta} \hat{L}_{\beta\alpha} = |V_\Gamma| + \sum_{(\alpha,\beta) \in E_\Gamma} \frac{1}{T_\alpha T_\beta} \ . \tag{7}$$

Then the simplified entropy becomes

$$H = \frac{|V_\Gamma|}{2} - \frac{|V_\Gamma|}{4} - \sum_{(\alpha,\beta)\in E_\Gamma} \frac{1}{4\,T_\alpha T_\beta} = \frac{|V_\Gamma|}{4} - \sum_{(\alpha,\beta)\in E_\Gamma} \frac{1}{4\,T_\alpha T_\beta} \quad . \tag{8}$$

As a result, the approximated complexity of the supergraph depends on two factors. The first is the order of supergraph, i.e. the number of nodes of the supergraph. The second is the degree of the nodes of the supergraph.

Finally, by adding together the two contributions to the code-length, the overall code-length is

$$\mathcal{L}(\mathcal{G},\Gamma) = LL(\mathcal{G}|\Gamma) + LL(\Gamma) = \tag{9}$$

$$-\frac{1}{|\mathcal{G}|} \sum_{G_i\in\mathcal{G}} \sum_{a\in V_i} \log\{ \sum_{\alpha\in V_\Gamma} K_a^i \exp[\mu \sum_{b\in V_i} \sum_{\beta\in V_\Gamma} D_{ab}^i M_{ab} s_{b\beta}^i] \} + \frac{|V_\Gamma|}{4} - \sum_{(\alpha,\beta)\in E_\Gamma} \frac{1}{4\,T_\alpha T_\beta} \quad .$$

Unfortunately, due to the mixture structure, the direct estimation of the supergraph structure M from the above code-length criterion is not tractable in closed-form. For this reason, we resort to using the expectation maximization algorithm.

4   Expectation-Maximization

Having developed our computational model which poses the problem of learning the supergraph as that of minimizing the code-length, in this section, we provide a concrete algorithm to locate the supergraph structure using our code-length criterion using expectation-maximisation. With the above likelihood function and the code-length developed in the previous section, Figueiredo and Jain's formulation of EM[14] involves maximizing

$$\Lambda^{(n+1)}(\mathcal{G}|\Gamma, \mathcal{S}^{(n+1)}) = \frac{1}{|\mathcal{G}|} \sum_{G_i\in\mathcal{G}} \sum_{a\in V_i} \sum_{\alpha\in V_\Gamma} Q_{a\alpha}^{i,(n)} \{\ln K_a^i + \mu \sum_{b\in V_i} \sum_{\beta\in V_\Gamma} D_{ab}^i M_{\alpha\beta}^{(n)} s_{b\beta}^{i,(n+1)}\}$$

$$-\frac{|V_\Gamma|}{4} + \sum_{(\alpha,\beta)\in E_\Gamma} \frac{1}{4\,T_\alpha^{(n)} T_\beta^{(n)}} \quad . \tag{10}$$

The expression above can be simplified since the first term under the curly braces contributes a constant amount. Based on this observation, the critical quantity in determining the update direction is

$$\hat{\Lambda}^{(n+1)} = \tag{11}$$

$$\frac{1}{|\mathcal{G}|} \sum_{G_i\in\mathcal{G}} \sum_{a\in V_i} \sum_{\alpha\in V_\Gamma} \sum_{b\in V_i} \sum_{\beta\in V_\Gamma} Q_{a\alpha}^{i,(n)} D_{ab}^i M_{\alpha\beta}^{(n)} s_{b\beta}^{i,(n+1)} - \frac{|V_\Gamma|}{4} + \sum_{(\alpha,\beta)\in E_\Gamma} \frac{1}{4\,T_\alpha^{(n)} T_\beta^{(n)}} \quad .$$

In order to optimize our weighted code-length criterion, we use graduated assignment [10] to update both the assignment matrices $\mathcal{S}$ and the structure of the supergraph, i.e. the supergraph adjacency matrix M. The updating process is realized by computing the derivatives of $\hat{\Lambda}^{(n+1)}$, and re-formulating the underlying discrete assignment problem as a continuous one using softmax.

In the **maximization step**, we have two parallel iterative update equations. The first update mode involves softening the assignment variables, while the second aims to modify the edge structure in the supergraph. Supergraph edges that are unmatchable disappear by virtue of having weak connection weights and cease to play any significant role in the update process. Experiments show that the algorithm appears to be numerically stable and appears to converge uniformly.

To update the assignment matrices, we commence by computing the partial derivative of the weighted code-length function in Equation (11) with respect to the elements of the assignment matrices, which gives

$$\frac{\partial \hat{\Lambda}^{(n+1)}}{\partial s_{b\beta}^{i,(n+1)}} = \frac{1}{|\mathcal{G}|} \sum_{a \in V_i} \sum_{\alpha \in V_\Gamma} Q_{a\alpha}^{i,(n)} D_{ab}^i M_{\alpha\beta}^{(n)} \quad . \tag{12}$$

To ensure that the assignment variables remain constrained to lie within the rage [0,1], we adopt the soft-max update rule

$$s_{a\alpha}^{i,(n+1)} \leftarrow \exp[\frac{1}{\tau} \frac{\partial \hat{\Lambda}^{(n+1)}}{\partial s_{a\alpha}^{i,(n+1)}}] / \sum_{\alpha' \in V_\Gamma} \exp[\frac{1}{\tau} \frac{\partial \hat{\Lambda}^{(n+1)}}{\partial s_{a\alpha'}^{i,(n+1)}}] \quad . \tag{13}$$

The value of the temperature τ in the update process has been controlled using a slow exponential annealing schedule of the form suggested by Gold and Rangarajan[10]. Initializing τ^{-1} with a small positive value and allowing it to gradually increase, the assignment variable $s_{a\alpha}^{i,(n+1)}$ corresponding to the maximum $\frac{\partial \hat{\Lambda}^{(n+1)}}{\partial s_{a\alpha}^{i,(n+1)}}$ approaches 1 while the remainder approach 0.

The partial derivative of the weighted code-length function in Equation (11) with respect to the elements of the supergraph adjacency matrix is equal to

$$\frac{\partial \hat{\Lambda}^{(n+1)}}{\partial M_{\alpha\beta}^{(n)}} = \frac{1}{|\mathcal{G}|} \sum_{G_i \in \mathcal{G}} \sum_{a \in V_i} \sum_{b \in V_i} Q_{a\alpha}^{i,(n)} D_{ab}^i s_{b\beta}^{i,(n+1)} - \frac{1}{(T_\alpha^{(n)})^2} \sum_{(\alpha,\beta') \in E_\Gamma} \frac{1}{4\,T_{\beta'}^{(n)}} \quad . \tag{14}$$

The soft-assign update equation for the elements of the supergraph adjacency matrix is

$$M_{\alpha\beta}^{(n+1)} \leftarrow \exp[\frac{1}{\tau} \frac{\partial \hat{\Lambda}^{(n+1)}}{\partial M_{\alpha\beta}^{(n)}}] / \sum_{(\alpha',\beta') \in E_\Gamma} \exp[\frac{1}{\tau} \frac{\partial \hat{\Lambda}^{(n+1)}}{\partial M_{\alpha'\beta'}^{(n)}}] \quad . \tag{15}$$

Recall that in Section 3 we discussed the encoding of the sample graphs, and chose to use the average of *Shannon-Fano code*. We can now elucidate that the reason for this choice is that as the number of the sample graphs increases, for instance in the limit as the size of the graph sample-set $\mathcal{G}$ increases, i.e. $N \to \infty$, the sum of permuted adjacency matrices of the sample graphs might dominate

Fig. 1. (a)Example images in the COIL dataset. (b)Example images in the toys dataset.

the magnitude of the second term in Equation (14). Thus the update algorithm might induce a complete supergraph that is fully connected. Hence, we choose to use its average rather than its sum.

In the **expectation step** of the EM algorithm, we compute the *a posteriori* correspondence probabilities for the nodes of the sample graphs to the nodes of the supergraph. Applying Bayes rule, the *a posteriori* correspondence probability for the nodes of the sample graph G_i at iteration $n + 1$ are given by

$$Q_{a\alpha}^{i,(n+1)} = \frac{\exp[\sum\limits_{b \in V_i} \sum\limits_{\beta \in V_\Gamma} D_{ab}^i M_{\alpha\beta}^{(n)} s_{b\beta}^{i,(n)}]\pi_\alpha^{i,(n)}}{\sum\limits_{\alpha' \in V_\Gamma} \exp[\sum\limits_{b \in V_i} \sum\limits_{\beta \in V_\Gamma} D_{ab}^i M_{\alpha'\beta}^{(n)} s_{b\beta}^{i,(n)}]\pi_{\alpha'}^{i,(n)}} . \tag{16}$$

In the above equation, $\pi_{\alpha'}^{i,(n)} = \langle Q_{a\alpha'}^{i,(n)} \rangle_a$, where $\langle\ \rangle_a$ means average over a.

5   Experiments

In this section, we report experimental results aimed at demonstrating the utility of our proposed generative model on real-world data. We use images from two datasets for experiments. The first dataset is the COIL which consists of images of 4 objects, with 72 views of each object from equally spaced directions over $360°$. We extract corner features from each image and use the detected feature points as nodes to construct sample graphs by Delaunay triangulation. The second dataset is a dataset consisting of views of toys, and contains images of 4 objects with 20 different views of each object. For this second dataset, the feature keypoints used to construct Delaunay graphs are extracted using the SIFT detector. Some example images of the objects from these two datasets are given in Figure 1.

The first part of our experimental investigation aims to validate our super-graph learning method. We test our proposed algorithm on both of the two datasets and in order to better analyze our method, we initialize the supergraph in our EM algorithm with different structures. For the COIL dataset, we initialize the supergraph structure with the median graph, i.e. the sample graph with the largest *a posteriori* probability from the supergraph. On the other hand, to initialize the structure of the supergraph in the toys dataset, we match pairs of graphs from a same object using the discrete relaxation algorithm [12]. Then we concatenate(merge) the common structures over for the sample graphs from a same object to form an initial supergraph. The initial supergraph constructed in this way preserves more of the structural variations present in the set of sample graphs. The median graph, on the other hand, captures more of the common salient information. We match the sample graphs from the two datasets against their supergraphs both using graduated assignment[10] and initialize the assignment matrices in our algorithm with the resulting assignment matrices. Using these settings, we iterate the two steps of the EM algorithm 30 times, and observe how the complexity of the supergraph, the average log-likelihood of the sample graphs and the overall code-length vary with iteration number. Figures 2 and 3 respectively shows the results for the COIL and toys datasets illustrated in Figure 1.

From Figure 2(a) it is clear that the von-Neumann entropy of the supergraph increases as the iteration number increases. This indicates that the supergraph structure becomes more complex with an increasing number of iterations. Figure 2(b) shows that the average of the log-likelihood of the sample graphs increases during the iterations. Figure 2(c) shows that the overall-code length decreases and gradually converges as the number of iterations increases. For the toys dataset, the von-Neumann entropy in Figure 3(a) shows an opposite trend and decreases as the number of iterations increases. The reason for this is that the initial supergraph we used for this dataset, i.e. the concatenated supergraph, accommodates too much structural variation from the sample graphs. The reduction of the von-Neumann entropy implies some trivial edges are eliminated or relocated. As a result the supergraph structure both condenses and simplifies with increasing iteration number. Although the complexity of the graphs behaves differently, the average of the likelihood of the graphs in Figure 3(b) and the overall-code length in Figure 3(c) exhibit a similar behaviour to those for the COIL dataset. In other words, our algorithm behaves in a stable manner both increasing the likelihood of sample graphs and decreasing the overall code-length on both datasets.

Our second experimental goal is to evaluate the effectiveness of our learned generative model for classifying out-of-sample graphs. From the COIL dataset, we aim 1) to distinguish images of cats from pigs on the basis of their graph representations and 2) distinguish between images of different types of bottle. For the toys dataset, on the other hand, we aim to distinguish between images of the four objects. To perform these classification tasks, we learn a supergraph for each object class from a set of samples and use Equation (3) to compute the *a posteriori* probabilities for each graph from a separate (out-of-sample) test-set. The class-label of the test graph is determined by the class of the supergraph

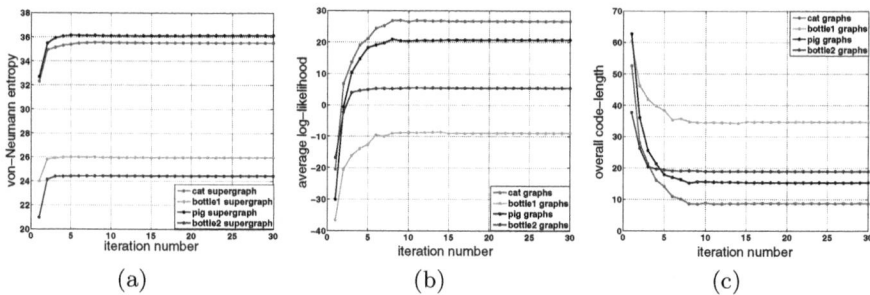

Fig. 2. COIL dataset: (a)variation of the complexity of the supergraph, encoded as von-Neumann entropy, during iterations, (b) variation of average log-likelihood of the sample graphs during iterations and (c) variation of the overall code-length during iterations.

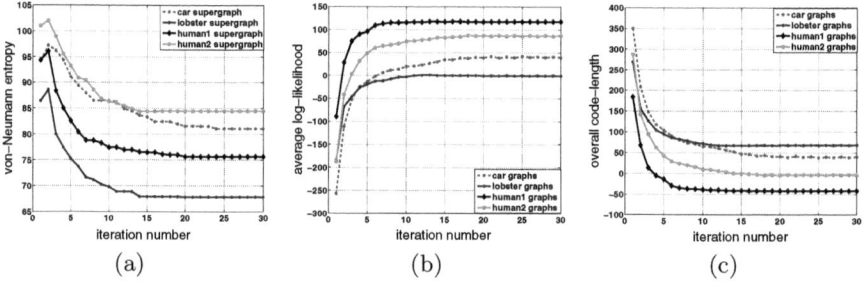

Fig. 3. Toy dataset: (a)variation of the complexity of the supergraph, encoded as von-Neumann entropy, during iterations, (b) variation of the average log-likelihood of the sample graphs during iterations and (c) variation of the overall code-length during iterations.

which gives the maximum *a posteriori* probability. The classification rate is the fraction of correctly identified objects computed using 10-fold cross validation. To perform the 10-fold cross validation for the COIL dataset, we index the 72 graphs from a same object according to their image view direction from $0°$ to $360°$, and in each fold we select 7 or 8 graphs that are equally spaced over the angular interval as test-set, and the remainder are used as as sample-set for training. The similar applies for the toys dataset. For comparison, we have also investigated the results obtained using two alternative constructions of the supergraph. The first of these is the median graph or concatenated graph used to initialize our algorithm. The second is the supergraph learned without taking its complexity into account, which means, this supergraph is learned by maximizing the likelihood function of the sample graphs given in equation (4). Table 1 shows the classification results obtained with the three different supergraph constructions. From the three constructions, it is the supergraphs learned using the MDL principle that achieve the highest classification rates on all the three classification tasks.

Table 1. Comparison of the classification results. The bold values are the average classification rates from 10-fold cross validation, followed by their standard error.

Classification Rate	cat & pig	bottle1 & bottle2	four objects (Toys)
learned supergraph(by MDL)	**0.824** ± 0.033	**0.780** ± 0.023	**0.763** ± 0.026
median graph/concatenated graph	**0.669** ± 0.052	**0.651** ± 0.023	**0.575**± 0.020
learned supergraph	**0.807** ± 0.056	**0.699** ± 0.029	**0.725** ± 0.022

6 Conclusion

In this paper, we have presented an information theoretic framework for learning a generative model of the variations in sets of graphs. The problem is posed as that of learning a supergraph. We provide a variant of EM algorithm to demonstrate how the node correspondence recover and supergraph structure estimation can be couched in terms of minimizing a description length criterion. Empirical results on real-world dataset support our proposed method by a) validating our learning algorithm and b) showing that our learned supergraph outperforms two alternative supergraph constructions. Our future work will aim to fit a mixture of supergraph to data sampled from multiple classes to perform graph clustering.

References

1. Friedman, N., Koller, D.: Being Bayesian about network structure.A Bayesian approach to structure discovery in Bayesian networks. Machine Learning, 95–125 (2003)
2. Christmas, W.J., Kittler, J., Petrou, M.: Modeling compatibility coefficient distribution. Image and Vsion Computing 14, 617–625 (1996)
3. Bagdanov, A.D., Worring, M.: First order Gaussian graphs for efficient structure classification. Pattern Recognition 36, 1311–1324 (2003)
4. Luo, B., Hancock, E.R.: A spectral approach to learning structural variations in graphs. Pattern Recognition 39, 1188–1198 (2006)
5. Torsello, A., Hancock, E.R.: Learning shape-classes using a mixture of tree-unions. IEEE PAMI 28, 954–967 (2006)
6. Luo, B., Hancock, E.R.: Structural graph matching using the EM alogrithm and singular value decomposition. IEEE PAMI 23, 1120–1136 (2001)
7. Rissanen, J.: Modelling by Shortest Data Description. Automatica, 465–471 (1978)
8. Passerini, F., Severini, S.: The von-neumann entropy of networks. arXiv:0812.2597 (2008)
9. Grunwald, P.: Minimum Description Length Tutorial. Advances in Minimum Description Length: Theory and Applications (2005)
10. Gold, S., Rangarajan, A.: A Graduated Assignment Algorithm for Graph Matching. IEEE PAMI 18, 377–388 (1996)
11. Wilson, R.C., Zhu, P.: A study of graph spectra for comparing graphs and trees. Pattern Recognition 41, 2833–2841 (2008)
12. Wilson, R.C., Hancock, E.R.: Structural matching by discrete relaxation. IEEE PAMI 19, 634–648 (1997)
13. Han, L., Wilson, R.C., Hancock, E.H.: A Supergraph-based Generative Model. In: ICPR, pp. 1566–1569 (2010)
14. Figueiredo, M.A.T., Jain, A.K.: Unsupervised learning of finite mixture models. IEEE PAMI 24, 381–396 (2002)

Information-Geometric Graph Indexing from Bags of Partial Node Coverages

Francisco Escolano, Boyan Bonev, and Miguel A. Lozano

University of Alicante
{sco,boyan,malozano}@dccia.ua.es

Abstract. In a previous work we have uncovered some of the most informative spectral features (Commute Times, Fiedler eigenvector, Perron-Frobenius eigenvector and Node Centrality) for graph discrimination. In this paper we propose a method which exploits information geometry (manifolds and geodesics) to characterize graphlets with covariance matrices involving the latter features. Once we have the vectorized covariance matrices in the tangent space each graph is characterized by a population of vectors in such space. Then we exploit bypass information-theoretic measures for estimating the dissimilarities between populations of vectors. We test this measure in a very challenging database (GatorBait).

1 Introduction

Graph indexing, as well as graph matching, building kernels between graphs and graph embedding, aims to discriminate graphs even when small perturbation (editions) arise. However, in the indexing case, such discrimination is typically addressed by computing a particular set of features (indices) characterizing the graph and then build a representation which is suitable for the discrimination task. Once such representation is available a query graph is processed for extracting their indices and then find the closer (more similar) graph in the database. Therefore, there are three elements to specify: (i) the indices, (ii) the representation, and (iii) the dissimilarity measure. Indices emerge from taking statistics. A significant approach is to considering the frequency of certain subgraphs, like in the *gIndex* approach [1]. Subgraph frequency implies solving graph-subgraph problems (wich are NP-hard) and subgraph-subgraph isomorphisms which are polinomial problems; in this approach is convenient to retain discriminative subgraphs which are typically the smallest ones. These discriminative subgraphs are useful for filtering false positives and to simplfy the number of graph-subgraph problems to solve, but since relationships between subgraphs is ignored, many false appear in practice. In the *summarization graph* approach [2] where the *bag of indexed subgraphs approach* is complemented by a complete graph encoding topological relationships between subgraphs; however, the main drawback is the high computational cost required to find such relationships. A successful alternative and less time-consuming alternative is to build indexes which characterizes the complete graph and all of its subgraphs [3]. However the latter method is

X. Jiang, M. Ferrer, and A. Torsello (Eds.): GbRPR 2011, LNCS 6658, pp. 52–61, 2011.
© Springer-Verlag Berlin Heidelberg 2011

constrained to trees (hierarchical structures) where subgraphs are sub-trees and indexes are built followin *nesting* approach; firstly the part of the spectrum of the complete tree is retained, and then the process considers recursively the spectrum of each of its subtrees. When considering general graphs, the nesting approach has inspired the concept of *node history* (a sequence of incrementally self-contained subgraphs emanating from each node) [4]. The latter approach, also related to spectral indexes, has been succeeded in quantifying the complexity of graphs because it captures global aspects of graphs. However, when tested in indexing problems it reported a poor performance [5]. In order to increase the discrimination performance we have explored two directions. The first one consists of elucidating what spectral features are the most discriminant among a reasonable catalog of them [6]. With a catalog of discriminative spectral features at hand, the second direction motivates this paper; it consists on exploring how these features can interact with a *set of partial coverages* instead of interacting with a *set of fulll node histories*. Specifying such interaction consists of defining two elements: a) bound the history a node to define a *partial node coverage*, and b) consider the statistical depedency between the spectral features describing each partial node coverage. Conceptually, a partial coverage is very close to the concept of *graphlet* whose spectrum has been recently described [7]: given the history of a node, we only consider the subgraph corresponding to a given order of expansion (the same order for all nodes); therefore the overlapping of all subgraphs defines a *coverage* of it. Regarding the quantification of the statistical dependency between spectral features we exploit covariances matrices and their vectorization in the tangent space as suggested in [8] and [9]. In this paper we compare two bypass information-theoretic dissimilarity measures between bags of subgraphs: multi-dimensional Henze-Penrose divergence and the total variance-kdp divergence. Although we use a bag of subgraphs, their topological dependencies are partially encoded since we use covariance matrices to describe node coverages and node coverages share more and more nodes as nodes are topologically closer in the original graph.

2   Subgraph Indexation

2.1   Partial Node Coverages

Let $G = (V, E)$ with $|V| = n$. Then the *history of a node* $i \in V$ is $h_i(G) = \{e(i), e^2(i)), \ldots, e^p(i)\}$ where: $e(i) \subseteq G$ is the *first-order expansion subgraph* given by i and all $j \sim i$, $e^2(i) = e(e(i)) \subseteq G$ is the *second-order expansion* consisting on $z \sim j : j \in V_{e(i)}, z \notin V_{e(i)}$, and so on until p cannot be increased. If G is connected $e^p(i) = G$, otherwise $e^p(i)$ is the connected component to which i belongs.(see [4]).

Every $h_i(G)$ defines a set of subgraphs $h_i(G) = \{e(i), e^2(i)), \ldots, e^p(i)\}$ where $e^l(i) \subseteq e^k(i)$ when $k > l$. If we select $k < p$ we obtain a $k-$order *partial node coverage* given by the subgraph $e^k(i)$. If we overlap the $k-$order partial coverages associated to all $i \in V$ we obtain a $k-$order *graph coverage*. However, as we must keep the number of free parameters in pattern recognition algorithms at

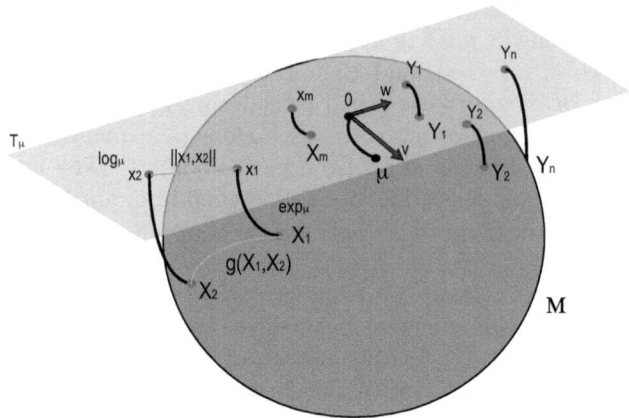

Fig. 1. Riemannian manifold (the sphere) and tangent space T_μ at point μ. Points in the tangent space are the de-projections (log) of their corresponding projections (exp) which lie in the Manifold. We show in different colors points corresponding to subgraphs of two different graphs: X_i and Y_j. We also show some examples of distances in the manifold (geodesics) $g(.,.)$ and in the tangent space $||.,.||$, and tangents u and w.

a minimum, herein we compute *the optimal order* for each node in the graph. To that end it is desirable to define a set of features and track their variability until a peak is detected in a sort of *structural scale-space*. However, this approach requires to compute the node history at least until a peak is found and it is quite computational demanding. Consequently, in this paper we will focus on setting experimentally a constant optimal order k^* for all graphs. In addition, in the information-geometry approach followed here (see [8][9]), the features are covariance matrices relying on spectral descriptors. More precisely, the features are vectorized covariances projected on a given tangent space (exponential chart).

2.2   Spectral Descriptors in Tangent Space

Consider $\Phi(i) = (f_1(i), \ldots, f_d(i))^T$ a vector of spectral descriptors of the partial node coverage $H = e^k(i) \subseteq G$ (commute times, Fiedler vector, Perron-Frobenius vector and node centrality). Such descriptors have been determined to be very informative for graph discrimination [6]. For commute times (CT) we consider both the Laplacian and the normalized Laplacian of H: the elements of the upper off-diagonal elements of the CT kernel are downsampled to select $m = |V_H|$ elements and they are normalized by m^2. Fiedler and Perron-Frobenius vectors have m elements by definition. Node centrality [11] is more selective than degree and it is related to the number of closed walks starting and ending at a each node. This measure is also normalized by m^2. For each partial coverage H we can compute the statistics of d spectral descriptors taking m samples. Such statistics can be easily encoded in a covariance matrix.

The set of $d \times d$ covariance matrices $\boldsymbol{X}_i = \frac{1}{n-1}\sum_{i=1}^{m}(\Phi(i) - \boldsymbol{\mu})(\Phi(i) - \boldsymbol{\mu})^T$, being $m = |V_H|$, lie in a Riemannian manifold $\mathcal{M}$ (see Fig. 1). For each $\boldsymbol{X} \in \mathcal{M}$

there exists a neighborhood which can be mapped to a given neighborhood in $\mathbb{R}^{d\times d}$. Such mapping is continuous bidirectional and one-to-one. As a Riemann manifold is differentiable, the derivatives at each $\boldsymbol{X}$ always exist, and such derivatives lie in the so called tangent space $T_{\boldsymbol{X}}$, which is a vector space in $\mathbb{R}^{d\times d}$. The tangent space at $\boldsymbol{T}_{\boldsymbol{X}}$ is endowed with an inner product $<.,.>_{\boldsymbol{X}}$ being $<\boldsymbol{u},\boldsymbol{v}>_{\boldsymbol{X}}= trace(\boldsymbol{X}^{-\frac{1}{2}}\boldsymbol{u}\boldsymbol{X}^{-1}\boldsymbol{v}\boldsymbol{X}^{-\frac{1}{2}})$. The tangent space is also endowed with an exponential map $\exp_{\boldsymbol{X}}: T_{\boldsymbol{X}} \to \mathcal{M}$ which maps a tangent vector $\boldsymbol{u}$ to a point $\boldsymbol{U} = \exp_{\boldsymbol{X}}(\boldsymbol{u}) \in \mathcal{M}$. Such mapping is one-to-one, bidirectional and continuously differentiable and maps u to the point reached by the unique geodesic (minimum-length curve connecting two points in the manifold) from $\boldsymbol{X}$ to $\boldsymbol{U}$: $g(\boldsymbol{X},\boldsymbol{U})$. The exponential map is only one-to-one in the neighborhood of $\boldsymbol{X}$ and this implies that the inverse mapping $\log_{\boldsymbol{X}}: \mathcal{M} \to T_{\boldsymbol{X}}$ is uniquely defined in a small neighborhood of $\boldsymbol{X}$. Therefore, we have the following mappings for going to the manifold and back (to the tangent space) respectively:

$$\exp_{\boldsymbol{X}}(\boldsymbol{u}) = \boldsymbol{X}^{\frac{1}{2}}\exp(\boldsymbol{X}^{-\frac{1}{2}}\boldsymbol{u}\boldsymbol{X}^{-\frac{1}{2}})\boldsymbol{X}^{\frac{1}{2}}, \log_{\boldsymbol{X}}(\boldsymbol{U}) = \boldsymbol{X}^{\frac{1}{2}}\log(\boldsymbol{X}^{-\frac{1}{2}}\boldsymbol{U}\boldsymbol{X}^{-\frac{1}{2}})\boldsymbol{X}^{\frac{1}{2}},$$

$$(1)$$

and $g^2(\boldsymbol{X},\boldsymbol{U}) =< \log_{\boldsymbol{X}}(\boldsymbol{U}),\log_{\boldsymbol{X}}(\boldsymbol{U}) >_{\boldsymbol{X}}= trace\left(\log^2(\boldsymbol{X}^{-\frac{1}{2}}\boldsymbol{U}\boldsymbol{X}^{-\frac{1}{2}})\right)$, where we take the matrix exponentiation and logarithm. The tangent space allows us to vectorize de result of the inverse mapping in order to work in a vector space with Euclidean distances which are approximations of the geodesics. In [8] the following orthonormal vectorization operator is proposed (off-diagonal elements are multiplied by $\sqrt{2}$):

$$vec_{\boldsymbol{X}}(u) = vec_{\boldsymbol{I}}(u)(\boldsymbol{X}^{-\frac{1}{2}}u\boldsymbol{X}^{-\frac{1}{2}}), vec_{\boldsymbol{I}}(u) = (u_{11}\sqrt{2}u_{12}\ldots u_{22}\sqrt{2}u_{23}\ldots u_{dd})^T.$$

$$(2)$$

2.3   Encoding Graphs in Tangent Space

Each graph X has $n_X = |V_X|$ partial coverages, one for each node. Therefore, we have n overlapped subgraphs H_{X_i} each one characterized by a covariance matrix $\boldsymbol{X}_i$ based on $m_{H_{X_i}} = |V_{H_{X_i}}|$ samples. Then, ech graph can be encoded by a population of n_X points in a manifold $\mathcal{M}$. For instance, another graph Y will be encoded by n_Y covariance matrices $\boldsymbol{Y}_j$ in the same manifold $\mathcal{M}$. In order to compare both populations we can map then back to a given tangent space. However we must determine what is the origin of such space. Let us denote by $\boldsymbol{Z}_k$ with $k = 1,\ldots,N$ (being $N = n_X + n_Y$) each covariance matrix coming from X or from Y. A fair selection of the tangent space origin is the Karcher mean defined as $\mu = \arg\min_{Z\in\mathcal{M}} d^2(\boldsymbol{Z}_k,\boldsymbol{Z})$. The Karcher mean can be obtained after few iterations of $\mu^{t+1} = \exp_{\mu^t}(\bar{\boldsymbol{X}}^t)$ where $\bar{\boldsymbol{X}}^t = \frac{1}{N}\sum_{k=1}^N \log_{\mu^t}(\boldsymbol{Z}_k)$. Once we have μ, we have an origin for the tangent space, and then we can project all matrices $\boldsymbol{Z}_k$ in such space (see Fig. 1) through $z_k = vec_{\mu}(\log_{\mu}(\boldsymbol{Z}_k))$. Now we characterize each graph by a population of vectors in the tangent space. The problem of determining their dissimilarity can be then reduced to compare the overlap of both populations. To that end, in the following section we introduce a couple of bypass information-theoretic (IT) measures which have proved to be effective in our previous work in the context of shape comparison [12].

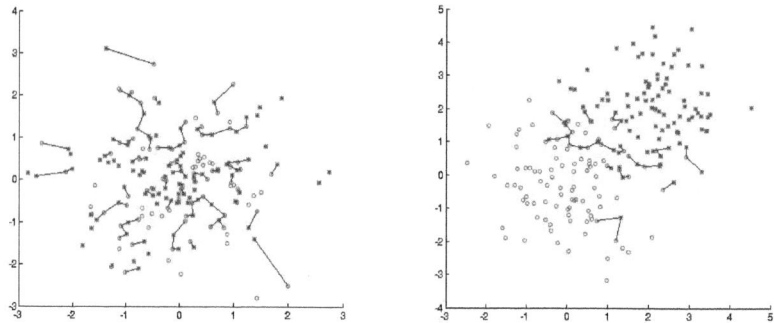

Fig. 2. Two examples of Friedman-Rafsky estimation of the Henze and Penrose divergence applied to samples drawn from two Gaussian densities. Left: the two densities have the same mean and covariance matrix ($D_{HP}(f\|g) = 0.5427$). Right: the two densities have different means ($D_{HP}(f\|g) = 0.8191$).

3   Bypass IT Dissimilarity Measures

3.1   Henze-Penrose Divergence

The Henze and Penrose divergence [13] between two distributions f and g is

$$D_{HP}(f\|g) = \int \frac{p^2 f^2(z) + q^2 g^2(z)}{pf(z) + qg(z)} dz \ , \qquad (3)$$

where $p \in [0,1]$ and $q = 1-p$. This divergence is the limit of the Friedman-Rafsky run length statistic [14], that in turn is a multi-dimensional generalization based on MST[1]s of the Wald-Wolfowitz test. The Wald-Wolfowitz statistic computes the divergence between two distributions f_X and g_O in $\mathcal{R}^d$, when $d = 1$, from two sets of n_x and n_o samples, respectively. First, the $n = n_x + n_o$ samples are ordered in ascending order and labeled as X and O according to their corresponding distribution. The test is based on the number of runs R, being a run a sequence of consecutive and equally labeled samples. The test is calculated as:

$$W = \frac{R - \frac{2n_o n_x}{n} - 1}{\left(\frac{2n_x n_o(2n_x n_o - n)}{n^2(n-1)}\right)^{\frac{1}{2}}} \ . \qquad (4)$$

The two distributions are considered similar if R is low and therefore W is also low. This test is consistent in the case that n_x/n_o is not close to 0 or ∞, and when $n_x, n_o \to \infty$. The Friedman-Rafsky test generalizes Eq. 4 to $d > 1$, due to the fact that the MST relates samples that are close in $\mathcal{R}^d$. Let $X = \{x_i\}$ and $O = \{o_i\}$ be two sets of samples drawn from f_X and g_O, respectively. The steps of the Friedman-Rafsky test are:

---

[1] Minimum-Spanning Tree.

1. Build the MST over the samples from both X and O.
2. Remove the edges that do not connect a sample from X with a sample from O.
3. The proportion of non-removed edges converges to 1 minus the Henze Penrose divergence (Eq. 3) between f_X and g_O.

See an example in Fig. 2.

3.2   Total Variation k-dP Divergence

The main drawback of both the Henze-Penrose and the Leonenko's-based divergences is the high temporal cost of building the underlying data structures (e.g. MSTs). This computational burden is due to the calculation of distances. A new entropy estimator recently developed by Stowell and Plumbley overcomes this problem [15]. They proposed an entropy estimation algorithm that relies on data spacing without computing any distance. This method is inspired by the data partition step in the k-d tree algorithm. Let X be a d-dimensional random variable, and $f(x)$ its pdf. Let $A = \{A_j | j = 1, \ldots, m\}$ be a partition of X for which $A_i \cap A_j = \emptyset$ if $i \neq j$ and $\bigcup_j A_j = X$. Then, we can approximate $f(x)$ in each cell as:

$$f_{A_j} = \frac{\int_{A_j} f(x)}{\mu(A_j)}, \quad , \tag{5}$$

where $\mu(A_j)$ is the d-dimensional volume of A_j. If $f(x)$ is unknown and we are given a set of samples $X = \{x_1, \ldots, x_n\}$ from it, being $x_i \in \mathcal{R}^d$, we can approximate the probability of $f(x)$ in each cell as $p_j = n_j/n$, where n_j is the number of samples in cell A_j. Thus,

$$\hat{f}_{A_j}(x) = \frac{n_j}{n\mu(A_j)} , \tag{6}$$

being $\hat{f}_{A_j}(x)$ a consistent estimator of $f(x)$ as $n \to \infty$. Then, to obtain the entropy estimation for A we have

$$\hat{H} = \sum_{j=1}^{m} \frac{n_j}{n} \log\left(\frac{n}{n_j}\mu(A_j)\right) . \tag{7}$$

The partition is created recursively following the data splitting method of the k-d tree algorithm. At each level, data is split at the median along one axis. Then, data splitting is recursively applied to each subspace until an uniformity stop criterion is satisfied. The aim of this stop criterion is to ensure that there is an uniform density in each cell in order to best approximate $f(x)$. The chosen uniformity test is fast and depends on the median. The distribution of the median of the samples in A_j tends to a normal distribution that can be standardized as:

$$Z_j = \sqrt{n_j}\frac{2med_d(A_j) - min_d(A_j) - max_d(A_j)}{max_d(A_j) - min_d(A_j)}, \tag{8}$$

where $med_d(A_j)$, $min_d(A_j)$ and $max_d(A_j)$ are the median, minimum and maximum, respectively, of the samples in cell A_j along dimension d. An improbable value of Z_j, that is, $|Z_j| > 1.96$ (the 95% confidence threshold of a standard normal distribution) indicates significant deviation from uniformity. Non-uniform cells should be divided further. An additional heuristic is included in the algorithm in order to let the tree reach a minimum depth level: the uniformity test is not applied until there are less than $\sqrt{n}$ data points in each partition, that is, until the level

$$L_n = \left\lceil \frac{1}{2} \log_2(n) \right\rceil \tag{9}$$

is reached. Then, our k-d partition based divergence (k-dP divergence) [12] follows the spirit of the *total variation* distance, but may also be interpreted as a L1-norm distance. The total variation distance between two probability measures P and Q on a σ-algebra F^2 is given by:

$$sup\{|P(X) - Q(X)| : X \in F\} \ . \tag{10}$$

In the case of a finite alphabet, the total variation distance is

$$\delta(P, Q) = \frac{1}{2} \sum_x |P(x) - Q(x)| \ . \tag{11}$$

Let $f(x)$ and $g(x)$ be two distributions, from which we draw a set X of n_x samples and a set O of n_o samples, respectively. If we apply the partition scheme of the k-d partition algorithm to the set of samples $X \bigcup O$, the result is a partition A of $X \bigcup O$, being $A = \{A_j | j = 1, \ldots, p\}$. For $f(x)$ and $g(x)$ the probability of any cell A_j is respectively given by

$$f(A_j) = \frac{n_{x,j}}{n_x} = f_j, \ \ g(A_j) = \frac{n_{o,j}}{n_o} = g_j \tag{12}$$

where $n_{x,j}$ is the number of samples of X in cell A_j and $n_{o,j}$ is the number of samples of O in the cell A_j. Since the same partition A is applied to both sample sets, and considering the set of cells A_j a finite alphabet, we can compute the k-dP *total variation divergence* between $f(x)$ and $g(x)$ as:

$$D_{kdP}(f||g) = \frac{1}{2} \sum_{j=1}^{p} |f_j - g_j| \ . \tag{13}$$

The latter divergence satisfies $0 \le D(f||g) \le 1$. The minimum value $D(O||X) = 0$ is obtained when all the cells A_j contain the same proportion of samples from X and O. By the other hand, the maximum value $D(O||X) = 1$ is obtained when all the samples in any cell A_j belong to the same distribution. We show in Fig. 3 two examples of divergence estimation using Eq. 13.

---

[2] A σ-algebra over a set X is a non-empty collection of subsets of X (including X itself) that is closed under complementation and countable unions of its members.

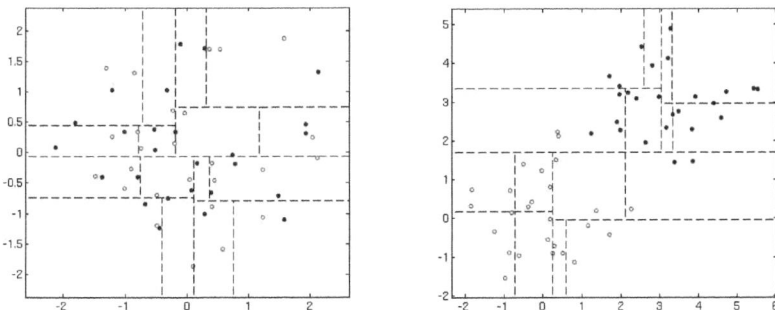

Fig. 3. Two examples of divergence estimation applied to samples drawn from two Gaussian densities. Left: both densities have the same mean and covariance matrix ($D(f||g) = 0.24$). Right: the two densities have different means. Almost all the cells contain samples obtained from only one distribution ($D(f||g) = 0.92$).

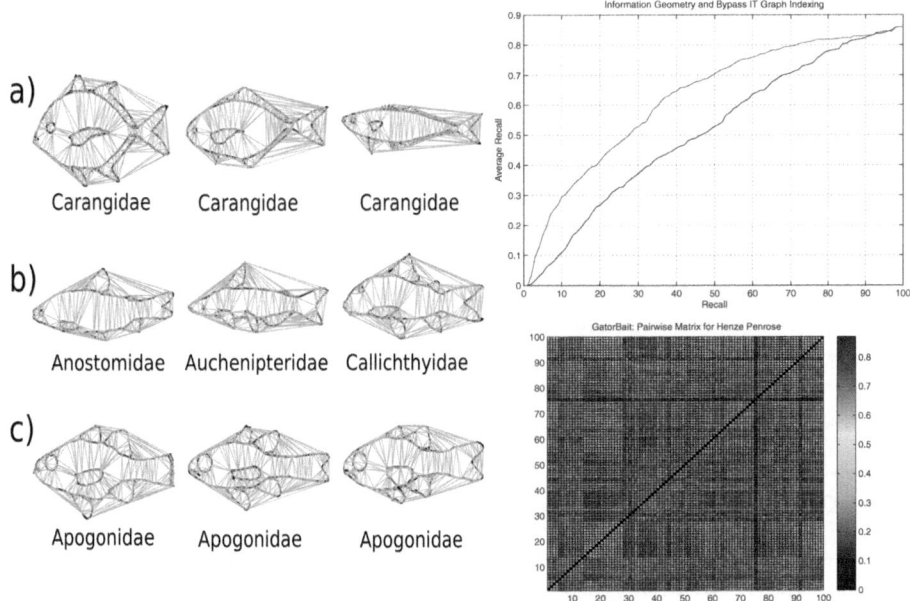

Fig. 4. Examples of the Gator database (left), average recall curves (top-right) and pairwise dissimilarity matrix for the Henze-Penrose divergence (bottom-right)

4   Experiments

In order to test both the proposed indexing approach we have chosen a challening database, the *GatorBait_100*[3] ichthtyology database. GatorBait has 100 shapes representing fishes from 30 different classes. We have extracted Delaunay graphs from their shape quantization (Canny algorithm followed by contour decimation).Since the classes are associated to fish genus and not to species, we find high intraclass variability in many cases – see a) in Fig. 4-left where the corresponding class has 8 species. There are also very similar species from different classes (row b)) and few homogeneous clases (row c)). There are 10 classes with one species (not included in the analysis and performance curves), 11 with $1 - 3$ individuals, 5 with $4 - 6$ individuals and only 4 classes with more than 6 species. We show the average retrieval-recall results for both bypass IT measures in Fig. 4-right: Henze-Penrose divergence computed via MSTs outperforms total variation. In all cases, the order of all partial node coverages is set to $k = 5$ which has been obtained experimentally. The number of feature descriptors considered is $d = 5$: commute times (from the normalized and un-normalized Laplacian), Perron-Frobenius, Fiedler vector and node centrality vector.

5   Conclusions

The main contribution of this paper is the characterization of graphs in terms of bags of features in a tangent space, and the use of this methodology, in combination with bypass IT dissimilarity measures for graph indexing. Features come from the inverse projection of covariance matrices in the tangent space defined by a common Karcher mean. Covariance matrices rely on several spectral subgraph descriptors which have been qualified as very informative in our previous work. We have tested the approach in a very hard graph database (GatorBait). One limitation of the present approach is the automatic selection of the order of the subgraphs (partial node coverages). We have designed an strategy for definining saliency in a structural scale space but it is out of the scope of this paper. In a near future we will contribute with new experiments considering automatic order selection.

References

[1]   Yan, X., Yu, P.S., Han, J.: Graph Indexing: A Frequent Structure-based Approach. In: Proc. of SIGMOD 2004 (2004)

[2]   Zou, L., Chen, L., Zhang, H., Lu, Y.S., Lou, Q.: Summarization graph indexing: Beyond frequent structure-based approach. In: Haritsa, J.R., Kotagiri, R., Pudi, V. (eds.) DASFAA 2008. LNCS, vol. 4947, pp. 141–155. Springer, Heidelberg (2008)

[3]   Shokoufandeh, A., Macrini, D., Dickinson, S., Siddiqi, K., Zucker, S.W.: Indexing Hierarchical Structures Using Graph Spectra. IEEE Trans. on PAMI 27(7), 1125–1140 (2005)

---

[3] http://www.cise.ufl.edu/~anand/publications.html

[4]   Escolano, F., Lozano, M.A., Hancock, E.R., Giorgi, D.: What is the complexity of a network? The heat flow-thermodynamic depth approach. In: Hancock, E.R., Wilson, R.C., Windeatt, T., Ulusoy, I., Escolano, F. (eds.) SSPR&SPR 2010. LNCS, vol. 6218, pp. 286–295. Springer, Heidelberg (2010)

[5]   Escolano, F., Giorgi, D., Hancock, E.R., Lozano, M.A., Falcidieno, B.: Flow Complexity: Fast Polytopal Graph Complexity and 3D Object Clustering. In: Torsello, A., Escolano, F., Brun, L. (eds.) GbRPR 2009. LNCS, vol. 5534, pp. 253–262. Springer, Heidelberg (2009)

[6]   Bonev, B., Escolano, F., Giorgi, D., Biasotti, S.: High-dimensional spectral feature selection for 3D object recognition based on reeb graphs. In: Hancock, E.R., Wilson, R.C., Windeatt, T., Ulusoy, I., Escolano, F. (eds.) SSPR&SPR 2010. LNCS, vol. 6218, pp. 119–128. Springer, Heidelberg (2010)

[7]   Kondor, R., Shervashidze, N., Borgwardt, K.M.: The Graphlet Spectrum Proc. of ICML 2009 (2009)

[8]   Tuzel, O., Porikli, F., Meer, P.: Pedestrian Detection via Classification on Riemannian Manifolds. IEEE Trans. on PAMI 30(10), 1–15 (2008)

[9]   Pennec, X., Fillard, P., Ayache, N.: A Riemannian Framework for Tensor Computing. IJCV 38(1), 41–66 (2006)

[10] Qiu, H., Hancock, E.R.: Clustering and Embedding Using Commute Times. IEEE Trans. on PAMI 29(11), 1873–1890 (2007)

[11] Estrada, E., Rodrguez-Velázquez, J.A.: Subgraph centrality in complex networks. Physical Review E 71, 56103 (2005)

[12] Escolano, F., Lozano, M.A., Bonev, B., Suau, P.: Bypass Information-Theoretic Shape Similarity from Non-rigid Points-based Alignment. In: Proc. of CVPR Workshops, NORDIA (2010)

[13] Henze, N., Penrose, M.: On the multi-variate runs test. Annals of statistics 27, 290–298 (1999)

[14] Friedman, J.H., Rafsky, L.C.: Mutivariate Generalization of the Wald-Wolfowitz and Smirnov Two-Sample Tests. Annals of Statistics 7(4), 697–717 (1979)

[15] Stowell, D., Plumbley, M.D.: Fast Multidimensional Entropy Estimation by K-d Partitioning. IEEE Signal Processing Letters 16(6), 537–540 (2009)

Maximum Likelihood for Gaussians on Graphs

Brijnesh J. Jain and Klaus Obermayer

Berlin Institute of Technology, Germany
jbj@cs.tu-berlin.de

Abstract. We show that extending the Gaussian distribution to the domain of graphs corresponds to truncated Gaussian distributions in Euclidean spaces. Based on this observation, we derive a maximum likelihood method for estimating the parameters of the Gaussian on graphs. In conjunction with a naive Bayes classifier, we applied the proposed approach to image classification.

1   Introduction

In probability theory and machine learning, the Gaussian distribution is often used as a first approximation for vector-valued random variables that cluster around a single mean. In addition, it forms the basic building block for Gaussian mixture models. Gaussian mixture models in conjunction with the maximum likelihood method form a popular technique for smooth approximation of arbitrarily shaped densities.

In many applications, however, the data we want to learn about are often represented by attributed graphs, including point patterns, sequences, and trees as special cases. Examples of such data include scene models in computer vision, protein structures in bioinformatics, chemical compounds in computational chemistry, and XML documents in computational linguistics.

Our long-term goal is to investigate under which conditions and to which extent can Gaussian mixture models in conjunction with the maximum likelihood method be used for approximation of arbitrary densities on attributed graphs. Clearly, the first step to approach this problem consists in providing a framework that allows us to emulate Gaussian distributions on attributed graphs.

Attempts at developing probabilistic models on graphs aim at inferring statistical variations of structural models from their individual primitives, i.e. from vertices, edges, and/or their attributes [1,5,7,14,16,17,19,20]. As an alternative, the proposed approach regards attributed graphs as events of a probability space that can then be lifted locally to a Euclidean space to infer the statistical variation of individual primitives.

This contribution presents a framework for adapting the Gaussian distribution to attributed graphs such that the parameters can be fitted by the maximum likelihood method in a feasible way. For this, we represent graphs as points of some Riemannian orbifold [8]. Since an orbifold looks locally like a Euclidean space almost everywhere, we can project the Gaussian distribution to the graph space. Lifting the Gaussian on graphs back to the Euclidean space yields a truncated

X. Jiang, M. Ferrer, and A. Torsello (Eds.): GbRPR 2011, LNCS 6658, pp. 62–71, 2011.
© Springer-Verlag Berlin Heidelberg 2011

Gaussian. For estimating the parameters of a truncated Gaussian distribution, we maximize the log-likelihood. Since no closed solution of the log-likelihood exists, we present a gradient-based update scheme. A key problem is that determining the gradients of the log-likelihood involves possibly intractable approximations of integrals. In experiments, we ignored the integrals and obtained reasonable results on three benchmark datasets compiled by [13].

2 Graph Orbifolds

In this section, we introduce attributed graphs and represent them as point of some orbifold. Most of this presentation including proofs of statements and claims is based on the structure space formalism proposed by [8].

Representation of Graphs. Let $\mathbb{E}$ be a d-dimensional Euclidean space. An *attributed graph* $X = (V, E, \alpha)$ consists of a set V of *vertices*, a set $E \subseteq V \times V$ of *edges*, and an *attribute function* $\alpha : V \times V \to \mathbb{E}$, such that $\alpha(i, j) \neq \mathbf{0}$ for each edge and $\alpha(i, j) = \mathbf{0}$ for each non-edge. Attributes $\alpha(i, i)$ of vertices i may take any value from $\mathbb{E}$.

For simplifying the mathematical treatment, we assume that all graphs are of order n, where n is chosen to be sufficiently large. Graphs of order less than n, say $m < n$, can be extended to order n by including isolated vertices with attribute zero. For practical issues, it is important to note that limiting the maximum order to some arbitrarily large number n and extending smaller graphs to graphs of order n are purely technical assumptions to simplify mathematics. For pattern recognition problems, these limitations should have no practical impact, because neither the bound n needs to be specified explicitly nor an extension of all graphs to an identical order needs to be performed. When applying the theory, all we actually require is that the graphs are finite.

A graph X is completely specified by its *matrix representation* $\mathbf{X} = (\mathbf{x}_{ij})$ with elements $\mathbf{x}_{ij} = \alpha(i, j)$ for all $1 \leq i, j \leq n$. Let $\mathcal{X} = \mathbb{E}^{n \times n}$ be the Euclidean space of all $(n \times n)$-matrices with elements from $\mathbb{E}$ and let Π^n be the set of all $(n \times n)$-permutation matrices. For each $\mathbf{P} \in \Pi^n$ we define a mapping

$$\gamma_{\mathbf{P}} : \mathcal{X} \to \mathcal{X}, \quad \mathbf{X} \mapsto \mathbf{P}^{\mathsf{T}} \mathbf{X} \mathbf{P}.$$

Then $\mathcal{G} = \{\gamma_{\mathbf{P}} : \mathbf{P} \in \Pi^n\}$ is a finite group acting on $\mathcal{X}$. For $\mathbf{X} \in \mathcal{X}$, the *orbit* of $\mathbf{X}$ is the set defined by $[\mathbf{X}] = \{\gamma(\mathbf{X}) : \gamma \in \mathcal{G}\}$. The quotient set

$$\mathcal{X}_\mathcal{G} = \mathcal{X}_G = \{[\mathbf{X}] : \mathbf{X} \in \mathcal{X}\}$$

consisting of all orbits is a *graph orbifold*. Its *orbifold chart* is the surjective continuous mapping

$$\pi : \mathcal{X} \to \mathcal{X}_\mathcal{G}, \quad \mathbf{X} \mapsto [\mathbf{X}]$$

that projects each matrix representation $\mathbf{X}$ to its orbit $[\mathbf{X}]$.

Suppose that $\mathbf{X}$ is a matrix representation of some attributed graph X. Then the orbit $[\mathbf{X}]$ consists of all possible matrices that represent X. By identifying the attributed graphs X with the orbits $[X]$, we can regard graphs as point of the graph orbifold $\mathcal{X}_\mathcal{G}$. The orbifold chart $\pi : \mathcal{X} \to \mathcal{X}_\mathcal{G}$ projects matrices $\boldsymbol{X}$ to the graphs X they represent.

For notational convenience, we identify $\mathcal{X}$ with $\mathbb{E}^N$, where $N = n^2$ and consider vector- rather than matrix representations of graphs. We obtain a *vector representation* $\boldsymbol{x}$ of graph X by concatenating the columns of a matrix $\boldsymbol{X}$ representing X. We write $\boldsymbol{x} \in X$ if $\boldsymbol{x} \in \mathcal{X}$ projects to $X \in \mathcal{X}_\mathcal{G}$ via the orbifold chart $\pi(\boldsymbol{x}) = X$.

Intrinsic Metric. The *intrinsic metric* of a graph orbifold $\mathcal{X}_\mathcal{G}$ is of the form

$$d(X, X') = \min \left\{ \|\boldsymbol{x} - \boldsymbol{x}'\| \ : \ \boldsymbol{x} \in X, \boldsymbol{x}' \in X' \right\},$$

where $\|\cdot\|$ is the Euclidean distance on $\mathcal{X}$. We call a pair $(\boldsymbol{x}, \boldsymbol{x}') \in X \times X'$ with $\|\boldsymbol{x} - \boldsymbol{x}'\| = d(X, X')$ an *optimal alignment* of X and X'. Note that the intrinsic metric is not a artificial construction for analytical purposes but rather is based on a generalized concept of maximum common subgraph and therefore appears in different guises as a common choice of proximity measure for graphs [3,6,18].

Orbifold Functions. Suppose that $\mathcal{X}_\mathcal{G}$ is a graph orbifold with orbifold chart $\pi : \mathcal{X} \to \mathcal{X}_\mathcal{G}$. An *orbifold function* is a mapping of the form $f : \mathcal{X}_\mathcal{G} \to \mathbb{R}$. The *lift* of f is a function $\tilde{f} : \mathcal{X} \to \mathbb{R}$ satisfying $\tilde{f} = f \circ \pi$. The lift $\tilde{f}$ is invariant under group actions of $\mathcal{G}$, that is $\tilde{f}(\boldsymbol{x}) = \tilde{f}(\gamma(\boldsymbol{x}))$ for all $\gamma \in \mathcal{G}$.

Fundamental Domains. For $\boldsymbol{x} \in \mathcal{X}$, we define the *isotropy group* of $\boldsymbol{x}$ as the set o all elements of $\mathcal{G}$ that fix $\boldsymbol{x}$, that is $\mathcal{G}_{\boldsymbol{x}} = \{\gamma \in \mathcal{G} \ : \ \gamma\boldsymbol{x} = \boldsymbol{x}\}$. The isotropy group of $\boldsymbol{x}$ is trivial if $\mathcal{G}_{\boldsymbol{x}} = \{\mathrm{id}\}$. Obviously, if the isotropy group of $\boldsymbol{x}$ is trivial, then the isotropy group of $\gamma\boldsymbol{x}$ is trivial for all $\gamma \in \mathcal{G}$. We say a structure $X \in \mathcal{X}_\mathcal{G}$ has trivial isotropy group if the isotropy group of any of its vector representations is trivial.

A *fundamental domain* of $\mathcal{G}$ in $\mathcal{X}$ is a closed subset $\mathcal{D} \subset \mathcal{X}$ with $\mathcal{X} = \bigcup_{\gamma \in \mathcal{G}} \gamma\mathcal{D}$ and $\mathrm{int}(\gamma_1 \mathcal{D}) \cap \mathrm{int}(\gamma_2 \mathcal{D}) = \emptyset$ for all $\gamma_1, \gamma_2 \in \mathcal{G}$ with $\gamma_1 \neq \gamma_2$. Suppose that $\boldsymbol{x} \in \mathcal{X}$ is a vector representation with trivial isotropy group. A *Dirichlet fundamental domain* of $\boldsymbol{x}$ is a fundamental domain satisfying

$$\mathcal{D}_{\boldsymbol{x}} = \left\{ \boldsymbol{y} \in \mathcal{X} \ : \ \|\boldsymbol{x} - \boldsymbol{y}\| \leq \|\boldsymbol{x} - \gamma\boldsymbol{y}\|, \ \gamma \in \mathcal{G} \right\}.$$

Let $\mathcal{D}$ be a fundamental domain. We call an injective mapping $\psi : \mathcal{X}_\mathcal{G} \to \mathcal{D}$ with $\pi(\psi(X)) = X$ a *lift* of $\mathcal{X}_\mathcal{G}$ into $\mathcal{D}$. Note that two different lifts of $\mathcal{X}_\mathcal{G}$ into the same fundamental domain are equal almost everywhere.

The Disintegration Formular. We assume that $(\mathcal{X}, \|\cdot\|)$ is the Euclidean space and $(\mathcal{X}, \mathfrak{B}, \lambda)$ is the Lebesgue-Borel measure space. The action of the finite group $\mathcal{G}$ on $\mathcal{X}$ induces the measure space $(\mathcal{X}_\mathcal{G}, \mathfrak{B}_\mathcal{G}, \lambda_\mathcal{G})$, where

$$\mathfrak{B}_\mathcal{G} = \left\{ \mathcal{B} \subset \mathcal{X}_\mathcal{G} \ : \ \pi^{-1}(\mathcal{B}) \in \mathfrak{B} \right\}$$

and $\lambda_\mathcal{G}$ is the induced quotient measure. Then for any λ-integrable function f on $\mathcal{X}$ the following important disintegration formula holds [4]

$$\int_{\mathcal{X}} f(\boldsymbol{x})\lambda(d\boldsymbol{x}) = \int_{\mathcal{X}_{\mathcal{G}}} f^*(X)\lambda_{\mathcal{G}}(dX),$$

where f^* is the unique function defined by

$$f^*(\pi(\boldsymbol{x})) = \frac{1}{|\mathcal{G}|} \sum_{\gamma \in \mathcal{G}} f(\gamma \boldsymbol{x}).$$

Thus, we can express the disintegration formular as

$$\int_{\mathcal{X}_{\mathcal{G}}} f^*(X)\lambda_{\mathcal{G}}(dX) = \int_{\mathcal{X}} \frac{1}{|\mathcal{G}|} \sum_{\gamma \in \mathcal{G}} f(\gamma \boldsymbol{x})\lambda(d\boldsymbol{x})$$

$$= \frac{1}{|\mathcal{G}|} \sum_{\gamma \in \mathcal{G}} \int_{\mathcal{X}} f(\gamma \boldsymbol{x})\lambda(d\boldsymbol{x}) = \int_{\mathcal{X}} f(\boldsymbol{x})\lambda(d\boldsymbol{x}).$$

For our purposes it is useful to reduce integration on $\mathcal{X}_{\mathcal{G}}$ to fundamental domains. For any fundamental domain $\mathcal{D}$ of $\mathcal{G}$ in $\mathcal{X}$, we have

$$\int_{\mathcal{X}_{\mathcal{G}}} f^*(X)\lambda_{\mathcal{G}}(dX) = \int_{\mathcal{D}} f(\boldsymbol{x})\lambda_{\mathcal{D}}(d\boldsymbol{x}) = |\mathcal{G}| \int_{\mathcal{D}} f(\boldsymbol{x})\lambda(d\boldsymbol{x}),$$

where $\lambda_{\mathcal{D}}$ is the measure defined by some lift $\psi : \mathcal{X} \to \mathcal{D}$. The integral is well-defined, because two lifts into $\mathcal{D}$ are equal almost everywhere.

3   Quotient Gaussians on Graphs

This section extends Gaussian distributions from vector-valued to graph-valued random variables.

A *Gaussian radial function* on the Euclidean space $\mathcal{X} = \mathbb{R}^D$ is of the form

$$\phi(\boldsymbol{x}|a, \boldsymbol{c}, \sigma) = a \cdot \exp\left(-\frac{\|\boldsymbol{x} - \boldsymbol{c}\|^2}{2\sigma^2}\right),$$

where the parameter $a > 0$ controls the *height*, $\boldsymbol{c} \in \mathcal{X}$ is the *center*, and $\sigma > 0$ is the *width* of ϕ. We obtain a *Gaussian distribution* with *mean* $\boldsymbol{c}$ and variance σ^2, if the height a of the radial function ϕ satisfies $a = \left((2\pi)^{D/2}\sigma^D\right)^{-1}$.

In order to extend Gaussian distributions to quotient spaces $\mathcal{X}_{\mathcal{G}}$, we set

$$\phi(X|C, \sigma) = \frac{1}{(2\pi)^{D/2}\sigma^D} \exp\left(-\frac{d(X, C)^2}{2\sigma^2}\right),$$

where $C \in \mathcal{X}_{\mathcal{G}}$ is the center and σ is the width. We define the *quotient Gaussian distribution* on $\mathcal{X}_{\mathcal{G}}$ by

$$f(X) = \frac{1}{a(C, \sigma)} \cdot \phi(X|C, \sigma),$$

where

$$a(C, \sigma) = \int_{X_G} \phi(X|C, \sigma) \lambda_{\mathcal{G}}(dX).$$

is the height that scales f to a density on $\mathcal{X}_{\mathcal{G}}$. The lift $\tilde{\phi}$ of ϕ is of the form

$$\tilde{\phi}(\boldsymbol{x}|C, \sigma) = \frac{1}{(2\pi)^{D/2} \sigma^D} \exp\left(-\frac{\min_{\boldsymbol{c} \in C} \|\boldsymbol{x} - \boldsymbol{c}\|^2}{2\sigma^2}\right)$$

$$= \max_{\boldsymbol{c} \in C} \frac{1}{(2\pi)^{D/2} \sigma^D} \exp\left(-\frac{\|\boldsymbol{x} - \boldsymbol{c}\|^2}{2\sigma^2}\right),$$

giving rise to the lift $\tilde{f} = \tilde{\phi}(\boldsymbol{x}|C, \sigma)/a(C, \sigma)$ of f. This shows that the quotient Gaussian distribution can be viewed as the pointwise maximum of a set of Gaussian radial functions on $\mathcal{X}$ with identical width and height, but distinct centers.

Without loss of generality, we can assume that the center of a quotient Gaussian distribution C has trivial isotropy group in some subspace $\mathcal{U}_{\mathcal{G}}$ of $\mathcal{X}$. Then all graphs $X \in \mathcal{X}_{\mathcal{G}}$ with nonzero density lie also in $\mathcal{U}_{\mathcal{G}}$. In this case, we indentify $\mathcal{X}_{\mathcal{G}}$ with $\mathcal{U}_{\mathcal{G}}$. Now suppose that $\boldsymbol{c}$ is an arbitrary vector representation that projects to C. Truncating the lift $\tilde{f}$ to the fundamental Dirichlet domain $\mathcal{D}_{\boldsymbol{c}}$ of $\boldsymbol{c}$ yields

$$\tilde{f}^t(\boldsymbol{x}) = \begin{cases} \tilde{f}(\boldsymbol{x}) & : \quad \boldsymbol{x} \in \mathcal{D}_{\boldsymbol{c}} \\ 0 & : \quad \boldsymbol{x} \notin \mathcal{D}_{\boldsymbol{c}} \end{cases}.$$

Let ψ lift $\mathcal{X}_{\mathcal{G}}$ into $\mathcal{D}_{\boldsymbol{c}}$. From $f(X) = \tilde{f}^t(\psi(X))$ for all $X \in \mathcal{X}_{\mathcal{G}}$ follows that we can restrict lifts $\tilde{f}$ of quotient Gaussian distributions to truncated Gaussian distributions on a fundamental Dirichlet domain. The height $a(C, \sigma)$ of the truncated Gaussian can be interpreted as the probability of being in the fundamental Dirichlet domain of $\boldsymbol{c}$. When restricting to $\mathcal{D}_{\boldsymbol{c}}$, we can rewrite the lift $\tilde{\phi}(\boldsymbol{x}|C, \sigma)$ and height $a(C, \sigma)$ by

$$\tilde{\phi}(\boldsymbol{x}|\boldsymbol{c}, \sigma) = \frac{1}{(2\pi)^{D/2} \sigma^D} \exp\left(-\frac{\|\boldsymbol{x} - \boldsymbol{c}\|^2}{2\sigma^2}\right)$$

$$a(\boldsymbol{c}, \sigma) = \int_{\mathcal{D}_{\boldsymbol{c}}} \tilde{\phi}(\boldsymbol{x}|\boldsymbol{c}, \sigma) \lambda(d\boldsymbol{x}),$$

giving rise to express the truncated Gaussian as $\tilde{f}^t(\boldsymbol{x}) = \tilde{\phi}(\boldsymbol{x}|\boldsymbol{c}, \sigma)/a(\boldsymbol{c}, \sigma)$. Note that we have $a(\boldsymbol{c}, \sigma) \leq 1$, where equality $a(\boldsymbol{c}, \sigma) = 1$ means that the truncated Gaussian is actually a Gaussian distribution on $\mathcal{D}_{\boldsymbol{c}} = \mathcal{X}$. In this case, $\mathcal{G}$ is the trivial group and the quotient space $\mathcal{X}_{\mathcal{G}}$ coincides with $\mathcal{X}$. In what follows, we assume that $\mathcal{G}$ is non-trivial, that is $a(\boldsymbol{c}, \sigma) < 1$.

The center $\boldsymbol{c}$ and the squared width σ^2 of a truncated Gaussian $\tilde{f}^t(\boldsymbol{x})$ on $\mathcal{D}_{\boldsymbol{c}}$ are not the expectation $\mathbb{E}[\boldsymbol{x}]$ and variance $\mathbb{V}[\boldsymbol{x}]$ of this distribution. The expectation and variance can be obtained from the center and squared width plus an adjustment for the truncation on the distribution

$$\mathbb{E}[\boldsymbol{x}] = \boldsymbol{c} + \delta_{\mathbb{E}}(\boldsymbol{c}, \sigma) \tag{1}$$

$$\mathbb{V}[\boldsymbol{x}] = \sigma^2 + \delta_{\mathbb{V}}(\boldsymbol{c}, \sigma) \tag{2}$$

Intuitively, we expect that the adjustment $\delta_{\mathbb{E}}(\boldsymbol{c}, \sigma)$ shifts the expectation $\mathbb{E}[\boldsymbol{x}]$ into the appropriate tail of the distribution. A closed form solution of $\mathbb{E}[\boldsymbol{x}]$ and $\mathbb{V}[\boldsymbol{x}]$ are unknown (at least to the authors). However, note that the goal is to make inferences on the center and width of a quotient Gaussian distribution rather than inferring its expectation and variance.

4   Maximum-Likelihood

The General Case. Let $(\mathcal{X}_{\mathcal{G}}, \mathfrak{B}_{\mathcal{G}}, \lambda_{\mathcal{G}})$ be the quotient measure space induced by the group action of $\mathcal{G}$ on the measure space $(\mathcal{X}, \mathfrak{B}, \lambda)$. We assume that all measures under consideration are complete. A *parametric model* for $\mathcal{X}_{\mathcal{G}}$ with parameter space $\Theta_{\mathcal{G}}$ is a family $\mathcal{F} = \{f_{\Theta} : \Theta \in \Theta_{\mathcal{G}}\}$ of probability density functions on $\mathcal{X}_{\mathcal{G}}$ with respect to the quotient measure $\lambda_{\mathcal{G}}$.

Suppose that $\mathcal{S} = \{X_1, \ldots, X_N\} \subseteq \mathcal{X}_{\mathcal{G}}$ is a sample of N independent and identically distributed structures coming from some unknown distribution $f_{\theta^*} \in \mathcal{F}$. The goal is to estimate the unknown parameter θ^* using the maximum likelihood method. The likelihood is defined by

$$\mathcal{L}(\theta \,|\, \mathcal{S}) = \prod_{i=1}^{N} f_{\theta}(X_i)$$

The maximum likelihood estimate of θ^* is defined by the value $\hat{\theta}$ that maximizes the likelihood $\mathcal{L}(\theta \,|\, \mathcal{S})$. Instead of $\mathcal{L}(\theta \,|\, \mathcal{S})$, it is sometimes more convenient to maximize the log-likelihood

$$\ell(\theta) = \ln \mathcal{L}(\theta \,|\, \mathcal{S}) = \sum_{i=1}^{N} \ln f_{\theta}(X_i).$$

Quotient Gaussians. We apply the maximum likelihood to the case, where the samples are drawn from a quotient Gaussian distribution with unknown center C^* and width σ^*. The parametric model is then of the form

$$\mathcal{F} = \{f_{\Theta} : \Theta = (C, \sigma), C \in \mathcal{X}_{\mathcal{G}}, \sigma \geq 0\}.$$

For a given $\Theta = (C, \sigma)$, we first choose an arbitrary vector representation $\boldsymbol{c} \in C$ with trivial isotropy group and replace Θ by $\boldsymbol{\theta} = (\boldsymbol{c}, \sigma)$ and the quotient Gaussian f_{Θ} by a truncated Gaussian $\tilde{f}_{\theta}^{t}(\boldsymbol{x})$ defined on the fundamental domain $\mathcal{D}_{\boldsymbol{c}}$.

Suppose that ψ lifts $\mathcal{X}_{\mathcal{G}}$ into $\mathcal{D}_{\boldsymbol{c}}$. Then $\tilde{\mathcal{S}} = \{\boldsymbol{x_1}, \ldots, \boldsymbol{x_N}\} \subseteq \mathcal{D}_{\boldsymbol{c}}$ with $\boldsymbol{x}_i = \psi(X_i)$ is a lift of the sample $\mathcal{X}$. The set $\tilde{\mathcal{S}}$ is uniquely defined if $\tilde{\mathcal{S}} \subseteq \text{int}(\mathcal{D}_{\boldsymbol{c}})$. We lift the parametric model of $\mathcal{X}_{\mathcal{G}}$ to a parametric model of $\mathcal{X}$ by defining

$$\tilde{\mathcal{F}} = \left\{\tilde{f}_{\boldsymbol{\theta}}^{t} : \boldsymbol{\theta} = (\boldsymbol{c}, \sigma) \in \mathcal{X} \times \mathbb{R}_+, \tilde{f}_{\boldsymbol{c}, \sigma}^{t} \circ \psi = f_{\pi(\boldsymbol{c}), \sigma}\right\}.$$

The lifted log-likelihood to be maximized is then of the form

$$\tilde{\ell}(\boldsymbol{\theta}) = \sum_{i=1}^{N} \ln \tilde{f}_{\boldsymbol{\theta}}^{t}(\boldsymbol{x}_i).$$

Maximizing the Likelihood. Since the truncated Gaussian $\tilde{f}_{\theta}^t$ is differentiable on $\mathcal{D}_c$ with respect to θ, the gradient of the log-likelihood exists and is of the form

$$
\nabla_{\theta}\,\tilde{\ell}(\theta) = \begin{pmatrix} \nabla_c\,\tilde{\ell}(\theta) \\ \nabla_{\sigma}\,\tilde{\ell}(\theta) \end{pmatrix} = \begin{pmatrix} \dfrac{N}{\sigma^2}\left\{ \delta_{\mathbb{E}}(c,\sigma) + c - \dfrac{1}{N}\displaystyle\sum_{i=1}^{N} x_i \right\} \\[3ex] \dfrac{N}{\sigma^3}\left\{ \delta_{\mathbb{V}}(c,\sigma) + \sigma^2 - \dfrac{1}{N}\displaystyle\sum_{i=1}^{N} \|x_i - c\|^2 \right\} \end{pmatrix}
$$

Recall that $\delta_{\mathbb{E}}(c,\sigma)$ and $\delta_{\mathbb{V}}(c,\sigma)$ are the adjustments defined in (1) and (2). Setting the derivatives to zero and solving the equations accordingly yields

$$
c = \frac{1}{N}\sum_{i=1}^{N} x_i - \delta_{\mathbb{E}}(c,\sigma) \tag{3}
$$

$$
\sigma^2 = \frac{1}{N}\sum_{i=1}^{N} \|x_i - c\|^2 - \delta_{\mathbb{V}}(c,\sigma). \tag{4}
$$

Note that equations (3) and (4) are not closed-form solutions for the center c and width σ, because the adjustments $\delta_{\mathbb{E}}(c,\sigma)$ and $\delta_{\mathbb{V}}(c,\sigma)$ involve integrals depending on both, c and σ.

Equation (3) shows that the maximum likelihood estimate of the center is an estimate of the unknown expectation $\mathbb{E}[x]$ by the sample mean plus an adjustment from the expectation to the center. Similarly, from equation (4) follows that the maximum likelihood estimate of the squared variance is an estimate of the unknown variance $\mathbb{V}[x]$ plus an adjustment from the variance to the width.

In principle, the adjustments $\delta_{\mathbb{E}}(c,\sigma)$ and $\delta_{\mathbb{V}}(c,\sigma)$ can be approximated using Monte Carlo integration. In a practical setting, however, this approach turns out to be computationally too intensive, because it requires numerous NP-hard distance calculations. For this reason, we sacrifice exactness of the solution for the sake of computational efficiency and ignore both adjustments $\delta_{\mathbb{E}}(c,\sigma)$ and $\delta_{\mathbb{V}}(c,\sigma)$. Then we obtain

$$
c = \frac{1}{N}\sum_{i=1}^{N} x_i \tag{5}
$$

$$
\sigma^2 = \frac{1}{N}\sum_{i=1}^{N} \|x_i - c\|^2. \tag{6}
$$

By definition of the Dirichlet fundamental domain $\mathcal{D}_c$, the sample vectors x_i are optimally aligned against the center c. If (x_i, c) are optimal alignments, then the graph $C = \pi(c)$ is a sample Frechet mean of the sample graphs $X_i = \pi(x_i)$, that is

$$
C = \arg\min_{Y \in \mathcal{X}_g} \sum_{i=1}^{N} d(X_i, Y)^2.
$$

An efficient method to approximate a sample Frechet mean is the incremental arithmetic mean method proposed by [9]

$$c(1) = x_1$$

$$c(i) = \frac{i-1}{i}c(i-1) + \frac{1}{i}x_i, \qquad 1 < i \leq N,$$

where $(c(i-1), x_i)$ is an optimal alignment for all $i \in \{2, \ldots, N\}$ and x_1 is an arbitrarily chosen vector representation that projects to sample graph $X_1 \in \mathcal{S}$. We obtain the maximum likelihood estimate of the center by setting $\hat{c} = c(N)$ and projecting to $\hat{C} = \pi(\hat{c})$. Finally, we use the maximum likelihood estimate $\hat{c}$ to determine the maximum likelihood estimate of the width according to

$$\hat{\sigma}^2 = \frac{1}{N}\sum_{i=1}^{N}\|x_i - \hat{c}\|^2,$$

where $(\hat{c}, x_i)$ are optimal alignments.

5   Experiments

We conducted first experiments to investigate the performance and behavior of the proposed approximation of the maximum likelihood method for graphs. For this purpose, we considered image classification problems.

Data. We selected the letter, grec, and fingerprint data sets from the IAM graph database repository [13]. Each data set is divided into a training, validation, and a test set. Table 1 provides a summary of the main characteristics of the data sets. For further details we refer to [13].

Methods. For each of the three data sets, we assumed that the conditional probability density functions $p(X|y_i)$ of class y_i is a quotient Gaussian with center C_i and width σ_i for all $i \in \{1, \ldots, K\}$. We considered two cases: (1) the widths σ_i may differ class-wise, and (2) the classes have common width $\sigma_i = \sigma$. In both cases, we used the maximum likelihood estimates of the parameters of the quotient Gaussians and then applied a naive Bayes (bayes$_1$, bayes$_2$) classifier. We compared the proposed maximum likelihood based classifiers against the following methods: (1) k-nearest neighbor (knn), (2) the similarity kernel in conjunction with the SVM (sk-svm) proposed by [15], (3) the family of Lipschitz embeddings in conjunction with SVM (le-svm) also proposed by [15], and (4) the learning graph quantization methods (lgq, lgq2.1) proposed by [10].

Experimental Setup. For each data set, we used the training and validation set for parameter estimation of the bayes, rbf, and knn classifier. To obtain the classification accuracy, we used the corresponding test set. For calculating graph distances, we applied the extended Bron-Kerbosch algorithm [11] with clique selection and maximum number of $10\,|V_Z|$ recursive calls, where V_Z is the vertex set of the underlying association graph.

Table 1. Summary of main characteristics of the data sets. The tiny numbers in parentheses show the size of the training, validation, and test set, respectively.

data set	#(classes)	avg(nodes)	max(nodes)	avg(edges)	max(edges)
letter (750, 750, 750)	15	4.7	8	3.1	6
grec (286, 286, 528)	22	11.5	24	11.9	29
fingerprint (500, 300, 2000)	4	8.3	26	14.1	48

Table 2. Classification accuracy (in %)

	knn	sk-svm	le-svm	lgq	lgq2.1	bayes$_1$	bayes$_2$
letter	82.0	79.1	92.5	81.5	85.7	80.4	81.3
grec	96.8	94.9	96.8	86.2	92.6	80.3	89.9
fingerprint	80.0	41.0	82.8	79.9	81.5	78.1	79.2

Results. Table 2 summarizes the results. Results of sk-svm and le-svm were taken from [15] and of lgq and lgq2.1 from [12]. Since both, sk-svm and le-svm, refer to a family of related methods rather than a single method, Table 2 presents the best result over all methods of the le-svm family for each data set. In doing so, the comparison is optimistically biased towards sk-svm and le-svm. In contrast to [10], the lgq variants use a single prototype for each class.

Despite neglecting the adjustments $\delta_{\mathbb{E}}$ and $\delta_{\mathbb{V}}$, the results show that the approximated versions of the maximum likelihood method used by the bayes classifier work reasonably well. The generalization capability of the bayes classifier improves on all data sets when using the same width for each class. Compared to the other classifiers, the Bayes classifier is of limited capability. This finding is in line with results for vectorial representations [2]. In contrast to the other classifiers, however, the Bayes classifier is extremely fast to train and use for classification, scales well with number of training patterns and requires less memory than, in particular, knn and the svm-based methods.

This results together with the improved results of lgq and lgq2.1 using more than one prototype per class [10] suggest to extend the proposed approach to mixture models of quotient Gaussians. Mixture models of quotient Gaussian in turn theoretically justify more powerful approaches such as robust soft learning graph quantization [12].

6 Conclusion

The quotient Gaussian on attributed graphs emulates the Gaussian distribution by clustering around a single mean. Geometrically, the quotient Gaussian on graphs reduces to a truncated Gaussian on vectors. Via the truncated Gaussian, we can estimate the center and width of the quotient Gaussian using the maximum likelihood method. The benefit of this approach is that it forms the mathematical foundation for estimating the parameters of other distributions on graphs using the maximum likelihood method. This includes, in particular, mixture models of quotient Gaussians, which form the theoretical basis of robust soft learning graph quantization as proposed by [12].

References

1. Bagdanov, A.D., Worring, M.: First order Gaussian graphs for efficient structure classification. Pattern Recognition 36, 1311–1324 (2003)
2. Caruna, R., Niculescu-Mizil, A.: An empirical comparison of supervised learning algorithms. In: ICML (2006)
3. Cour, T., Srinivasan, P., Shi, J.: Balanced graph matching. In: NIPS (2006)
4. Eaton, M.L.: Group Invariance Applications in Statistics. In: Institute of Mathematical Statistics and American Statistical Association (1989)
5. Friedman, N., Koller, D.: Being bayesian about network structure. Machine Learning 50(1–2), 95–125 (2003)
6. Gold, S., Rangarajan, A.: Graduated Assignment Algorithm for Graph Matching. IEEE Transactions on PAMI 18, 377–388 (1996)
7. Hong, P., Huang, T.S.: Spatial pattern discovery by learning a probabilistic parametric model from multiple attributed relational graphs. Journal of Discrete Applied Mathematics (2004)
8. Jain, B., Obermayer, K.: Structure Spaces. Journal of Machine Learning Research 10, 2667–2714 (2009)
9. Jain, B.J., Obermayer, K.: Algorithms for the sample mean of graphs. In: Jiang, X., Petkov, N. (eds.) CAIP 2009. LNCS, vol. 5702, pp. 351–359. Springer, Heidelberg (2009)
10. Jain, B., Srinivasan, S.D., Tissen, A., Obermayer, K.: Learning Graph Quantization. In: Hancock, E.R., Wilson, R.C., Windeatt, T., Ulusoy, I., Escolano, F. (eds.) SSPR&SPR 2010. LNCS, vol. 6218, pp. 109–118. Springer, Heidelberg (2010)
11. Jain, B., Obermayer, K.: Extending Bron Kerbosch for the Maximum Weight Clique Problem. arXiv:1101.1266v1 (2011)
12. Jain, B., Obermayer, K.: Generalized Learning Graph Quantization. In: Jiang, X., Ferrer, M., Torsello, A. (eds.) GbRPR 2011. LNCS, vol. 6658, pp. 122–131. Springer, Heidelberg (2011)
13. Riesen, K., Bunke, H.: IAM graph database repository for graph based pattern recognition and machine learning. In: da Vitoria Lobo, N., Kasparis, T., Roli, F., Kwok, J.T., Georgiopoulos, M., Anagnostopoulos, G.C., Loog, M. (eds.) S+SSPR 2008. LNCS, vol. 5342, pp. 287–297. Springer, Heidelberg (2008)
14. Solé-Ribalta, A., Serratosa, F.: A Structural and Semantic Probabilistic Model for Matching and Representing a Set of Graphs. In: Torsello, A., Escolano, F., Brun, L. (eds.) GbRPR 2009. LNCS, vol. 5534, pp. 164–173. Springer, Heidelberg (2009)
15. Riesen, K., Bunke, H.: Graph Classification by Means of Lipschitz Embedding. IEEE Transactions on Systems, Man, and Cybernetics 39(6), 1472–1483 (2009)
16. Sanfeliu, A., Serratosa, F., Alquezar, R.: Second-Order Random Graphs for modelling sets of Attributed Graphs and their application to object learning and recognition. IJPRAI 18(3), 375–396 (2004)
17. Serratosa, F., Alquezar, R., Sanfeliu, A.: Function-Described Graphs for modelling objects represented by attributed graphs. Pattern Recognition 36(3), 781–798 (2003)
18. Umeyama, S.: An eigendecomposition approach to weighted graph matching problems. IEEE Transactions on PAMI 10(5), 695–703 (1988)
19. Wong, A.K.C., Constant, J., You, M.L.: Random Graphs Syntactic and Structural Pattern Recognition. World Scientific, Singapore (1990)
20. Zhang, D.Q., Chang, S.F.: Learning Random Attributed Relational Graph for Part-based Object Detection. Columbia University ADVENT Technical Report, #212-2005-6 (2005)

Towards Performance Evaluation of Graph-Based Representation

Salim Jouili and Salvatore Tabbone

LORIA UMR 7503 - University of Nancy 2
BP 239, 54506 Vandoeuvre-lès-Nancy Cedex, France
{salim.jouili,tabbone}@loria.fr

Abstract. Graphs give a universal and flexible framework to describe the structure and relationship between objects. They are useful in many different application domains like pattern recognition, computer vision and image analysis. In the image analysis context, images can be represented as graphs such that the nodes describe the features and the edges describe their relations. In this paper we, firstly, review the graph-based representations commonly used in the literature. Secondly, we discuss, empirically, the choice of a graph-based representation on three different image databases and show that the representation has a real impact on the method performances and experimental results in the literature on graph performance evaluation for similarity measures should be considered carefully.

1   Introduction

Graphs are very flexible data structures that offer a great capacity of abstraction. In pattern recognition, graph-based representation has been widely used this last decade for several types of images such graphic symbols [6], shape [21], ancient documents [7], etc. The major advantage of graphs is the explicit representation of the relational configuration between the different primitives of an object. Typically, this representation is invariant to several types of changes (rotation, translation ...). In addition, through the graph-based representation, one can transform a recognition problem to a graph matching problem. In the literature, we distinguish several methods of extracting graphs from images. Generally, nodes correspond to salient features found in the image, and edges describe relationships that can link these characteristics. Naturally, salient features and their relations depend on the type of the considered images.

In this paper we review the graph-based representations commonly used in the literature. The first family contains the graph based on the point of interest extracted from the image. The second family of graphs is based on the layout of regions included in an image. In the third family, we introduce the skeleton graph as a graph-based representation of images. Then, in the last family, we describe the graph extraction using the spatial information between image primitives. Finally, using three different databases, we evaluate the performance of these methods. We show experimentally that the graph-based representation has a real

X. Jiang, M. Ferrer, and A. Torsello (Eds.): GbRPR 2011, LNCS 6658, pp. 72–81, 2011.
© Springer-Verlag Berlin Heidelberg 2011

impact on the recognition rate. Moreover, following the chosen representation, one approach can be better or not than others. In this perspective, we can say that experimental results in the literature on graph performance evaluation for similarity measures should be considered carefully.

2 Graph-Based Representation

2.1 Graph Based on Points of Interest

The notion of interest points was introduced by Moravec [10]. It allows to locate the points where the image signal is rich in information. In an image, interest points are usually corners, junctions or points of high textural variations. They are widely used in the literature for image matching. Among the algorithms for detecting interest points, we distinguish the algorithm of Harris [5].

As part of the graph-based representation of images, interest points provide important information on the location of information-rich areas in the image. In the literature, several studies have used the interest points to construct graphs. In [14,22], the authors applied the Delaunay triangulation on these points to extract the graph representation of an image. Briefly, let $\mathcal{PI}$ the set of the interest point of an image I, the triangulation of $\mathcal{PI}$ defines a set of triangles $\mathcal{T} = \{T_1, ..., T_n\}$ such that:

- The vertices of the triangles are the points of $\mathcal{PI}$.
- $\forall p \in \mathcal{PI}, \forall i \in [1, n]$, then p is a vertice of T_i or $p \notin T_i$
- $\mathcal{T} = \{T_1, ..., T_n\}$ is a partition of the image I.

2.2 Region Adjacency Graph

The region adjacency graph (RAG) was introduced by Rosenfeld [15]. It consists of modeling the adjacency relations between regions in a segmented image. A RAG is a planar graph representing the image regions by nodes and adjacency relations between these regions by edges. Thus, as illustrated in Figure 1 two nodes of the graph are connected by an edge if their corresponding regions are adjacent. In the literature, several formal definitions of RAG coexist. The major difference between these definitions is related to the orientation of the graph.

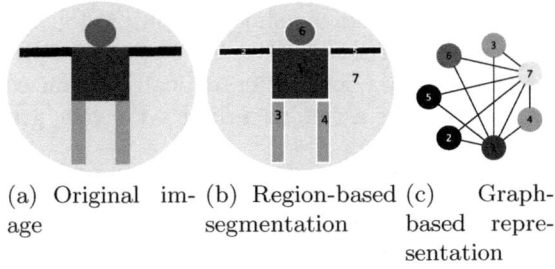

(a) Original image

(b) Region-based segmentation

(c) Graph-based representation

Fig. 1. An example of a graph-based representation

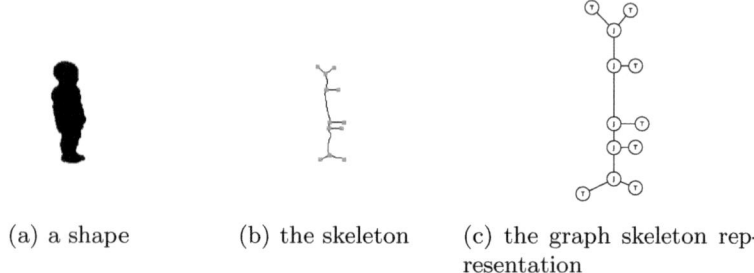

(a) a shape   (b) the skeleton   (c) the graph skeleton representation

Fig. 2. An example of graph skeleton representation

2.3 Skeleton Graph

In shape recognition, the skeleton of a shape is a thin version of that shape. It is composed by a set of thinned curves and lines which are equidistant from the shape's boundaries[1]. The thickness of a shape skeleton is one pixel.

This shape representation is used in several works based on graphs. Generally, the skeleton graph results in the classification of skeleton pixels into three classes; junction points, branch points and endpoints [4,16,21]. These sets of points are defined as follows:

- A junction point in a skeleton is an intersection between at least three branches.
- A branch point is a point belonging to a branch.
- An endpoint is an endpoint of a branch, which is not connected to any other branches.

Thus, in a skeleton graph $G = (V, E)$, junctions and endpoints form the set of nodes V, and the branch point sets correspond to the edges (i.e. E) of the graph. Figure 2(b) illustrates the skeletonization of the shape of figure 2(a), the characteristic points are marked by green squares. The skeleton graph is shown in Figure 2(c). In this graph we have assumed that each node is labelled with a letter describing the point type (T for endpoints and J for junction points). Indeed, the graph is often enriched by other information such as branch length, morphological distance, the curvature variance of the branches [16].

Note that this method of skeleton graph extraction is not unique. In the literature, we can also distinguish the shock graph [20,18], which considers the skeleton of a shape as a set of shocks. Briefly, a shock graph is an abstraction that decomposes a shape into a set of hierarchically organized primitive parts. Thus, a graph of shocks can be a tree or a directed acyclic graph.

2.4 Spatial Relation Graph

With some types of images, spatial relations [2] between different parts of an image are important for better representation and thus better recognition

---

[1] Each point on the skeleton is equidistant from at least two opposite points on the boundary.

performance. In this sense, the spatial relation graphs use spatial relationships for image representation. Specifically, a spatial relation graph, generally, is a directed graph where the nodes represent the primitive components of the image and edges represent the spatial relationships between these components. In [17,23], the authors use graphs to represent graphical symbols. They start by breaking down each symbol in different families of primitives (circles, endpoints ...). Then, they determine the spatial relationships between each family. The graphs are then constructed by storing in the nodes the primitive characteristics (using local descriptors) and the spatial relationships in the edges.

The main difference between this kind of graph-based representation and the previous representations is the fact that in a spatial relation graph the edge must be labelled. Indeed, the edge labels must indicates the spatial relation between the adjacent nodes.

3   Discussion

In the previous section, we presented four major families of graph-based representation commonly used in the context of the pattern recognition. We note that in each family, we can found several graph extraction methods that can differ by the information we put into the graph. We can say that the representation and the content of a graph depend on the image types and the needs of the recognition systems. For example, RAG extraction methods used for ancient document images [7] are not suitable for graphical symbols [6]. Moreover, the definition of a region in an image depends on the objectives and techniques considered. In this sense, a region is sometimes defined as a pixel set (contiguous or non-contiguous pairs) with similar values. A region can also be defined as a set of features (components) belonging to the same semantic family. Several other definitions exist, but these various definitions do not limit the use of graphs. Indeed, the RAG representation can be adapted to all these definitions. This can be also applied to other families of graphs (points of interest, spatial relations or skeleton graphs) for which the primitives can have various definitions. Indeed, we can define several strategies for extracting graphs from each image type, and this justifies the high capacity of the graph-based representation. This is can be also enhanced by adding labels to nodes and edges. These labels provide an additional means to better represent the images. Concretely, the specification of labels defines suitably the relations between image primitives. For example, in the spatial relation graphs, the labels of the edges represent explicitly the topological layout of primitives.

These four families of graph-based representation are not the unique possibilities to represent one image by a graph. In fact, a graph-based representation can be carried out manually with the help (totally or partially) of a user. Moreover, one can consider the combination of two or more families of representation. For example, it is possible to combine the graphs of points of interest with the graphs of spatial relations, by adding spatial information into the edges.

However, this wide range to represent and define graphs poses a great challenge which can be summarized into the following question:

– Does the choice of a graph-based representation depend on the image type?

Intuitively, the answer to the this question is *"yes"*. To argue our response, let us consider the following simple hypothesis; *A graph should define the best representation of a considered image* i.e, a graph must contain the salient features of an image. Since these features depend on the image type, hence the graph necessarily depends also on the image type. Therefore, the choice of a graph representation and the information we put into the graph should be guided by the image type. This response is partially because many graph-based representation techniques can be appropriate for a given application context. In this perspective, we consider in the following an empirical study of performance techniques where the choice of the best graph extraction method is set from the experimental results.

4   Empirical Impact of the Graph-Based Representation

In the previous section, we have considered, intuitively, that the choice of a graph-based representation technique depends on the image type. Consequently, the use of a suitable technique involves good recognition results. Contrarily, an unsuitable technique for extracting graph involves poorer results. In this section, we study empirically the impact of the graph extraction technique on the recognition results. To do this, we consider three image data-sets: the first data-set contains binary image of shapes, the second one consists of logo images, and the third is a collection of ancient document images. In the following we examine the impact of the graph-based representation technique for each database.

4.1   Data

– **Shape database:** This database is provided by the LEMS Laboratory at Brown University[2]. It contains 216 binary shapes divided into 18 classes of 12 shapes (Fig. 3). Each class of the database contains 12 transformations of each shape, occlusion, rotation and scaling.

Fig. 3. Samples from the Shape database [19]

– **Logo database:** This database [1] (see Figure 3) consists of binary images of trademark-logos (see Figure 4). The graph database used in our experiments consists of 80 graphs, with 10 classes and 8 graphs per class.

Fig. 4. Samples from the Logo database

---

[2] http://www.lems.brown.edu/

Fig. 5. Samples of ancient ornamental letters images

- **Ornamental letters database:** The ornamental letters database[3], also called *lettrines* extracted from documents of the fifteenth and sixteenth century. They correspond to images widely used in books and very reused over time as initials in the beginning of chapters or paragraphs (see Figure 5 for examples). In this paper we used a subset from this database that consists of 280 images, 4 classes and 70 images per class.

4.2 Experimental Setup

For each database, except Shape, three different graph representations are used. These representations are:

RAG : This representation consists of representing each image by a region adjacency graph. In fact, a graph is extracted from a region-based segmentation [3] of the considered image. The nodes of the graph represent the regions and the edges describe their adjacency relationships. This representation is not used for the Shape database because the adjacency between regions is not a discriminating feature. Indeed, all images are binary and consist of two regions.

H+D : Here, graphs are extracted by the Delaunay triangulations on the detecting points of interest using Harris algorithm [5].

Skel : The representation consists of a skeleton graph such that the junctions and endpoints form the nodes and the branch points correspond to the edges.

The impact of the graph-based representations is studied in a classification context. We make use of the k-nearest neighbours classifier because it can be applied directly to the graph domain using any graph similarity measure without any further adaptations. The number of nearest neighbours for k-nearest neighbour classifier is set to 3. For all databases, in order to obtain reliable results, we randomly reshuffle test and training sets. Thirty percent of the samples were randomly assigned to the training set and the remaining samples were used as a test set.

To deepen our study we use six graph similarity measures. For the sake of completeness, we briefly describe these methods in the following:

Jouili [8] : this method combines a node signatures extraction with an optimal assignment method for approximating the graph edit distance. Concretely, the authors use the node signatures to consider the graph edit distance as an instance of an assignment problem which can be solved by the Hungarian method.

---

[3] Provided by the CESR - University of Tours on the context of the ANR Navidomass project http://l3iexp.univ-lr.fr/navidomass/

Robles-Kelly [14] : this method uses a spectral method to represent graphs by strings, and then the similarity of graphs is measured according to the edit distance of strings in a probabilistic framework.

Lopresti [9] : this method introduces the paradigm of graph probing. This technique consist on using a probe into graphs to determine some particular information. The measure of similarity between two graphs is an L1 norm distance of two corresponding vectors. For the construction of vectors, Lopresti present three classes of construction each one led by a question, Class0: "*How many vertices with degree n are present in graph G = (V,E)?*", Class 1: "*How many vertices with in-degree m and out-degree n are present in G?*", Class2 : "*How many vertices labelled as **att** are present in G?*". The use of such class depends on the type of graph.

Therefore, for each graph, a representative vector is computed. Concretely, for the Class0, let $G = (V,E)$ be an undirected graph, the vector associated to G is: $PR(G) \equiv (n_0, n_1, n_2, ...)$ where $n_i = |\{ v$ in $V \mid \deg(v) = i\}|$. So the distance between two graphs is $L_1(PR1, PR2)$. In the remainder, we keep the Class0 for the set of experiments.

Papadopoulos [12] : this method introduces the degree sequence of a graph, i.e. the non-increasing sequence of the degrees of vertices in a graph. This degree sequence of a graph is used to compute the distance between two graphs. The distance between two graphs consists of the minimum number of primitive operations which are required such that the two graphs have the same degree sequence.

Neuhaus [11] : this technique uses the A*-Beamsearch algorithm to compute a sub-optimal solution of the graph edit distance. Indeed, they authors use a search tree to represent the optimization problem, such that the root node represents the starting point, inner nodes correspond to partial solutions, and leaf nodes to complete solutions.

Riesen [13] : this approach considers the approximation of the graph edit distance as an instance of an assignment problem. The method computes the edit distance between two graphs based on a bipartite graph matching by means of the Hungarian algorithm and provides sub-optimal edit distance results.

Note that, Neuhaus and Riesen methods need the optimum edit cost functions as parameters. In this paper, the edit costs are computed with the same protocol proposed in [11,13]. We extract a subset, called validation set, from each data set. Then we determine the edit costs that are optimal in the validation sets. We use the same edit cost functions for both algorithms.

Let us recall that the objective of this experiment is to study the impact of graph-based representation and not the performance of graph similarity measures. In this perspective, to answer to our previous question on the link between the graph representation and the image type and indirectly the performance of each graph similarity measure vs the graph-based representation, it is interesting that a graph similarity measure is evaluated on different graph-based representations for each database.

4.3   Results

In Table 1, we present the classification accuracy rates (k-nn classification) performed on the three images databases for each graph similarity measure. Each
line shows the classification results using a measure of distance and different
graph-based representations for each database. We can note that in all lines
of the table, the results are different from one representation to another. This
shows that the choice of the type of graph representation has an impact on the
performance of the classification. By considering the Shape database, the skeleton graph representation provides the best results for almost all graph similarity
measures (four methods among six). Considering the Logo and the Lettrine
databases, it is clear that the region adjacency graph is the best representation for all the similarity measures. That is, the region topology description (for
the Logo and the Lettrine databases) is more discriminate than the skeleton
representation.

From these results, we see clearly that the impact of the graph representation
techniques can highly affect the classification accuracy. For instance, for the
Lettrine database using of the Papadopoulos [12] measure, the difference in the
accuracy rate between the skeleton graphs and the RAGs is around 21% which
is very important. An other interesting point is on the ranking position of these
similarity methods. For instance we can remark that rank of one method may be
different following the graph-based representation. For example with the Logo

Table 1. Classification accuracy. The edit cost line corresponds to the costs used as
parameters for the Neuhaus and Riesen methods, i.e. Edit costs = (node cost, edge
cost).

	Shape		Logo			Lettrine		
	H+D	Skel	RAG	H+D	Skel	RAG	H+D	Skel
Jouili	56,60%	**61.29%**	**93,40%**	76,13%	75,47%	**81.04%**	68.32%	69.92%
Robles-Kelly	57,89%	**67,52%**	**82,03%**	69,33%	71,27%	**72.01%**	59.54%	63.33%
Lopresti	**35,99%**	34,72%	**91,37%**	81,83%	80,93%	**79.82%**	60.85%	58.78%
Papadopoulos	**46,22%**	40,16%	**94,67%**	84,40%	81,87%	**80.00%**	65.43%	62.29%
Riesen	30,06%	**47,73%**	**85,57%**	70,49%	67,93%	**41.87%**	39.25%	40.54%
Neuhaus	30,53%	**46,44%**	**86,70%**	73,93%	69,20%	**42.29%**	40.09%	41.67%
Edit costs	(1.8,0.5)	(0.9,0.5)	(0.3,0.1)	(1.5,0.9)	(1.1,0.1)	(0.3,0.1)	(1.8,0.9)	(1,0.5)

Table 2. Recommendations

	Shape	Logo	Lettrine
Jouili [8]	skeleton	RAG	RAG
Robles-Kelly [14]	skeleton	RAG	RAG
Lopresti [9]	H+D	RAG	RAG
Papadopoulos [12]	H+D	RAG	RAG
Riesen [13]	skeleton	RAG	RAG
Neuhaus [11]	skeleton	RAG	RAG

dataset, the Lopresti[9] method is at the rank 3 for the RAG representation and rank 2 for the skeleton and interest points representations. In this perspective one has to consider carefully the comparative experimental results provided in the literature because following the graph representation the rank between the compared methods could be partially different. We can also consider that a similarity measure approach is robust if its rank is not affected by the graph representation (for example [14] for the shape, [12] for the logo and [8] for the lettrine).

From these results, we deduce also a set of recommendations (Table 2) which consist of the best graph representation for each database and each graph similarity measure. For instance, for the Shape database, we recommend to use the skeleton graph to represent the images with the Jouili [8], Robles-Kelly [14], Riesen [13] or Neuhaus [11] similarity measures. Moreover, if the Lopresti [9] or Papadopoulos [12] approaches are used to classify the Shape database, we recommend the graph representation based on the interest points. For the two other databases the RAG seems to be the best choice whatever the measure of similarity.

5 Conclusion

In this paper we have investigated how the graph structure can be used to represent the images. We have reviewed the graph-based representations commonly used in the literature. We classified these representations on four families: graph based on points of interest, region adjacency graph, skeleton graph and spatial relation graph. From our empirical studies, we have concluded that the choice of a graph-based representation technique depends on the image type. That is, the choice of a graph representation technique can highly affect the classification accuracy and the ranking position of a similarity measure method. Therefore, we can say that experimental results in the literature on graph performance evaluation for similarity measures should be considered carefully.

References

1. Doermann, D.S., Rivlin, E., Weiss, I.: Logo Recognition. Technical Report CS-TR-3145, University of Maryland, College Park, College Park, MD (1993)
2. Egenhofer, M.J., Shariff, A.R.B.M.: Metric details for natural-language spatial relations. ACM Transactions on Information Systems 16(4), 295–321 (1998)
3. Felzenszwalb, P.F., Huttenlocher, D.P.: Efficient graph-based image segmentation. International Journal of Computer Vision 59(2) (September 2004)
4. Goh, W.-B.: Strategies for shape matching using skeletons. Computer Vision and Image Understanding 110(3), 326–345 (2008)
5. Harris, C., Stephens, M.: A combined corner and edge detection. In: Proc. 4th Alvey Vision Conf., pp. 189–192 (1988)
6. Lladòs, J., Martí, E., Villanueva, J.J.: Symbol recognition by error-tolerant subgraph matching between region adjacency graphs. IEEE Transactions on Pattern Analysis and Machine Intelligence 23(10), 1137–1143 (2001)

7. Jouili, S., Coustaty, M., Tabbone, S., Ogier, J.-M.: Navidomass: Structural-based approaches towards handling historical documents. In: ICPR, pp. 946–949 (2010)
8. Jouili, S., Mili, I., Tabbone, S.: Attributed graph matching using local descriptions. In: Blanc-Talon, J., Philips, W., Popescu, D., Scheunders, P. (eds.) ACIVS 2009. LNCS, vol. 5807, pp. 89–99. Springer, Heidelberg (2009)
9. Lopresti, D.P., Wilfong, G.T.: A fast technique for comparing graph representations with applications to performance evaluation. International Journal on Document Analysis and Recognition 6(4), 219–229 (2003)
10. Moravec, H.: Towards automatic visual obstacle avoidance. In: Proceedings of the 5th International Joint Conference on Artificial Intelligence, p. 584 (August 1977)
11. Neuhaus, M.: Bridging the gap between Graph edit distance and Kernel Machines. PhD thesis, University of Bern (2006)
12. Papadopoulos, A.N., Manolopoulos, Y.: Structure-based similarity search with graph histograms. In: Proceedings of International Workshop on Similarity Search (DEXA IWOSS 1999), pp. 174–178 (September 1999)
13. Riesen, K., Bunke, H.: Approximate graph edit distance computation by means of bipartite graph matching. Image Vision Comput. 27(7), 950–959 (2009)
14. Robles-Kelly, A., Hancock, E.R.: Graph edit distance from spectral seriation. IEEE Transactions on Pattern Analysis and Machine Intelligence 27(3), 365–378 (2005)
15. Rosenfeld, A.: Adjacency in digital pictures. Information and Control 26(1), 24–33 (1974)
16. Ruberto, C.D.: Recognition of shapes by attributed skeletal graphs. Pattern Recognition 37(1), 21–31 (2004)
17. Santosh, K., Wendling, L., Lamiroy, B.: Using spatial relations for graphical symbol description. In: International Conference on Pattern Recognition, pp. 2041–2044. IEEE Computer Society, Los Alamitos (2010)
18. Sebastian, T.B., Klein, P.N., Kimia, B.B.: Recognition of shapes by editing their shock graphs. IEEE Transactions on Pattern Analysis and Machine Intelligence 26(5), 550–571 (2004)
19. Sharvit, D., Chan, J., Tek, H., Kimia, B.B.: Symmetry-based indexing of image databases. Journal of Visual Communication and Image Representation 9, 366–380 (1998)
20. Siddiqi, K., Shokoufandeh, A., Dickinson, S.J., Zucker, S.W.: Shock graphs and shape matching. International Journal of Computer Vision 35(1), 13–32 (1999)
21. Torsello, A., Hancock, E.R.: A skeletal measure of 2d shape similarity. Computer Vision and Image Understanding 95(1), 1–29 (2004)
22. Wilson, R.C., Hancock, E.R., Luo, B.: Pattern vectors from algebraic graph theory. IEEE Transactions on Pattern Analysis and Machine Intelligence 27(7), 1112–1124 (2005)
23. Xiaogang, X., Sun, Z., Peng, B., Jin, X., Liu, W.: An online composite graphics recognition approach based on matching of spatial relation graphs. International Journal on Document Analysis and Recognition 7(1), 44–55 (2004)

Measuring the Distance of Generalized Maps

Camille Combier[1,2], Guillaume Damiand[1,2], and Christine Solnon[1,2]

[1] Université de Lyon, CNRS
[2] Université Lyon 1, LIRIS, UMR5205, F-69622, France

Abstract. Generalized maps are widely used to model the topology of nD objects (such as images) by means of incidence and adjacency relationships between cells (vertices, edges, faces, volumes, ...). In this paper, we define a first error-tolerant distance measure for comparing generalized maps, which is an important issue for image processing and analysis. This distance measure is defined by means of the size of a largest common submap, in a similar way as a graph distance measure may be defined by means of the size of a largest common subgraph. We show that this distance measure is a metric, and we introduce a greedy randomized algorithm which allows us to efficiently compute an upper bound of it.

1 Introduction

Generalized maps are widely used to model the topology of nD objects subdivided in cells (*e.g.*, vertices, edges, faces, volumes, ...) by means of incidence and adjacency relationships between these cells. In 2D, they are an extension of planar graphs, and a generalization for higher dimensions. In particular, generalized maps are very well suited to model 2D and 3D images, and there exist efficient algorithms for extracting maps from images [DBF04, Dam08]. In [DDLHJ+09], we have defined two basic comparison tools, *i.e.*, map isomorphism (which involves deciding if two generalized maps are equivalent) and submap isomorphism (which involves deciding if a copy of a pattern generalized map may be found in a target generalized map), and we have proposed efficient polynomial time algorithms for solving these two problems. However, these decision algorithms cannot be used to quantify the distance between two generalized maps as soon as there is no inclusion relation between them.

In this paper, we define a first error-tolerant distance measure for comparing generalized maps, which is an important issue for image processing and analysis. This distance measure is defined by means of the size of a largest common submap, in a similar way as a graph distance measure is defined by means of the size of a largest common subgraph in [BS98]. In Section 2, we briefly recall definitions related to generalized maps. In Section 3, we define the distance measure and we show that it is a metric distance. In Section 4, we describe a greedy randomized algorithm which allows us to efficiently compute an approximation of this distance measure. In Section 5, we give first experimental results showing that our algorithm is able to compute good approximations of the distance.

X. Jiang, M. Ferrer, and A. Torsello (Eds.): GbRPR 2011, LNCS 6658, pp. 82–91, 2011.

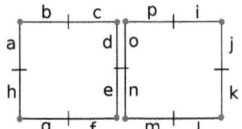

	a	b	c	d	e	f	g	h	i	j	k	l	m	n	o	p
α_0	h	c	b	e	d	g	f	a	p	k	j	m	l	o	n	i
α_1	b	a	d	c	f	e	h	g	j	i	l	k	n	m	p	o
α_2	a	b	c	o	n	f	g	h	i	j	k	l	m	e	d	p

Fig. 1. Example of 2G-map: consecutive darts separated with a little segment are 0-sewn (*e.g.*, b and c); consecutive darts separated with a dot are 1-sewn (*e.g.*, a and b); parallel darts are 2-sewn (*e.g.*, d and o); a is 2-free because $\alpha_2(a) = a$.

2   Generalized Maps and (Sub)Map Isomorphism

Let us first recall basic definitions on generalized maps, which are a generalization of combinatorial maps. We refer the reader to [Lie94] for more details.

Definition 1. *(n G-map) Let $n \geq 0$. An n-dimensional generalized map (or n G-map) is defined by a tuple $M = (D, \alpha_0, \ldots, \alpha_n)$ such that (i) D is a finite set of darts; (ii) $\forall i \in [0, n]$, α_i is an involution[1] on D; and (iii) $\forall i, j \in [0, n]$ such that $i + 2 \leq j$, $\alpha_i \circ \alpha_j$ is an involution.*

We will say that a dart $d \in D$ is *i-free* whenever $d = \alpha_i(d)$, and that it is *i-sewn* to $d' \in D$ whenever $d = \alpha_i(d')$ and $d \neq d'$. Fig. 1 displays an example of 2G-map which describes an object composed of two adjacent faces.

Map isomorphism checks for the equivalence of nG-maps. It has been defined in [Lie94]. Submap isomorphism checks for the inclusion of two nG-maps. It has been defined in [DDLHJ+09] for combinatorial maps. Definition 2 extends this definition to generalized maps (see Fig. 2 for an example).

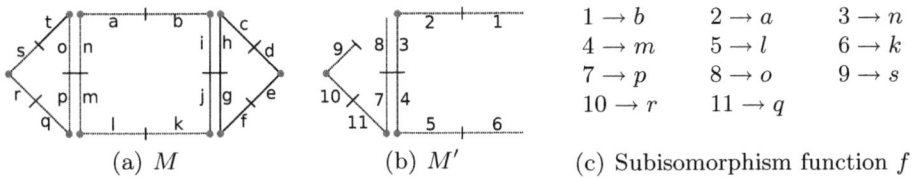

(a) M (b) M' (c) Subisomorphism function f

$$1 \to b \qquad 2 \to a \qquad 3 \to n$$
$$4 \to m \qquad 5 \to l \qquad 6 \to k$$
$$7 \to p \qquad 8 \to o \qquad 9 \to s$$
$$10 \to r \qquad 11 \to q$$

Fig. 2. Submap isomorphism example: f is a subisomorphism function from M' to M

Definition 2. *(n G-map subisomorphism) Let $M = (D, \alpha_0, \ldots, \alpha_n)$ and $M' = (D', \alpha'_0, \ldots, \alpha'_n)$ be two n G-maps. M is subisomorphic to M', denoted $M \sqsubseteq M'$ if there exist an injective function $f : D \to D'$ such that $\forall d \in D$, and $\forall i \in [0, n]$:*

- *if d is i-sewn, then $f(\alpha_i(d)) = \alpha'_i(f(d))$;*
- *if d is i-free, then either $f(d)$ is i-free, or $f(d)$ is i-sewn with a dart which is not matched by f to another dart of D, i.e., $\forall d_k \in D, f(d_k) \neq \alpha'_i(f(d))$.*

---

[1] An involution f on D is a bijective mapping from D to D such that $f = f^{-1}$.

3   Definition of a Distance Measure for Generalized Maps

Submap isomorphism may be used to decide of generalized map inclusion, but it cannot be used to quantify the distance between two nG-maps which are not related by an inclusion relationships. In this section, we propose to quantify this distance by means of the largest generalized map included into the two generalized maps to be compared. This definition may be viewed as an extension of the graph distance measure based on maximum common subgraphs defined in [BS98].

Let us first define generalized map sizes and maximum common submaps.

Definition 3. *(size of an nG-map) The size of an nG-map $M = (D, \alpha_0, \ldots, \alpha_n)$ is denoted $|M|$ and is defined by its number of darts, i.e., $|M| = |D|$.*

Definition 4. *(maximum common submap) Let M and M' be two nG-maps. A maximum common submap of M and M', denoted $mcs(M, M')$, is an nG-map such that $mcs(M, M') \sqsubseteq M$, $mcs(M, M') \sqsubseteq M'$; and $|mcs(M, M')|$ is maximal.*

Note that $mcs(M, M')$ is not necessarily unique as there may exist several common submaps which have a same size. A maximum common submap basically corresponds to the intersection of two nG-maps, and its size may be used to quantify their similarity: the larger a maximum common submap, the more similar the two nG-maps. To define a distance, we normalize this size with respect to the size of the largest of the two generalized maps, as proposed in [BS98] for graphs.

Definition 5. *(distance between two nG-maps) Let M_1 and M_2 be two nG-maps. The distance between M_1 and M_2 is defined by:*

$$d(M_1, M_2) = 0 \qquad\qquad if\ |M_1| = |M_2| = 0;$$
$$d(M_1, M_2) = 1 - \frac{|mcs(M_1, M_2)|}{max(|M_1|, |M_2|)} \quad otherwise.$$

Theorem 1. *Let $n \geq 1$. The distance d is a metric on the set $\mathcal{M}$ of all nG-maps so that the following properties hold:*

1. *Non-negativity: $\forall M_1, M_2 \in \mathcal{M}, d(M_1, M_2) \geq 0$;*
2. *Isomorphism of indiscernibles:*
    $\forall M_1, M_2 \in \mathcal{M}, d(M_1, M_2) = 0$ iff M_1 and M_2 are isomorphic;
3. *Symmetry: $\forall M_1, M_2 \in \mathcal{M}, d(M_1, M_2) = d(M_2, M_1)$;*
4. *Triangle inequality: $\forall M_1, M_2, M_3 \in \mathcal{M}, d(M_1, M_3) \leq d(M_1, M_2) + d(M_2, M_3)$.*

Proof. Properties 1, 2, and 3 are direct consequences of Def. 5.
Let us denote $m_{ij} = mcs(M_i, M_j)$, and $S_{ij} = max(|M_i|, |M_j|)$, and let us show Property 4 by considering separately the two following cases.
  (Case 1): $d(M_1, M_2) + d(M_2, M_3) \geq 1$.
In this case, the triangle inequality trivially holds as $d(M_1, M_3) \leq 1$.
  (Case 2): $d(M_1, M_2) + d(M_2, M_3) < 1$.

In this case, let us first show that there exists at least one dart d of M_2 which belongs both to m_{12} and m_{23} or, in other words, that the sum of the number of darts of m_{12} and m_{23} is strictly greater than the number of darts of M_2 so that at least one dart of M_2 belongs to the two common submaps, $i.e.$,

$$|M_2| < |m_{12}| + |m_{23}| \tag{1}$$

This inequation can be proven by considering all possible order relations between nG-map sizes. For example, if $|M_1| \geq |M_3| \geq |M_2|$, then:

(Case 2) $\Leftrightarrow 1 - \frac{|m_{12}|}{|M_1|} + 1 - \frac{|m_{23}|}{|M_3|} < 1$ (by Def. 5, and as $S_{12} = |M_1|$ and $S_{23} = |M_3|$)

$\Leftrightarrow |M_3| < \frac{|M_3|}{|M_1|}|m_{12}| + |m_{23}|$ (by multiplying by $|M_3|$)

$\Rightarrow |M_3| < |m_{12}| + |m_{23}|$ (as $\frac{|M_3|}{|M_1|} < 1$)

$\Rightarrow |M_2| < |m_{12}| + |m_{23}|$ (as $|M_3| \geq |M_2|$).

Ineq. (1) can be proven in a very similar way for the five other possible order relations between nG-map sizes.

Ineq. (1) shows that the sum of the sizes of the two common submaps m_{12} and m_{23} is always strictly greater than the size of M_2 so that there are at least $|m_{12}| + |m_{23}| - |M_2|$ darts that both belong to m_{12} and m_{23}. Therefore, the nG-map $mcs(m_{12}, m_{23})$ is a common submap of M_1, M_2, and M_3 which has at least $|m_{12}| + |m_{23}| - |M_2|$ darts. This nG-map gives a lower bound on the size of the maximum common submap of M_1 and M_3, $i.e.$,

$$|m_{13}| \geq |m_{12}| + |m_{23}| - |M_2| \tag{2}$$

Let us use this lower bound to show that the triangle inequality holds. When developing the triangle inequality w.r.t. Def. 5, it becomes:

$$|m_{13}| \geq \frac{S_{13}}{S_{12}}|m_{12}| + \frac{S_{13}}{S_{23}}|m_{23}| - S_{13} \tag{3}$$

Let us prove (3) by considering all order relations between nG-map sizes:

(Case 2.1): $|M_1| \geq |M_2| \geq |M_3|$ so that $S_{13} = |M_1|$, $S_{12} = |M_1|$, $S_{23} = |M_2|$. Ineq. (3) becomes $|m_{13}| \geq |m_{12}| + \frac{|M_1|}{|M_2|}|m_{23}| - |M_1|$. As $|m_{13}| \geq |m_{12}| + |m_{23}| - |M_2|$ (Ineq. (2)), we have to show that $|m_{23}| - |M_2| \geq \frac{|M_1|}{|M_2|}|m_{23}| - |M_1|$, $i.e.$, $|m_{23}| \leq |M_2|$ (as $|M_2| - |M_1| < 0$). This inequality trivially holds by Def. 4.

(Case 2.2): $|M_2| \geq |M_1| \geq |M_3|$ so that $S_{13} = |M_1|$, $S_{12} = |M_2|$, $S_{23} = |M_2|$. Ineq. (3) becomes $|m_{13}| \geq \frac{|M_1|}{|M_2|}|m_{12}| + \frac{|M_1|}{|M_2|}|m_{23}| - |M_1|$. As $\frac{|M_1|}{|M_2|} \leq 1$, Ineq. (2) implies that $|m_{13}| \geq \frac{|M_1|}{|M_2|}(|m_{12}| + |m_{23}| - |M_2|)$. Therefore, Ineq. (3) holds.

(Case 2.3): $|M_1| \geq |M_3| \geq |M_2|$ so that $S_{13} = |M_1|$, $S_{12} = |M_1|$, $S_{23} = |M_3|$. Ineq. (3) becomes $|m_{13}| \geq |m_{12}| + \frac{|M_1|}{|M_3|}|m_{23}| - |M_1|$. As $|m_{13}| \geq |m_{12}| + |m_{23}| - |M_2|$ (Ineq. (2)), we have to show that $|m_{23}| - |M_2| \geq \frac{|M_1|}{|M_3|}|m_{23}| - |M_1|$, $i.e.$, $|m_{23}| \leq |M_3|\frac{|M_2| - |M_1|}{|M_3| - |M_1|}$ (as $|M_3| - |M_1| < 0$). This inequality trivially holds by Def. 4 because $\frac{|M_2| - |M_1|}{|M_3| - |M_1|} \geq 1$.

The 3 others cases can be obtained by inverting M_1 and M_3. $\square$

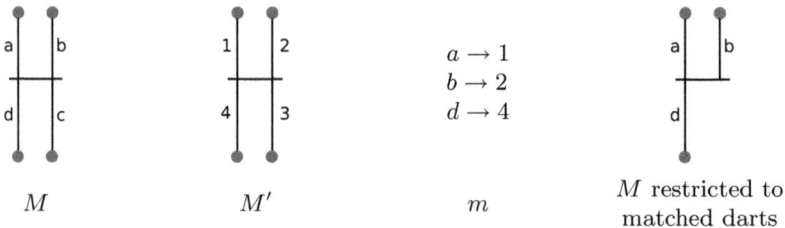

Fig. 3. Example of matching $m : M \to M'$ which does not induce a consistent submap. Constraint $\alpha_i \circ \alpha_j(d) = \alpha_j \circ \alpha_i(d)$ of Def. 1 is not satisfied in M restricted to $\{a, b, d\}$ as $\alpha_2(\alpha_0(a)) = d$ and $\alpha_0(\alpha_2(a)) = b$.

4   Algorithm for Approximating the nG-Map Distance

Computing the distance between two nG-maps basically involves computing their maximum common submap. We have given in [DDLHJ+09] a polynomial time algorithm for deciding if there exists a submap isomorphism between two connected generalized maps (such that there exists a path of i-sewn darts between every pair of darts). This problem becomes $\mathcal{NP}$-complete as soon as the pattern nG-map is not connected. As the maximum common submap of two nG-maps is not necessarily connected, computing a maximum common submap is at least as hard as deciding of submap isomorphism between two non connected nG-maps and, therefore, the maximum common submap problem is $\mathcal{NP}$-hard.

In this section, we describe a polynomial-time algorithm which efficiently compute an upper bound of $d(M, M')$. The idea is to build a submap which is common to M and M' and which is (hopefully) as large as possible. To compute a common submap, we actually build a dart matching which induces a submap, in a rather similar way as we compute a common subgraph by building a vertex matching which induces a subgraph. More precisely, a dart matching is a function $m : D \to D' \cup \{\epsilon\}$ such that, for each dart $d \in D$, if $m(d) = \epsilon$ then d is not matched, otherwise d is matched to $m(d) \in D'$. Symmetrically, a dart $d' \in D'$ is said to be not matched if $\forall d_k \in D, m(d_k) \neq d'$, otherwise d' is said to be matched with the dart $d \in D$ such that $m(d) = d'$. A matching induces a submap which is obtained by removing from M all non matched darts. Obviously, a matching must preserves involutions so that two darts i-sewn must be matched to two darts which are also i-sewn. A matching must also induce a consistent submap which satisfies constraint (iii) of Def. 1, as illustrated in Fig. 3. More precisely, Def. 6 defines consistent matchings.

Definition 6. (consistent dart matching) Let $M = (D, \alpha_0, \ldots, \alpha_n)$ and $M' = (D', \alpha'_0, \ldots, \alpha'_n)$ be two nG-maps. A matching $m : D \to D' \cup \{\epsilon\}$ is consistent if

1. it matches different darts of M to different darts of M', i.e.,
   $\forall d_1, d_2 \in D, d_1 \neq d_2 \Rightarrow m(d_1) = \epsilon \vee m(d_2) = \epsilon \vee m(d_1) \neq m(d_2)$;
2. it preserves all involutions between matched darts i.e.,
   $\forall d \in D$ such that $m(d) \neq \epsilon$, $\forall i \in [0, n]$:

- *if d is i-free and m(d) is i-sewn then $\alpha_i'(m(d))$ is not matched;*
- *if d is i-sewn and m(d) is i-free then $\alpha_i(d)$ is not matched;*
- *if both d and m(d) are i-sewn then $m(\alpha_i(d)) = \alpha_i'(m(d))$ or both $\alpha_i(d)$ and $\alpha_i'(m(d))$ are not matched.*

3. *matched darts satisfy constraint (iii) of Def. 1, i.e.,*
$\forall d \in D$ *such that* $m(d) \neq \epsilon$, $\forall i,j \in [0,n]$ *such that* $i + 2 \leq j$, *we have*
$(\alpha_i(d) \neq d \wedge m(\alpha_i(d)) \neq \epsilon \wedge \alpha_j(d) \neq d \wedge m(\alpha_j(d)) \neq \epsilon) \Rightarrow m(\alpha_i(\alpha_j(d))) \neq \epsilon.$

Definition 7. *(submap induced by a matching) Let* $M = (D, \alpha_0, \ldots, \alpha_n)$ *and* $M' = (D', \alpha_0', \ldots, \alpha_n')$ *be two nG-maps. The submap induced by a consistent matching* $m : D \to D' \cup \{\epsilon\}$ *is* $M_{\downarrow m} = (D'', \alpha_0'', \ldots, \alpha_n'')$ *such that*

- $D'' = \{d \in D \mid m(d) \neq \epsilon\};$
- $\forall d \in D'', \forall i \in [0,n], if \alpha_i(d) \in D''$ *then* $\alpha_i''(d) = \alpha_i(d);$ *otherwise* $\alpha_i''(d) = d.$

Proposition 1. *Let M and M' be two nG-maps and m be a consistent matching from M to M'. $M_{\downarrow m}$ is an nG-map such that $M_{\downarrow m} \sqsubseteq M$ and $M_{\downarrow m} \sqsubseteq M'$.*

Proof. Cond. 3 of Def. 6 ensures that $M_{\downarrow m}$ is a consistent nG-map which satisfies constraints of Def. 1. $M_{\downarrow m}$ is a submap of M by Def. 7. $M_{\downarrow m}$ is a submap of M' because Cond. 1 of Def. 6 ensures that darts of M are matched to different darts of M' and because Cond. 2 and 3 of Def. 6 ensure that all involutions are preserved by m. □

Algorithm 1 describes a polynomial-time algorithm which computes a consistent matching m in an incremental way. It starts from an empty matching. At each iteration, it chooses a couple (d, d') which is candidate to be added to m (line 5) according to a choice procedure described in Section 4.1. When adding this couple of darts to m, some other new couples must also be added in order to ensure its consistency. This propagation step is done line 6 and is described in Section 4.2. If propagation detects an inconsistency, then another couple of darts must be chosen (repeat loop lines 4-7); otherwise, m is updated with respect to the results of the propagation step (line 9) and the set of candidate couples is updated (line 10) as described in Section 4.3. When no more couple of darts can be consistently added to m, an upper bound of the distance is computed with respect to m (line 11). Note that the size of $M_{\downarrow m}$ is actually equal to the number of couples of darts in m.

4.1   Choice of a Couple of Darts (Line 5)

At each iteration, we choose a couple of darts $(d, d') \in Cand$ to be added to m. This choice is done in a greedy way, by choosing a couple of darts which maximizes the number of preserved involutions, *i.e.*, a couple (d, d') such that $|\{i \in [0,n] | m(\alpha_i(d)) = \alpha_i'(d'))\}|$ is maximal. The goal is to favor the choice of darts which are sewn to already matched darts: the more sewn, the better.

Algorithm 1. APPROXD(M, M')

Input: two nG-maps $M = (D, \alpha_0, \ldots, \alpha_n)$ and $M' = (D', \alpha'_0, \ldots, \alpha'_n)$
Output: returns an upper bound of $d(M, M')$

1 Let m be the set of matched darts; initialize m to $\emptyset$
2 $Cand \leftarrow D \times D'$
3 **while** $Cand \neq \emptyset$ **do**
4 　**repeat**
5 　　choose $(d, d') \in Cand$ and remove (d, d') from $Cand$
6 　　$m' \leftarrow$ propagate(m, d, d')
7 　**until** m' is consistent **or** $Cand = \emptyset$;
8 　**if** m' is consistent **then**
9 　　$m \leftarrow m'$
10 　　update $Cand$ with respect to m'

11 **return** $1 - \frac{|m|}{max(|M|, |M'|)}$

There usually exist several couples which maximize preserved involutions. We have defined two different ways of breaking ties, which are experimentally compared in Section 5. The first way to break ties (denoted *Rand*) consists in randomly choosing a couple within the set of candidate couples which maximize preserved involutions. The second way to break ties (denoted *Deg*) is defined with respect to degrees and co-degrees of incident cells. Indeed, each dart d is incident to $n + 1$ i-cells (one in each dimension): the 0-cell corresponds to the vertex incident to d, the 1-cell to the edge, the 2-cell to the face, the 3-cell to the volume, etc. Each i-cell (with $0 \leq i < n$) has a degree, which is the number of incident $(i + 1)$-cells (*e.g.*, the degree of a vertex (resp. edge, face) is the number of incident edges (resp. faces, volumes)). Each i-cell (with $0 < i \leq n$) also has a co-degree, which is the number of incident $(i - 1)$-cells (*e.g.*, the co-degree of an edge (resp. face, volume) is the number of incident vertices (resp. edges, faces)). Hence, when several couples of candidate darts preserve the same number of involutions, *Deg* break ties by choosing the couple of darts which maximizes the number of i-cells with the same degrees and co-degrees, thus favoring the choice of darts which have a similar neighborhood. Further ties are randomly broken.

4.2 Propagation of a Couple of Darts to Ensure Consistency (Line 6)

Adding (d, d') to m may lead to an inconsistent matching which does not satisfies Cond. 3 of Def. 6 (such as the one displayed in Fig. 3). Hence, we don't simply add (d, d') to m but compute the whole set m' of couples of darts that must be added to m so that the matching is consistent according to Def. 6. This propagation is done recursively for each new couple of darts added to m'. It either stops when the matching has been completed to a consistent matching, or when an inconsistency has been detected (when it is not possible to consistently add (d, d') to m). In this latter case, the repeat loop (lines 4-7) must be re-iterated to choose another couple of darts to be added to m.

4.3 Update of *Cand* with Respect to m' (Line 10)

A consistent matching m must satisfy the 3 conditions of Def. 6. Condition 3 is ensured by the propagation step. Conditions 1 and 2 are ensured by removing from the set *Cand* every couple (d_k, d'_k) which does not satisfy them. More precisely, we remove from *Cand* every couple (d_k, d'_k) such that d_k or d'_k are already matched to another dart in m' (Cond. 1 of Def. 6), or such that there exists a couple of matched darts $(d, d') \in m'$ such that d is i-sewn with d_k whereas d' is not i-sewn with d'_k for some dimension i (Cond. 2 of Def. 6).

5   First Experimental Results

ApproxD (described in Algorithm 1) greedily computes a consistent matching m for two nG-maps M and M'. This matching induces a common submap $M_{\downarrow m}$ which can be used to compute an upper bound of the distance between M and M', corresponding to $1 - \frac{|m|}{max(|M|,|M'|)}$ (as $|m| = |M_{\downarrow m}|$).

ApproxD is not deterministic as ties are randomly broken when choosing a couple of darts (line 5). Therefore, we can run it several times and finally return the smallest computed upper bound. Let us denote *Rand(k)* (resp. *Deg(k)*) the algorithm which runs ApproxD k times and uses the *Rand* (resp. *Deg*) procedure to break ties when choosing a couple of darts (line 5). Note that for each different run of ApproxD, we enforce the algorithm to choose a different couple of darts when choosing the first couple: for *Rand(k)* we randomly order the $|M| \cdot |M'|$ possible couples of darts whereas for *Deg(k)* we order them by decreasing order with respect to the number of i-cells with same (co-)degrees (see Section 4.1)[2].

We compare *Rand(k)* and *Deg(k)* on a synthetic benchmark of randomly generated couples of 3G-maps. As we cannot compute the exact distance, we have generated couples of 3G-maps for which we know an upper bound on the distance by construction: we randomly generate a first 3G-map M with 2000 darts; we then extract a submap of M which has $p\%$ of its 2000 darts; we finally generate a new 3G-map M', starting from the submap of M by randomly adding (and sewing) new darts to it until it has 2000 darts. We have generated different couples of 3G-maps, with different values of p, denoted $(p\%)$. By construction, we know that the distance for an instance $(p\%)$ is lower or equal to $1 - p/100$. However, this is just a bound as the parts of M and M' which do not belong to the $p\%$ common part may also have some common patterns.

Table 1 compares $Deg(1000)$ and $Rand(1000)$ on nine $(p\%)$ instances, with p ranging from 10 to 90. For each instance, we report average results on 50 runs. On these instances, $Deg(1000)$ always computes a smaller upper bound than $Rand(1000)$. The difference is more particularly sensible for larger values of p. In particular, the bound computed by $Deg(1000)$ is twice as small as the one

---

[2] We have also performed experiments where ApproxD chooses the next couple of darts (line 5) completely randomly (*i.e.*, independently from the number of preserved involutions), or where ApproxD does not perform the propagation step (line 6). In both cases, the computed matchings are much poorer and induce very bad approximations of the distance so that we do not report results of these variants.

Table 1. Comparison of $Deg(1000)$ and $Rand(1000)$ on $(p\%)$ instances. Each line first gives the value of p and the upper bound on d corresponding to p ($i.e.$, $1 - p/100$). Then, for each algorithm, it successively gives the best computed upper bound within a limit of 1000 matching constructions, and the number of matching constructions needed to reach this upper bound (average and standard deviation on 50 runs).

	$Deg(1000)$			$Rand(1000)$		
p	Upper bound		#constructions	Upper bound		#constructions
	avg	(sdv)	avg	(sdv)	avg	(sdv)	avg	(sdv)
10 ($d \le 0.9$)	0.309	(0.0038)	330	(245)	0.324	(0.0031)	498	(276)
20 ($d \le 0.8$)	0.274	(0.0030)	406	(281)	0.311	(0.0150)	467	(303)
30 ($d \le 0.7$)	0.258	(0.0028)	459	(211)	0.293	(0.0204)	435	(279)
40 ($d \le 0.6$)	0.238	(0.0022)	451	(280)	0.275	(0.0204)	570	(271)
50 ($d \le 0.5$)	0.209	(0.0025)	469	(194)	0.252	(0.0324)	502	(289)
60 ($d \le 0.4$)	0.186	(0.0015)	645	(219)	0.229	(0.0388)	545	(261)
70 ($d \le 0.3$)	0.158	(0.0014)	362	(323)	0.212	(0.0545)	471	(277)
80 ($d \le 0.2$)	0.107	(0.0009)	271	(304)	0.155	(0.0493)	418	(282)
90 ($d \le 0.1$)	0.068	(0.0005)	94	(175)	0.127	(0.0749)	581	(251)

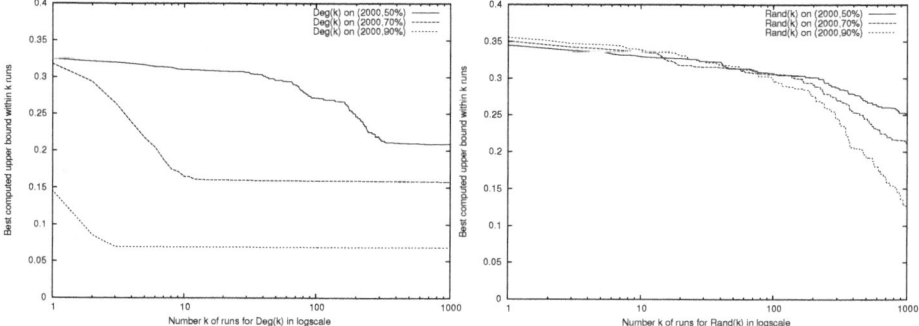

Fig. 4. Evolution of the upper bound w.r.t. the number of matching constructions (average on 50 runs) for instances (50%), (70%) and (90%): $Deg(k)$ on the left and $Rand(k)$ on the right

computed by $Rand(1000)$ when $p = 90$. Bounds computed by $Deg(1000)$ and $Rand(1000)$ are much smaller than $1 - p/100$ for small values of p, thus showing that the parts of M and M' which do not belong to the $p\%$ common part actually also share many sub-patterns. Note also that $Deg(1000)$ always computes upper bounds smaller than $1 - p/100$, whereas $Rand(1000)$ has not been able to compute an upper bound smaller or equal to $1 - p/100$ when $p = 90$. Standard deviations of bounds are also smaller for $Deg(1000)$ than for $Rand(1000)$.

Table 1 also gives the number of matching constructions needed to reach the best upper bound (within a limit of 1000 constructions). It shows us that Deg usually needs less constructions than $Rand$ (whereas it computes better approximations). Note also that the number of constructions needed by Deg to find the best upper bound decreases for large values of p. In particular, the best approximation is found in a hundred or so constructions when $p = 90\%$. Fig. 4 displays the evolution of the quality of the bound with respect to the number k of matching constructions for three different values of p. It shows us that the upper bound computed by $Deg(k)$ decreases quicker than the upper bound computed by $Rand(k)$, and that the higher p, the quicker it decreases.

Let us finally point out that matching constructions are rather quickly computed. For instance, on an Intel Xeon E5520 CPU with 16GB RAM, Algorithm 1 runs in less than 1 second (resp. 2s, 11s, 56s) when 3G-maps have 1000 (resp. 2000, 4000, and 8000) darts. Note that CPU times do not really depend on the size of the common part p.

6   Conclusion

We have introduced a first distance measure for comparing nG-maps. This distance is based on the size of the largest common submap and we have shown that it is a metric. We have described a polynomial time algorithm which is able to compute an approximation of this distance and very first experimental results have shown us that this algorithm uses relevant heuristics, which allow it to find better approximations, and that the computed approximations are always better than bounds obtained by construction. Further work will concern the evaluation of our distance measure on a real image retrieval application where images are modeled by 2G-maps. We also plan to extend our work to partial submaps, such that not all involutions are preserved by the matching, and more generally, to a nG-map edit distance which evaluates the distance between two generalized maps by means of the smallest number of edit operations needed to transform one nG-map into the other, in way similar to graph edit distances [Bun97].

References

[BS98]   Bunke, H., Shearer, K.: A graph distance metric based on the maximal common subgraph. PRL 19(3-4), 255–259 (1998)

[Bun97]   Bunke, H.: On a relation between graph edit distance and maximum common subgraph. PRL 18, 689–694 (1997)

[Dam08]   Damiand, G.: Topological model for 3d image representation: Definition and incremental extraction algorithm. CVIU 109(3), 260–289 (2008)

[DBF04]   Damiand, G., Bertrand, Y., Fiorio, C.: Topological model for two-dimensional image representation: definition and optimal extraction algorithm. CVIU 93(2), 111–154 (2004)

[DDLHJ+09] Damiand, G., De La Higuera, C., Janodet, J.-C., Samuel, E., Solnon, C.: Polynomial algorithm for submap isomorphism: Application to searching patterns in images. In: Torsello, A., Escolano, F., Brun, L. (eds.) GbRPR 2009. LNCS, vol. 5534, pp. 102–112. Springer, Heidelberg (2009)

[Lie94]   Lienhardt, P.: N-dimensional generalized combinatorial maps and cellular quasi-manifolds. International Journal of Computational Geometry and Applications 4(3), 275–324 (1994)

Aggregated Search in Graph Databases: Preliminary Results

Haytham Elghazel[1] and Mohand-Said Hacid[2]

[1] Université de Lyon, 69000, Lyon, France; Université Lyon 1,
Laboratoire GAMA, 69622 Villeurbanne
`haytham.elghazel@univ-lyon1.fr`
[2] Université de Lyon, 69000, Lyon, France; Université Lyon 1,
Laboratoire LIRIS, 69622 Villeurbanne
`mohand-said.hacid@univ-lyon1.fr`

Abstract. Graphs are widely used to model complicated data semantics in many applications (*e.g.* spacial databases, image databases,...). Querying graph databases is costly since it involves subgraph isomorphism testing, which is an NP-complete problem [7]. Most of the existing query processing techniques are based on the framework of filtering-and-verification to reduce computation costs. However, to the best f our knowledge, the problem of assembling graphs to provide an answer to a given query graph (*i.e.* information need) if it is not present in one single source, is not investigated. In this paper, we try to highlight the potential and the motivation that lies behind the graph aggregation query problem. We propose a first approach to support aggregated search in graph databases. We discuss the preliminary results from the algorithmic point of view.

Keywords: graph databases, query processing, aggregated search.

1 Introduction

Database research has been facing a new challenge raised by the emergence of massive, complex structural data, in the form of sequences, trees, and graphs [8]. Among all the complex structured data, graph is among the most sophisticated and general form of structure. Conceptually, any kind of data can be represented by graphs. Graphs have become increasingly important in modelling complicated structures and schemaless data in many application domains such as bioinformatics, chemistry, web, social networks, sensor networks and telecommunication, etc. For instance, graphs may represent molecular structures of chemical compounds in chemistry, the organization of entities for images in computer vision, the ER diagrams in database design, the UML diagrams in software engineering, and so on. In addition, Web sites, XML and RDF documents can also be modelled as graphs. Therefore, it is obvious that graph databases will become more used in the near future. As a result, the problem of processing query on graph databases, becomes important.

X. Jiang, M. Ferrer, and A. Torsello (Eds.): GbRPR 2011, LNCS 6658, pp. 92–101, 2011.
© Springer-Verlag Berlin Heidelberg 2011

In some cases, the success of an application significantly relies on the efficiency of the query processing system. The classical graph query problem can be described as follows: Given a graph database $\mathcal{D} = \{g_1, g_2, \ldots, g_n\}$ and a query graph q, find all the graphs $g_i \in \mathcal{D}$ such that q is a subgraph of g_i. It is inefficient to perform a sequential scan on the graph database and check whether q is a subgraph of g_i. Sequential scan is very costly because one has to not only access the whole graph database but also check subgraph isomorphism which is NP-complete. The problem of graph query processing has been tackled recently and a lot of methods has been proposed [3,11,13,14,1,7]. The underlying techniques have focused on quickly retrieving the graphs that are supposed to be the answer to the given query. Consequently, most of these methods follow the framework of filtering and verification which is based on feature-based indexing. First, the filtering step uses a feature-based index to eliminate part of the negative results and produces a candidate answer set. Then, the verification step verifies whether the query is indeed a subgraph of each candidate graph. Since the candidate answer set is in general much smaller than the entire graph database, query processing using the indexes is more efficient than the sequential scan approach. The existing filtering and verification techniques include *GraphGrep* [3], *gIndex* [11], *TreePi* [13] and *Tree+Δ* [14]. *GraphGrep* [3] is a path-based technique which tries to enumerate all existing paths in a database within a threshold length and index them. Then, it uses the index to identify every graph that contains all the paths in the query graph. After getting the candidate answer, it records all the embeddings of the query paths in each candidate. Rather than doing real subgraph isomorphism testing, it performs join operations on these embeddings to figure out the possible isomorphism mapping between the query graph and each graph in the candidate set. The main problem of *GraphGrep* is the following; the extraction of paths from graphs results in a large amount of structural information that is not preserved, and as a consequence it may affect the performance of the index's pruning capability. To address this problem, some recent work uses more complex and selective sub-structures as index features. Among them, *gIndex* [11] is proposed as a graph-based indexing approach which selects only frequent and discriminative subgraphs as index patterns and identifies the graphs in the database which contain those subgraphs. Then, *gIndex* performs naive subgraph isomorphism tests for final verification. It is reported in [11] that *gIndex* significantly outperforms *GraphGrep* in terms of index and candidate set size. Another example is given by the *TreePi* algorithm [13] which tries to index frequent and discriminative subtrees rather than subgraphs, as trees are easier to manipulate than graphs and can capture most information of the original database. *TreePi* also uses a new pruning technique based on the concept of *Center Distance Constraints* to further reduce the size of the candidate set. The basic idea is that if the query graph appears in a candidate graph, distances between pairs of features in query graph must be preserved in the candidate graph as well. At the end, a new subgraph isomorphism test algorithm is used to perform final verification by using the location information of the indexing structures. In order to determine whether the query graph q is subgraph isomorphic to a candidate graph g, the algorithm tries to reconstruct q, using a new canonical reconstruction form, by joining one by one subtrees in g that satisfy center distance constraint. Zhao et al. [14] have proposed $Tree+\Delta$ approach by extending $TreePi$ to achieve better pruning capability by adding a small number of discriminative graphs to the index structure. Another recent indexing technique is given by FG-$Index$ [1], for which both frequent subgraphs and edges are chosen as feature set. FG-$Index$ supports verification-free subgraph search. The underlying observation is the following: if the query graph is contained in some feature in the index, all graphs in the database which contain that feature must also contain the query graph, and such graphs can be returned as final results without additional verification. This technique becomes effective only when many large subgraphs are indexed, thus construction cost and storage overhead of FG-Index are much larger than other feature-based indexing techniques. More recently, in order to significantly reduce the verification costs, Shang et al. [7] have proposed a new efficient subgraph isomorphism testing algorithm ($QuickSI$) by improving the widely applied Ullman's algorithm [9]. Then, to accommodate $QuickSI$ in the filtering phase, they have developed a novel feature-based index technique ($Swift$-$Index$) which significantly outperforms the existing techniques such as $gIndex$ and FG-$Index$.

Although the graph query problem has been tackled in the last decade, no attention has been paid to the problem of assembling graphs in a sensed way to provide an answer of the given query graph q if (1) no single candidate graph turns out to be isomorphic to q after the verification step, or (2) additional answers to q are needed to be returned. For example, if we use the graph in Figure 1 as the query q, then among the 3 graphs ($\mathcal{D} = \{g_1, g_2, g_3\}$) in Figure 2, no one is isomorphic to q. However, as shown in Figure 3. a possible output to q could be given by the aggregation of fragments of the graphs g_2 and g_3, A real example for such a problem could be given by the semantic web RDF documents. Suppose an RDF query graph for a hotel which tries to retrieve a hotel description, some pictures, an address, contact information, a map, hotel reviews and so on. However, all this information is not necessarily in the same RDF document. A possible issue has to consider different RDF documents (and then their corresponding graphs) such as the one for yellow pages, the one for online travel agencies, the one for image database, etc.

In view of this context, our problem seems to have similar intention as the problem of *substructure similarity search* (also called *approximate graph matching problem*) which try to discover all the graphs that approximately contain the query graph when any match for the latter can be found in the graph database [12]. However, our work differs from substructure similarity search ones in terms of giving different exact solutions (instead of relaxed ones) to the query graph by assembling graphs as answers to the query such that the aggregated graph contains the query. The challenge in this scenario is to answer the following questions : (1) how we determine the participating graphs to the aggregation, and (2) how such an aggregation could be built.

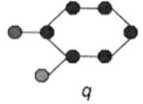

Fig. 1. A simple query graph

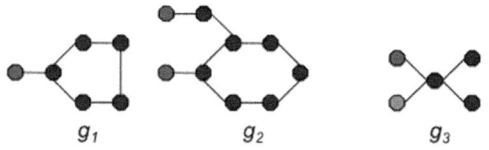

Fig. 2. A simple graph database

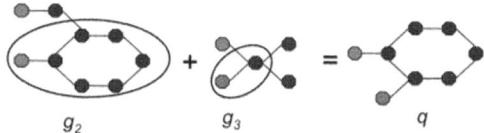

Fig. 3. An example of graph aggregation for querying graph database

The problems raised by aggregation in the context of documents are discussed in [5]. However, the paper does not address formal and algorithmic issues.

Motivated by this problem, we propose an approach intended to support the *graph aggregation* in the framework of query evaluation. This approach focuses on the *simple labeled graphs* (we refer to them as graphs in the rest of the paper). However, it can be easily extended to other kinds of graphs (for example *directed labeled graphs* in order to consider RDF and XML queries). Our design for data aggregation is targeted to supplement RDF query processing in such a way that query approximation (via data aggregation) will be supported. In our work, we are mainly interested in query aggregation in RDF databases.

2   Preliminaries

This section introduces the terminology used in this paper and formally define the problem.

Conceptually, any kind of data can be represented by graphs. In labeled graphs, vertices and edges represent entities and relationships, respectively. The attributes associated with entities and relationships are called labels.

More formally, a *labeled graph* g is defined as a 6-tuple $(V, E, L_v, L_e, F_v, F_e)$ where V is the set of vertices; $E \subseteq V \times V$ is the set of edges joining two distinct vertices; L_v is the set of vertex labels; L_e is the set of edge labels; F_v is a function $V \rightarrow L_v$ that assigns labels to vertices and F_e is a function $E \rightarrow L_e$ that assigns labels to edges. The vertex set and the edge set of a graph g are denoted by $V(g)$ and $E(g)$, respectively. Labeled graphs are generally classified, according to the direction of their edges, into two main classes: *directed labeled graphs* such as XML and RDF and *undirected labeled graph* such as social networks and chemical compounds. For example, the graphs shown in Figure 2 are undirected labeled graphs.

Definition 1 (Graph database). A graph database $\mathcal{D}$ is a collection of data graphs g_i where $\mathcal{D} = \{g_1, g_2, \ldots, g_n\}$.

Definition 2 (Candidate set). A candidate set $\mathcal{C}$ is a collection of data graphs from $\mathcal{D}$ that contain all the features appearing in the query graph q after the filtering phase.

Definition 3 (Non isomorphic set of graphs). A *non isomorphic set $\mathcal{N}$ of graphs* is a collection of data graphs from $\mathcal{C}$ that are not isomorphic to q after the verification phase. In other words, they are the graphs outside the final result of filtering and verification framework.

Definition 4 (Graph aggregation query problem). Given a non isomorphic set $\mathcal{N} = \{g_1, g_2, \ldots, g_m\}$ and a query graph q, the problem of *graph aggregation query* is to find different subsets $\mathcal{S} = \{g_1, g_2, \ldots, g_k\}$ from $\mathcal{N}$ (*i.e.* $k \leq m$) for which the joining of fragments (subgraphs) $P_{g_1}, P_{g_2}, \ldots, P_{g_k}$ from graphs $g_1, g_2, \ldots, g_k$ respectively, leads to q, that is $q = (P_{g_1} \bowtie P_{g_2} \bowtie \ldots \bowtie P_{g_k})$. Here, the semantics of the join operation is the one used in the example given figures 1, 2, and 3 to built a solution to the query of figure 1 by combining fragments stemming from graphs g_2 and g_3 of figure 2.

Although in this paper we focus on undirected labeled graphs, it is easy to extend our proposal to directed labeled graphs. In fact, the direction of an edge can simply be expressed by the label or by an additional flag. Thus the basic algorithm needs only small modifications [2].

3 Graph Aggregation for Query Processing Framework

In this section, we discuss the graph aggregated search approach we propose. As we said in the previous section, our approach builds on a *non isomorphic set $\mathcal{N}$* of graphs. This set contains all the graphs left out by the verification phase of the *filtering and verification framework*. In order to prepare this subset, we propose to perform any aforementioned technique [3,11,13,14,7].

Given a query graph q. In order to determine whether q is subgraph isomorphic to the aggregation of a graph subset $\mathcal{S} \subseteq \mathcal{N}$, our approach tries to reconstruct q from $\mathcal{S}$, using a four stages approach:

1. Initially select one graph g from $\mathcal{N}$ and set query graph q_1 to q,
2. Find the *maximum common subgraph* g_c between q_1 and g,

Algorithm 1. GraphAggregatedSearch(q, $\mathcal{N}$)

Require: q is a query graph; $\mathcal{N}$ is a non isomorphic graph set;
Ensure: $\mathcal{R}$ is a set of matched graphs subsets;

```
 1: for each g ∈ 𝒩 do
 2:    𝒮 := {g};
 3:    𝒯 := 𝒩 \ {g};
 4:    g_c := maximum_common_subgraph(q, g);
 5:    q_1 := query_generation(q, g_c);
 6:    while (q_1 ≠ ∅ or 𝒯 ≠ ∅) do
 7:       select a graph g' from 𝒯;
 8:       g_c := maximum_common_subgraph(q_1, g');
 9:       q_2 := query_generation(q_1, g_c);
10:       if (q_2 ≠ q_1) then
11:          𝒮 := 𝒮 ∪ {g'};
12:       end if
13:       𝒯 := 𝒯 \ {g'};
14:       q_1 := q_2;
15:    end while
16:    if (q_1 = ∅) then
17:       �R := �R ∪ 𝒮;
18:       𝒩 := 𝒩 \ 𝒮;
19:    else
20:       𝒩 := 𝒩 \ {g};
21:    end if
22: end for
23: return �R;
```

3. Generate a new query graph q_2 given by the subgraph of q_1 that does not exist in g_c (*i.e.* $q_2 = q_1 \setminus g_c$),
4. Repeat steps 2-3, for the not yet considered graphs in $\mathcal{N}$ using $q_1 = q_2$, until a mapping is found (*i.e.* $q_1 = \emptyset$) or no more graphs could be considered from $\mathcal{N}$.

If a mapping is found, the subset of graphs that contribute in this mapping will be added to the result set $\mathcal{R}$ and pruned from $\mathcal{N}$. Otherwise, only the initially considered graph g is withdrawn from $\mathcal{N}$. Finally, in order to find another subset of graphs which together contain q, we move on to the next graph in $\mathcal{N}$ and backtrack to the step 1. In this scenario, a graph query returns as output different disjoint subsets of $\mathcal{N}$.

The basic steps underlying the approach are given in Algorithm 1.

3.1 Maximum Common Subgraph Detection

Given a data graph g and a query graph q, the *maximum common subgraph* (MCS) problem consists in determining the largest induced subgraph of q isomorphic to a subgraph of g.

The detection of the MCS between g and q can be reduced to the problem of determining the *maximum clique*[1] in a *compatibility graph*, which is an NP-complete problem [6]. The compatibility graph (aka*association graph* and *modular product graph*) has the property that an MCS between the graphs g and q is equivalent to a maximum clique in their compatibility graph. Since clique-based algorithms seem to provide the most widely used approach to the MCS problem [6], we adopt it in our case.

In order to determine the compatibility graph G of g and q, the adjacency properties of these graphs is used. This graph is defined on the vertex set $V(g) \times V(q)$ with two vertices (u_i, v_i) and (u_j, v_j) being adjacent whenever $(u_i, u_j) \in E(g)$ and $(v_i, v_j) \in E(q)$, or $(u_i, u_j) \notin E(g)$ and $(v_i, v_j) \notin E(q)$. In the case of labeled graphs (as in our case), the definition of the compatibility graph is further restricted by requiring that the vertex and edge labels correspond according to some compatibility criteria.

Finally, to detect the maximum common subgraph between g and q, the maximum clique in the compatibility graph G has to be found. A recently proposed MaxCliqueDyn algorithm [4] for comparing large molecular structures and that is considerably faster than many other maximal clique algorithms will be adopted for this purpose.

The referenced resources in RDF triples have unique identifiers, IRIs.These identifiers can point to precise definitions of predicates or refer to specific concepts or objects. This means that the compatibility graph we can build in the case of RDF databases is not too large and hence the computation of maximum clique is not costly.

3.2   Query Generation

In this section, we discuss how to generate the new query graph q_1 once the maximum common subgraph g_c between a query graph q and a data graph g is found. The vertex set $V(q_1)$ and edge set $E(q_1)$ of q_1 are defined as:

- $V(q_1)$ is given by the vertices of q that are not in g_c and the vertices of q that are in g_c but adjacent to vertices that are not in g_c. Formally, $V(q_1) = V'(q_1) \cup V''(q_1)$ where:
  - $V'(q_1) = \{v_i | v_i \in V(q) \text{ and } v_i \notin V(g_c)\}$.
  - $V''(q_1) = \{v_i | (v_i, v_j) \in E(q) \text{ and } v_j \notin V(g_c)\}$.
- Two vertices v_i and v_j of $V(q_1)$ are connected by an edge if and only if the vertices are also connected in q. Formally, $(v_i, v_j) \in E(q_1)$ if $(v_i, v_j) \in E(q)$.

4   Performance Evaluation

In this section, we report first empirical results to evaluate the effectiveness and efficiency of our technique. Our experiments are conducted on synthetic

---

[1] A clique in a graph G is a subset of vertices in the graph such that each pair of vertices in the subset is connected by an edge in the graph G. A maximum clique is the largest such subset present in the graph.

Table 1. Performance Evaluation on Varying # Distinct Labels

# Distinct labels	Response time (s)	Aggregated output	Simple_QP output
5	0.0842	34.35	3.9
10	0.0619	6.85	0.54
15	0.0589	3.2	0.05

datasets. We generate a large number of graphs by using the synthetic graph data generator *GraphGen* [1]. The generator also allows us to specify various parameters such as the average graph density, graph size and the number of distinct vertex labels.

In these experiments, we focus on the capability and the scalability of our technique wrt the number of distinct vertex labels, which have a great impact on overall performance. It can be verified that (1) the less the number of distinct vertex labels, the harder the aggregated search problem is, since the size of the compatibility graph increases and then the maximum common subgraph detection became costly, and (2) the more the number of distinct vertex labels, the less the number of output solutions from a filtering-and-verification based approaches is and then the more interesting becomes aggregated search.

We use *GraphGen* to generate a set of graph datasets with different numbers of distinct labels, varying from 15 to 5. The average number of edges in the query graphs and data graphs are 10 and 30, respectively. There are 1000 data graphs and 100 queries in the considered experiments. The set of non isomorphic graphs $\mathcal{N}$, which is the input to our *graph aggregated search algorithm*, is given by a simple filtering-and-verification based approach that firstly enumerates the frequent subgraphs for filtering using the graph mining approach *gSpan* [10] and performs naive subgraph isomorphism tests for final verification.

Table 1 reports the overall performance of our aggregated technique in terms of average query response time and average output size on the 100 queries set. For an interesting assess of the results gained with our aggregated search, we propose to measure the advantage of our approach to find new output solutions compared to the classical subgraph isomorphism query approaches. For this purpose, the average output size of the simplest filtering-and-verification based approach is recorded and demonstrated in Table 1. We use *Simple_QP* to denote this simple approach for query processing.

As expected, performance of classical subgraph isomorphism query techniques, in terms of the number of output solutions, deteriorates with the increasing number of distinct vertex labels. When the number of distinct labels varies from 5 to 15, the number of query answers provided by *Simple_QP* decreases. In addition, it is noted that *Simple_QP* fails to return any result for many query graphs when the number of distinct labels reaches 10. However, the experiments confirm the ability of our technique to deliver new query answers that can improve the efficiency and the precision of a query processing system. For example, when the number of distinct labels is equal to 5, the average number of solutions provided by our approach is 8.807 times the numbers of solutions returned by *Simple_QP*. On the other hand, when the number of distinct labels drops from 15 to 5, the response time of aggregation slowly increased.

5   Conclusions

We have discussed aggregation in the context of data modeled as graphs. Aggregated search can be combined with query processing/approximation in order to provide greater flexibility in the querying of complex, irregular and semistructured data sets.

We introduced an algorithm dedicated to the task of computing fragments of graphs candidate to answer a user query. To this end, we built on feature-based indexing and verification to select graphs that will be subject to aggregation. Initial computational evaluations of this algorithm were provided. We are investigating the following issues for which we expect to get first feedback very quickly:

- Since the answers returned by our approach for a user query may vary as per the examining order of the non-isomorphic graphs, the goodness of the different answers is expected to be evaluated. In this context, we are working on the optimization of the examining order of graphs in the database in the sense to better improving the query processing in terms of runtime and answers quality.
- We are building a large scale RDF database together with a set of well defined and characterized queries. This work is led by master students and we expect to have these data set and queries ready for large scale experiments by the end of january.
- We are working on the design, prototyping and evaluation of a query interface. In practice, we expect that a visual query interface would be required, providing users with facilities for query specification. The idea is to provide the users with facilities to express RDQL queries and generate the corresponding specification in the format proposed in this paper. We plan to build on partial evaluation techniques used in coupling relational databases and logic-based programming languages.

Acknowledgment

This work is partially supported by ANR (Agence Nationale de la Recherche) project ANR-08-CORD-009 and by Rhne-Alpes Region, Cluster ISLE (Informatique, Signal, Logiciel Embarqué).

References

1. Cheng, J., Ke, Y., Ng, W., Lu, A.: Fg-index: towards verification-free query processing on graph databases. In: ACM SIGMOD International Conference on Management of Data, pp. 857–872 (2007)

2. Dreweke, A., Wörlein, M., Fischer, I., Schell, D., Meinl, T., Philippsen, M.: Graph-based procedural abstraction. In: Fifth International Symposium on Code Generation and Optimization (CGO), pp. 259–270 (2007)
3. Giugno, R., Shasha, D.: Graphgrep: A fast and universal method for querying graphs. In: International Conference on Pattern Recognition, pp. 112–115 (2002)
4. Janez, K., Dusanka, J.: An improved branch and bound algorithm for the maximum clique problem. MATCH Communications in Mathematical and in Computer Chemistry 58, 569–590 (2007)
5. Kopliku, A., Pinel-Sauvagnat, K., Boughanem, M.: Aggregated search: potential, issues and evaluation. Technical Report RT2009-4FR, IRIT (2009), http://www.irit.fr/PERSONNEL/SIG/kopliku/
6. Raymond, J.W., Willett, P.: Maximum common subgraph isomorphism algorithms for the matching of chemical structures. Journal of Computer-Aided Molecular Design 16(7), 521–533 (2002)
7. Shang, H., Zhang, Y., Lin, X., Yu, J.X.: Taming verification hardness: an efficient algorithm for testing subgraph isomorphism. In: International Conference on Very Large Data Bases, pp. 364–375 (2008)
8. Shasha, D., Wang, J.T.L., Giugno, R.: Algorithmics and applications of tree and graph searching. In: Proceedings of the twenty-first ACM SIGMOD-SIGACT-SIGART Symposium on Principles of Database Systems (PODS), pp. 39–52 (2002)
9. Ullmann, J.R.: An algorithm for subgraph isomorphism. Journal of ACM 23(1), 31–42 (1976)
10. Yan, X., Han, J.: gspan: Graph-based substructure pattern mining. In: International Conference on Data Mining, pp. 721–724 (2002)
11. Yan, X., Yu, P.S., Han, J.: Graph indexing: A frequent structure-based approach. In: ACM SIGMOD International Conference on Management of Data, pp. 335–346 (2004)
12. Yan, X., Yu, P.S., Han, J.: Substructure similarity search in graph databases. In: ACM SIGMOD International Conference on Management of Data, pp. 766–777 (2005)
13. Zhang, S., Hu, M., Yang, J.: Treepi: A novel graph indexing method. In: International Conference on Data Engineering, pp. 966–975 (2007)
14. Zhao, P., Yu, J.X., Yu, P.S.: Graph indexing: Tree + delta >= graph. In: International Conference on Very Large Data Bases, pp. 938–949 (2007)

Speeding Up Graph Edit Distance Computation through Fast Bipartite Matching

Stefan Fankhauser, Kaspar Riesen, and Horst Bunke

Institute of Computer Science and Applied Mathematics, University of Bern,
Neubrückstrasse 10, CH-3012 Bern, Switzerland
{bunke,fankhaus,riesen}@iam.unibe.ch

Abstract. In the field of structural pattern recognition graphs constitute a very common and powerful way of representing objects. The main
drawback of graph representations is that the computation of various
graph similarity measures is exponential in the number of involved nodes.
Hence, such computations are feasible for rather small graphs only. One
of the most flexible graph similarity measures is graph edit distance.
In this paper we propose a novel approach for the efficient computation of graph edit distance based on bipartite graph matching by means
of the Volgenant-Jonker assignment algorithm. Our proposed algorithm
provides only suboptimal edit distances, but runs in polynomial time.
The reason for its sub-optimality is that edge information is taken into
account only in a limited fashion during the process of finding the optimal node assignment between two graphs. In experiments on diverse
graph representations we demonstrate a high speed up of our proposed
method over a traditional algorithm for graph edit distance computation
and over two other sub-optimal approaches that use the Hungarian and
Munkres algorithm. Also, we show that classification accuracy remains
nearly unaffected by the suboptimal nature of the algorithm.

1 Introduction

Graph matching refers to the process of evaluating the structural similarity of
graphs. Numerous methods for graph matching have been proposed in the literature [1]. A prominent approach is to consider the spectral decomposition
of graphs (e.g. Singular Value Decomposition and Eigenvalues) rather than the
graphs themselves [2,3]. The basic idea is to represent graphs by the eigendecomposition of their structural matrices (i.e. adjacency or Laplacian matrix). The
resulting representation exhibits interesting properties for pattern recognition.
However, spectral methods are often sensitive to structural errors and do not allow arbitrary labels or attributes on the nodes and edges, although recent work
attempts to overcome those limitations [4]. In other approaches, artificial neural
networks, relaxation labeling techniques, and genetic algorithms have been used
to map the nodes of one graph to the nodes of another graph such that the edge
structure is preserved as accurately as possible [5,6,7]. Such algorithms perform
quite efficiently, but they are limited in that they are often applicable to special
classes of graphs only.

X. Jiang, M. Ferrer, and A. Torsello (Eds.): GbRPR 2011, LNCS 6658, pp. 102–111, 2011.

One of the most flexible methods for error-tolerant graph matching that is applicable to any kind of graphs is the graph edit distance [8,9]. The idea of graph edit distance is to define the dissimilarity of graphs by the amount of distortion that is needed to transform one graph into another. As a matter of fact, the edit distance of graphs has been used in the context of classification and clustering tasks in various applications[10,11].

The main drawback of graph edit distance is its computational complexity, which is exponential in the number of nodes of the involved graphs. Consequently, the application of edit distance is limited to graphs of rather small size. To render the matching of graphs less computationally demanding, a number of methods have been proposed. In some approaches, the basic idea is to perform a local search to solve the graph matching problem, that is, to optimize local criteria instead of global, or optimal ones [12,13]. In [14], a linear programming method for computing the edit distance of graphs with unlabeled edges is proposed. The method can be used to derive lower and upper edit distance bounds in polynomial time.

In this paper, we propose an efficient algorithm to speed up graph edit distance computation. The method is based on an assignment algorithm used to solve linear sum assignment problems. A similar approach is described in [15,16]. In the present paper, however, we use the algorithm of Volgenant and Jonker [17] to solve the assignment problem, which leads to a faster computation of suboptimal graph edit distance. In a series of experiments we analyse the impact of the chosen assignment algorithm on the whole procedure and demonstrate how the proposed method allows us to speed up the computation of graph edit distance substantially, while at the same time recognition accuracy is not much affected.

In Section 2, graph edit distance is introduced. In Section 3, the assignment algorithms used in this paper and their extension for computing graph edit distance are described. Section 4 gives some experimental results achieved with the proposed new method. Finally, in Section 5, we draw conclusions from this work.

2   Graph Edit Distance

The key idea of graph edit distance is to determine the minimal amount of distortion that is needed to transform one graph into another [8,9]. The considered distortions are given by insertions, deletions, and substitutions of nodes and edges. For two graphs – the source graph g_1 and the target graph g_2 – we delete some nodes and edges from g_1, relabel some of the remaining nodes and edges (substitutions) and insert some nodes and edges, such that g_1 is finally transformed into g_2. A sequence of edit operations that transform g_1 into g_2 is called an *edit path* between g_1 and g_2. To increase the power of this method, one can introduce a function that assigns a cost to each edit operation measuring the strength of the given distortion. The idea of such a cost function is that one can define whether or not an edit operation represents a strong modification of the underlying graph. Hence, between two similar graphs, there exists an inexpensive edit path, representing low cost operations, while for different graphs an edit path with high costs is needed. Consequently, the *edit distance* of two graphs is defined by the minimum cost edit path between two graphs. In the following we will denote a graph by $g = (V, E, \alpha, \beta)$, where V denotes a finite set of nodes, $E \subseteq V \times V$ a set of directed edges, $\alpha : V \rightarrow L_V$ a node labeling function assigning an attribute from a set L_V of node labels to each node, and $\beta : E \rightarrow L_E$ an edge labeling function. The substitution of a node u by a node v is denoted by $u \rightarrow v$, the insertion of u by $\varepsilon \rightarrow u$, and the deletion of u by $u \rightarrow \varepsilon$. A similar notation is used for edge substitution, insertion and deletion.

The edit distance can be computed by a tree search algorithm, where possible edit paths are iteratively explored, and the minimum-cost edit path can finally be retrieved from the search tree [8]. This method allows us to find the optimal edit path between two graphs. However, its drawback is its exponential time complexity, which makes the algorithm applicable to small graphs only. In this paper we propose another way of computing graph edit distance based on bipartite graph matching.

3   Bipartite Graph Matching by Assignment Algorithms

Standard tree search procedures for graph matching assign all nodes and edges of a graph to another graph by traversing some kind of search tree and minimizing the overall edit costs. If n and m denote the number of nodes of two graphs g_1 and g_2, the number of possible node assignments from g_1 to g_2 is given by $\frac{n!}{m!}$. Hence, the time complexity for this kind of algorithms is $O(n^m)$. However, the process of assigning nodes can be solved as a Linear Sum Assignment Problem instead, according to Def. 1.

Definition 1 (Linear Sum Assignment Problem (LSAP)). *Let us assume there are two sets U and V given, together with an $n \times n$ cost matrix $\mathbf{C} = (c_{ij})_{n \times n}$ of real numbers ($|U| = |V| = n$). The matrix elements $c_{ij} \geq 0$ correspond to the cost of assigning the i-th element of U to the j-th element of V. The assignment problem can be stated as finding a permutation $p = p_1, \ldots, p_n$ of the integers $1, 2, \ldots, n$ that minimizes $\sum_{i=1}^{n} c_{ip_i}$.*

Applied to the problem of graph edit distance computation, the sets U and V are the nodes of the graphs to be matched (V_1 and V_2) and the costs c_{ij} correspond to the node edit costs. We can expand the matrix to size $(n + m)$ to enable deletion or insertion of nodes. We define a cost matrix $C = (c_{ij})$ of dimension $(n + m) \times (n + m)$ such that entry c_{ij} corresponds to the cost of assigning the $i - th$ node of V_1 to the $j - th$ node of V_2 ($i \leq n, j \leq m$), the deletion of a node of V_1 ($i \leq n, j > m$), or the insertion of a node of V_2 ($i > n, j \leq m$).[1] For further details of setting up the cost matrix, we refer to [15].

---

[1] Note that, except for $c_{ij}, i \leq n, j \leq m$, all values except the diagonal of the according sub squares are set to infinity. This is because every node can only be inserted or deleted once. Costs on the diagonal of $c_{ij}, i > n, j > m$ are set to zero.

3.1 Assignment Algorithms

There are several algorithms known from the literature to solve the LSAP. In this paper, we consider three of them: HUNGARIAN [18], MUNKRES [19] and VOLGENANTJONKER [17].

- **Hungarian.** The first polynomial-time method for LSAP was the famous Hungarian algorithm by Kuhn [18] with time complexity $O(n^4)$. Nowadays, faster versions are available and the best time complexity for the Hungarian algorithm is $O(n^3)$ [20]. This is the algorithm used in the present paper.
- **Munkres.** A variation of the Hungarian algorithm was presented by Munkres [19]. He showed that his method requires at most $(11n^3 + 12n^2 + 31n)/6$ operations, where, however, some operations are "scan a line", thus leaving an $O(n^4)$ time complexity.
- **VolgenantJonker.** The shortest augmenting path algorithm by R. Jonker and T. Volgenant [17] has received most attention in the literature nowadays, due to its efficiency. The algorithm consists, like other shortest path algorithms, of three steps: a preprocessing method, used to find the first partial solution; a sparsification step to solve an instance with a reduced number of edges followed by a procedure that iteratively adds edges until an optimal solution is obtained; and a procedure to determine shortest paths. The preprocessing is the most important and time-consuming step of this algorithm. It consists of a column reduction process followed by two augmenting row reduction steps to find a first partial solution. The partial solution found is then improved by augmentation and completed through a smart implementation of a shortest paths algorithm in the style of Dijkstra[21,22]. Augmentation is done by constructing the auxiliary network graph and determining a minimal cost alternating path from an unassigned row to an unassigned column, which is then used to augment the solution. Although the time complexity of this algorithm ($O(n^3)$) is comparable to other LSAP algorithms, it is highly effective in practice. This is due to the fact that the partial solution found after preprocessing usually has a large number of assignments and only a few shortest paths are needed to complete it. Another benefit of this algorithm is its applicability to both sparse and dense graphs, and its insensitivity to the cost value range.

3.2 Bipartite Matching

We will refer to the method described in this section as BIPARTITE MATCHING. To compute the graph edit distance of two graphs, the procedure starts by generating a cost matrix containing the node edit costs as described earlier in this section. Included within those node edit costs are the edge edit costs of the adjacent edges for all node combinations. In the second step, the assignment algorithm computes the minimum cost node assignment.[2] Given the minimum cost node assignment, the implied edit operations of the edges are inferred, and

---

[2] We will refer to these costs as *implicitly* computed costs.

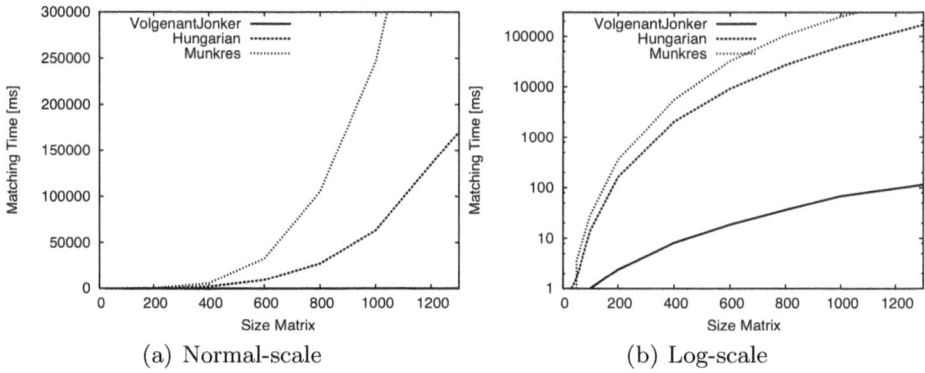

(a) Normal-scale   (b) Log-scale

Fig. 1. Comparison of the three assignment algorithms on randomly generated float matrices with size s and values $c_{i,j}$ between zero and one

the accumulated costs of the individual edit operations on both nodes and edges can be computed. Hence, the exact edit distance of the given assignment can be computed in linear time.[3] Note that there might exist other assignments with the same minimum node assignment cost, but possibly a smaller exact edit distance, as the explicit edge structure is only checked after the assignment algorithm has been applied. Hence, the node assignments and the implied edge assignments found by the algorithm need not correspond to an optimal solution, leading to a computed edit cost greater than or equal to the real graph edit distance.

Given this mapping These costs serve us as an approximate graph edit distance. The approximate edit distance values obtained by this procedure are equal to, or larger than, the exact distance values, since our suboptimal approach finds an optimal solution in a subspace of the complete search space.

4   Experimental Results

The purpose of the experiments described in this section is twofold. First, as BIPARTITE MATCHING uses a polynomial assignment algorithm to compute edit distance, we expect a crucial speed up when compared to exponential edit distance computation. This speed up is to be measured. Second, as our procedure comes up with a suboptimal solution only, we analyze the quality of the suboptimal edit distances making use of a k-nearest-neighbor classifier. The recognition rates obtained with this classifier allow us to compare the quality of the distances obtained from the different procedures.

4.1   Assignment Algorithms on Randomly Generated Float Matrices

Figure 1 shows the computation time of the assignment algorithms VOLGENANT-JONKER, HUNGARIAN, and MUNKRES on randomly generated matrices of floating point numbers with costs $c_{i,j} \in \,]0, 1[$ for $i, j \leq s$. The x-axis corresponds to the size

---

[3] We will refer to these costs as *explicitly* computed costs.

s of the matrices and the y-axis to the runtime [ms]. Assignments are computed for $s = 10, 50, 100, 200, \ldots, 1300$. For each size s and each algorithm, five test runs are made and the average value of the assignment runtime is recorded. In all test runs, the different assignment algorithms yield the same costs and the same assignments. For an assignment matrix with $s = 1300$ (which can be seen as mapping 1300 nodes of a graph g_1 to a graph g_2), VOLGENANTJONKER computes an assignment within 0.116, whereas HUNGARIAN needs 170,34 and MUNKRES 637,39, respectively. Algorithm VOLGENANTJONKER is by far the fastest algorithm and HUNGARIAN outperforms MUNKRES independent of size s.

4.2   Bipartite Matching on the IAM Data Sets

Table 1 characterizes the data sets used in our experiments[4]. The data consist of graphs representing line drawings (LETTER, DIGIT, GREC), proteins (AIDS, PROTEIN, MUTAGENICITY), and fingerprints. The parameters (costs of node and edge edit operations and value of k for the $k - NN$ classifier) are the same as reported in [23].

The recognition rates on the different data sets are based on $k - NN$ classification with graph edit distances and determined by the algorithm BIPARTITE MATCHING while using one of the given assignment algorithms. As reference method for an exact tree search method, we use the A* procedure as described in [8].[5] Comparing the results of the four algorithms on the LETTER data sets (Tab. 2), all methods achieve the same recognition rate (RR) on the LETTER LOW data set. The running time of the exact A* algorithm with 10,966 s is outperformed by the other procedures by a factor of 64 (MUNKRES), 69 (HUNGARIAN) and 79 (VOLGENANTJONKER). On the LETTER HIGH data set, the recognition rate increases from 89.87 to 90.13 by using A*. However, the computation time of the A* algorithm is with 12.3 hours disproportionately higher than the time of the VOLGENANTJONKERalgorithm with 146.2 seconds. An interesting fact is that for the LETTER MED data set higher recognition rates are obtained by the three suboptimal methods than by A*.

As explained in Sect. 3, the suboptimal distances are always larger than or equal to the true graph edit distances. This property is confirmed in Fig. 2, where a scatter plot of the suboptimal versus the true distances is shown.

For the following tests, we leave the A* algorithm beside because of its high computational complexity and concentrate on the other algorithms. Table 3 shows the results for the three other algorithms on the different data sets. In total, 5,937,500 matchings are computed. Time t_1 measures the whole assignment process, while time t_2 measures only the computation of the node mappings.[6]

---

[4] All data sets are available under www.iam.unibe.ch/fki/databases/iam-graph-database; see [23].

[5] Note that, for complexity reasons, exact distances are computed only for the LETTER data sets.

[6] Note that the considered assignment algorithm is also used in the first step described in Sect. 3.2 in order to find the minimum cost of matching the edges incident to a node.

Table 1. Data Sets

Data Set	Avg.#Nodes	Avg.#Edges	Tot.#Matchings
Letter Low	4.7	3.1	562,500
Letter Med	4.7	3.2	562,500
Letter High	4.7	4.5	562,500
Digit	8.9	7.9	1,000,000
GREC	11.5	12.2	375,000
AIDS	15.7	16.2	375,000
Protein	32.6	62.1	500,000
Fingerprint	5.4	4.4	1,000,000
Mutagenicity	30.3	30.8	1,000,000

Table 2. Evaluation of the classification accuracy and accumulated computation times (seconds) for the four algorithms on the Letter data sets

Data Set	A*		Munkres		Hungarian		VolgenantJonker	
	RR	$Time$	RR	$Time$	RR	$Time$	RR	$Time$
Letter Low	99.60	10,966.290	99.60	169.48	99.60	157.21	99.60	137.89
Letter Med	93.86	80,368.443	94.27	172.03	94.27	156.92	94.27	139.25
Letter High	90.13	44,459.416	89.87	190.28	89.87	178.15	89.87	146.20

Table 3. Recognition rates and running times (seconds) of Bipartite Matching using the different assignment algorithms, evaluated on the IAM data sets; t_2 is the time used to assign the nodes of two graphs and t_1 is the runtime of the whole procedure (including generation of the cost matrix and assignment of edges). The times shown are the accumulated runtimes for all matchings on each particular data set.

Data Set	Munkres			Hungarian Alg.			Volgenant-Jonker		
	RR	t_2	t_1	RR	t_2	t_1	RR	t_2	t_1
Letter Low	99.60	5.12	169.48	99.60	2.60	157.21	99.60	1.34	137.89
Letter Med	94.27	7.67	172.03	94.27	4.18	156.92	94.27	0.79	139.25
Letter High	89.87	8.51	190.28	89.87	4.15	178.15	89.87	1.50	146.20
Digit	96.75	417.22	1512.76	96.75	225.40	1046.85	96.75	24.82	636.70
GREC	97.73	22.44	307.59	97.73	10.88	267.86	97.73	2.11	224.49
AIDS	99.20	438.53	852.88	99.20	194.57	485.07	98.93	10.61	147.39
Protein	68.00	288.86	960.94	68.00	140.32	649.98	67.00	5.81	381.55
Fingerprint	63.20	63.03	200.58	63.25	36.05	150.51	62.95	4.28	94.27
Mutagenicity	68.30	14,223.93	16,345.59	68.30	7,411.11	9,017.22	67.60	59.92	880.70

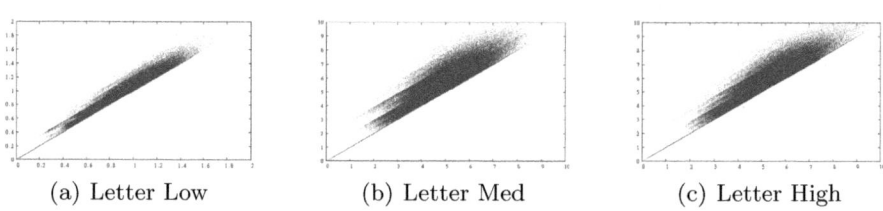

(a) Letter Low   (b) Letter Med   (c) Letter High

Fig. 2. Comparison of the edit distances computed by the exact algorithm (x-axis) and the suboptimal edit distances (y-axis)

In all cases, the three assignment algorithms yield the same (implicitly computed) costs. Figures 3 (a), (b) and (c) show the 100,000 computed suboptimal implicit edit costs (no explicit matching of edges according to Step 2 in Section

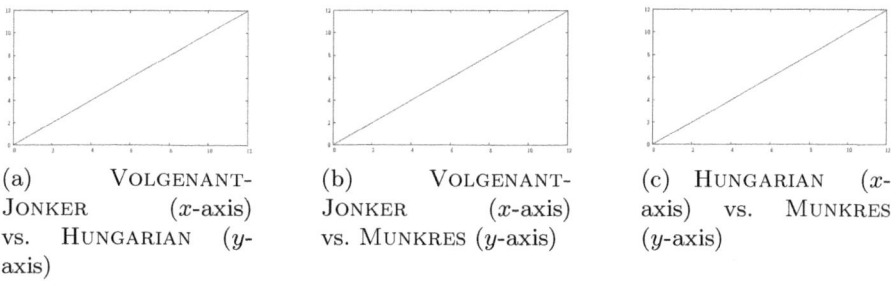

(a)   VOLGENANT-
JONKER   (x-axis)
vs.   HUNGARIAN   (y-
axis)

(b)   VOLGENANT-
JONKER   (x-axis)
vs. MUNKRES (y-axis)

(c) HUNGARIAN   (x-
axis)   vs.   MUNKRES
(y-axis)

Fig. 3. Comparison of the implicitly determined edit costs of the different algorithms

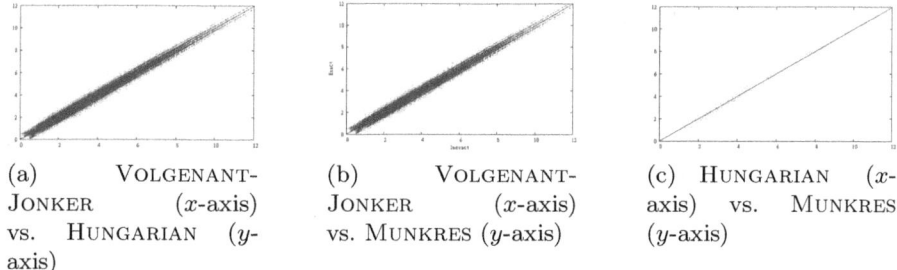

(a)   VOLGENANT-
JONKER   (x-axis)
vs.   HUNGARIAN   (y-
axis)

(b)   VOLGENANT-
JONKER   (x-axis)
vs. MUNKRES (y-axis)

(c) HUNGARIAN   (x-
axis)   vs.   MUNKRES
(y-axis)

Fig. 4. Comparison of the explicitly determined edit costs of the different algorithms

3.2 after assignment of nodes) on the FINGERPRINT data set, comparing the
values of VOLGENANTJONKER to those of HUNGARIAN and MUNKRES. As all
values lie on the straight $x = y$, we can conclude that both procedures yield
the same results. In some cases, however, there exist multiple assignments with
the same minimal cost, and the returned explicit assignment cost is no longer
the same (see Fig. 4). This has an effect on the recognition rates, e.g. on the
FINGERPRINT data set, where the recognition rate varies between 62.95, 63.15,
and 63.20 depending on the assignment algorithm used. Regarding the computation times, BIPARTITE MATCHING performs the fastest with the algorithm of
VOLGENANTJONKER independent of the data set used. Moreover, HUNGARIAN
is faster than BIPARTITE MATCHING using MUNKRES in all cases. For some data
sets, the time differences vary extremely. For example, computation of all node
assignments on the MUTAGENICITY data set is performed in 14,223 seconds by
MUNKRES, 7,411 by HUNGARIAN, and 60 by VOLGENANTJONKER.

5   Conclusion and Future Work

Graph edit distance is one of the most flexible and error-tolerant graph similarity
measures, which can be applied to a wide variety of problems. However, the
optimal comparison of two graphs is exponential in the number of nodes involved.
By reducing the matching process to an assignment problem, the complexity of
the procedure can be lowered to polynomial. As edges are considered implicitly

only, and the total edge costs used to transform a graph into another are applied after the node assignment has been found, the edit distance computed in this way is always greater than or equal to the exact graph edit distance, and hence suboptimal. For applications where it is not required to know the exact distances between two graphs, this procedure offers a versatile and fast alternative to the classical approach of graph edit distance computation.

For instance, a nearest-neighbor classifier to work properly does not need exact graph distances. Moreover, small distances, which are generally not much affected by our procedure, have more influence on the decision than large distances. In fact, experiments on a wide range of data sets show that by feeding the suboptimal edit distances into a k-nearest-neighbor classifier, the achieved classification accuracy computed by our novel procedure remains nearly the same, whereas the computation time is reduced dramatically. Among the assignment algorithms used in this paper, VOLGENANTJONKER performed better than HUNGARIAN or MUNKRES for all experiments. At the same time, recognition rates vary only slightly depending on the assignment algorithm actually used. In future work we will study the behavior of the proposed heuristic on other data sets, especially on non-labeled graphs.

References

1. Conte, D., Foggia, P., Sansone, C., Vento, M.: Thirty years of graph matching in pattern recognition. Int. Journal of Pattern Recognition and Artificial Intelligence 18(3), 265–298 (2004)
2. Luo, B., Wilson, R., Hancock, E.R.: Spectral embedding of graphs. Pattern Recognition 36(10), 2213–2223 (2003)
3. Wilson, R., Hancock, E.R.: Levenshtein distance for graph spectral features. In: Kittler, J., Petrou, M., Nixon, M. (eds.) Proc. 17th Int. Conf. on Pattern Recognition, vol. 2, pp. 489–492 (2004)
4. Lee, W.-J., Duin, R.P.: A labelled graph based multiple classifier system. In: Benediktsson, J.A., Kittler, J., Roli, F. (eds.) MCS 2009. LNCS, vol. 5519, pp. 201–210. Springer, Heidelberg (2009)
5. Christmas, W.J., Kittler, J., Petrou, M.: Structural matching in computer vision using probabilistic relaxation. IEEE Transactions on Pattern Analysis and Machine Intelligence 17(8), 749–764 (1995)
6. Scarselli, F., Gori, M., Tsoi, A.C., Hagenbuchner, M., Monfardini, G.: The graph neural network model. IEEE Transactions on Neural Networks 20(1), 61–80 (2009)
7. Cross, A., Wilson, R., Hancock, E.: Inexact graph matching using genetic search. Pattern Recognition 30(6), 953–970 (1997)
8. Bunke, H., Allermann, G.: Inexact graph matching for structural pattern recognition. Pattern Recognition Letters 1, 245–253 (1983)
9. Sanfeliu, A., Fu, K.S.: A distance measure between attributed relational graphs for pattern recognition. IEEE Transactions on Systems, Man, and Cybernetics (Part B) 13(3), 353–363 (1983)
10. Ambauen, R., Fischer, S., Bunke, H.: Graph edit distance with node splitting and merging and its application to diatom identification. In: Hancock, E.R., Vento, M. (eds.) GbRPR 2003. LNCS, vol. 2726, pp. 95–106. Springer, Heidelberg (2003)

11. Robles-Kelly, A., Hancock, E.R.: Graph edit distance from spectral seriation. IEEE Transactions on Pattern Analysis and Machine Intelligence 27(3), 365–378 (2005)
12. Boeres, M.C., Ribeiro, C.C., Bloch, I.: A randomized heuristic for scene recognition by graph matching. In: Ribeiro, C.C., Martins, S.L. (eds.) WEA 2004. LNCS, vol. 3059, pp. 100–113. Springer, Heidelberg (2004)
13. Sorlin, S., Solnon, C.: Reactive tabu search for measuring graph similarity. In: Brun, L., Vento, M. (eds.) GbRPR 2005. LNCS, vol. 3434, pp. 172–182. Springer, Heidelberg (2005)
14. Justice, D., Hero, A.: A binary linear programming formulation of the graph edit distance. IEEE Trans. on Pattern Analysis ans Machine Intelligence 28(8), 1200–1214 (2006)
15. Riesen, K., Bunke, H.: Approximate graph edit distance computation by means of bipartite graph matching. Image and Vision Computing 27(4), 950–959 (2009)
16. Raveaux, R., Burie, J.C., Ogier, J.M.: A graph matching method and a graph matching distance based on subgraph assignments. Pattern Recognition Letters 31(5), 394–406 (2010)
17. Jonker, R., Volgenant, T.: A shortest augmenting path algorithm for dense and sparse linear assignment problems. Computing 38, 325–340 (1987)
18. Kuhn, H.W.: The Hungarian method for the assignment problem. Naval Research Logistic Quarterly 2, 83–97 (1955)
19. Munkres, J.: Algorithms for the assignment and transportation problems. Journal of the Society for Industrial and Applied Mathematics 5, 32–38 (1957)
20. Burkard, R., Dell'Amico, M., Martello, S.: Assignment Problems. Society for Industrial and Applied Mathematics, Philadelphia (2009)
21. Dijkstra, E.W.: A note on two problems in connexion with graphs. Numerische Mathematik 1, 269–271 (1959)
22. Dantzig, G.B.: On the shortest route through a network. Management Science 6, 187–190 (1960)
23. Riesen, K., Bunke, H.: Graph Classification and Clustering based on Vector Space Embedding. World Scientific, Singapore (2010)

Two New Graph Kernels and Applications to Chemoinformatics

Benoit Gaüzère[1], Luc Brun[1], and Didier Villemin[2]

[1] GREYC UMR CNRS 6072,
[2] LCMT UMR CNRS 6507,
Caen, France
{benoit.gauzere,didier.villemin}@ensicaen.fr,
luc.brun@greyc.ensicaen.fr

Abstract. Chemoinformatics is a well established research field concerned with the discovery of molecule's properties through informational techniques. Computer science's research fields mainly concerned by the chemoinformatics field are machine learning and graph theory. From this point of view, graph kernels provide a nice framework combining machine learning techniques with graph theory. Such kernels prove their efficiency on several chemoinformatics problems. This paper presents two new graph kernels applied to regression and classification problems within the chemoinformatics field. The first kernel is based on the notion of edit distance while the second is based on sub trees enumeration. Several experiments show the complementary of both approaches.

Keywords: edit-distance, graph kernel, chemoinformatics.

1 Introduction

Chemoinformatics aims to predict or analyse molecule's properties through informational techniques. One of the major principle in this research field is the *similarity principle*, which states that two structurally similar molecules should have similar activities and properties. The structure of a molecule is naturally encoded by a labeled graph $G = (V, E, \mu, \nu)$, where the unlabeled graph (V, E) encodes the structure of the molecule while μ maps each vertex to an atom's label and ν characterizes a type of bond between two atoms (single, double, triple or aromatic).

A first family of methods introduced within the Quantitative Structure-Activity Relationship (QSAR) field is based on the correlation between molecule's descriptors such as the number of atoms and molecule's properties (e.g. molecule's boiling point). Vectors of descriptors may be defined from structural information [2], physical properties or biological activities and may be used within any statistical machine learning algorithm to predict molecule's properties. Such a scheme allows to benefit from the large set of tools available within the statistical machine learning framework. However, the definition of a vector from a molecule, ie. a graph, induces a loss of information. Moreover, for each application, the definition of a vectorial description of each molecule remains heuristic.

X. Jiang, M. Ferrer, and A. Torsello (Eds.): GbRPR 2011, LNCS 6658, pp. 112–121, 2011.
© Springer-Verlag Berlin Heidelberg 2011

A second family of methods, based on graph theory may be decomposed in two sub families. The first sub family [8], related to the data mining field, aims to discover sub graphs with a large difference of frequencies in a set of positive and negative examples. The second sub family [1], more related to the machine learning field, builds a structural description of each class of molecule so that the classification is conducted by mean of a structural matching between each prototype and a graph representation of an input molecule. Both sub families are however mainly restricted to the classification field.

Graph kernels can be understood as symmetric graph similarity measures. Using a semidefinite positive kernel, the value $k(G, G')$ where G, G' encode two input graphs corresponds to a scalar product between two vectors $\psi(G)$ and $\psi(G')$ in an Hilbert space (this space is only a Krein space if the kernel is non definite). Graph kernels provide thus a natural connection between structural pattern recognition and graph theory on one hand and statistical pattern recognition on the other hand. A large family of kernels is based on the definition of a bag of patterns for each graph and deduces graph similarity from the similarity between bags. Kashima [5] defines graph kernels based on the comparison of sets of walks extracted from each graph. Ramon and Gärtner [9] and Mahé [6] define kernels using an infinite set of tree patterns instead of walks. These methods improve the limited expressiveness of linear features such as walks hence providing a priori a more meaningful similarity measure. Instead of decomposing graphs into an infinite set of substructures (ie walks or trees), Shervashidze and Borgwardt[12] compute the kernel from the distribution of a predefined set of subgraphs, called *graphlets*. An other approach to the definition of graph kernels is proposed by Neuhaus and Bunke [7]. This approach aims to define definite positive kernels from the notion of edit distance. The main challenge of this approach is that the edit distance does not fulfill all requirements of a metric and hence does not readily lead to a definite positive kernel.

This paper presents two new kernels: A first kernel, presented in Section 2, combines graph edit distance and graph Laplacian kernel notions in order to obtain a definite positive graph kernel. A method to update efficiently this kernel is also proposed. Our second kernel, presented in Section 3, uses a different approach based on an explicit enumeration of subtrees within an acyclic unlabeled graph. The efficiency and complementarity of these two kernels is finally demonstrated in Section 4 through experiments.

2   Kernel from Edit Distance

An edit path between two graphs G and G' is defined as a sequence of operations transforming G into G'. Such a sequence may include vertex or edge addition, removal and relabeling. Given a cost function $c(.)$, associated to each operation, the cost of an edit path is defined as the sum of its elementary operation's costs. The minimal cost among all edit paths transforming G into G' is defined as the *edit distance* between both graphs. A high edit distance indicates a low similarity between two graphs while a small one indicates a strong similarity.

According to Bunke and Neuhaus[7], the computational cost of the exact edit distance grows exponentially with the size of the graphs. Such a property limits the computation of exact edit distance to small graphs. To overcome this problem, Bunke and Riesen[11] defined a method to compute a sub optimal edit distance. This method computes an approximate edit distance in $O(nv^2)$ where n and v are respectively equal to the number of nodes and to the maximal degree of both graph.

Unfortunately, edit distance doesn't define a metric and trivial kernels based on edit distance are not definite positive. Neuhaus and Bunke [7] proposed several method to overcome this important drawback, however the proposed kernels are not explicitly based on the minimization problem addressed by kernel methods. Such a minimization problem may be stated as follows: Given a kernel k and a dataset of graphs $D = \{G_1, \ldots, G_n\}$, the Gram matrix K associated to D is an $n \times n$ matrix defined by $K_{i,j} = k(G_i, G_j)$. Within the kernel framework, a classification or regression problem based on K may be stated as the minimization of the following formula on the set of real vectors of dimension n:

$$f^* = \arg\min_{f \in \mathbb{R}^n} CLoss(f, y, K) + f^t K^{-1} f \qquad (1)$$

where $CLoss(., ., .)$ denotes a given loss function encoding the distance between vector f and the vector of known values y.

As denoted by Steinke [14], the term $f^t K^{-1} f$ in equation 1 may be considered as a regularization term which counter balance the fit to data term encoded by the function $CLoss(., .)$. Therefore, the inverse of K (or its pseudo inverse if K is not invertible) may be considered as a regularization operator on the set of vectors of dimension n. Such vectors may be considered as functions mapping each graph of the database to a real value. Conversely, the inverse (or pseudo inverse) of any semi definite positive regularization operator may be considered as a kernel. We thus follow a kernel construction scheme recently introduced [1] which first builds a semi definite positive regularization operator on the set of functions mapping each graph $\{G_1, \ldots, G_n\}$ to a real value. The inverse, or pseudo inverse of this operator defines a kernel on the set $\{G_1, \ldots, G_n\}$.

In order to construct this regularization operator, let us define a $n \times n$ adjacency matrix W defined by $W_{ij} = e^{-\frac{d(G_i, G_j)}{\sigma}}$, where $d(., .)$ denotes the edit distance and σ is a tuning variable. The Laplacian of W is defined as $l = \Delta - W$ where Δ is a diagonal matrix defined by: $\Delta_{i,i} = \sum_{j=1}^{n} W_{i,j}$. Classical results from spectral graph theory [3] establish that l is a symmetric semi definite positive matrix whose minimal eigenvalue is equal to 0. Such a matrix is thus not invertible. To overcome this problem, Smola [13] defines the regularized Laplacian $\tilde{l}$ of W as $\tilde{l} = I + \lambda l$ where λ is a regularization coefficient. The minimal eigen value of $\tilde{l}$ is equal to 1 and the matrix $\tilde{l}$ is thus definite positive. Moreover, given any vector f, we have :

$$f^t \tilde{l} f = \|f\|^2 + \lambda \sum_{i,j=1}^{n} W_{ij}(f_i - f_j)^2 \qquad (2)$$

Intuitively, minimising equation 2, leads to build a vector f with a small norm which maps graphs with a small edit distance (and thus a strong weight) to close values. Such a constraint corresponds to the regularization term required by equation 1 in order to smoothly interpolate the test values y over the set of graphs $\{G_1, \ldots, G_n\}$. Our un normalized kernel, is thus defined as: $K_{un} = \tilde{l}^{-1}$.

Note that a regularized normalized Laplacian kernel may alternatively be considered by introducing the matrix $\tilde{L} = \Delta^{-\frac{1}{2}} \tilde{l} \Delta^{-\frac{1}{2}}$. We have in this case, for any vector f:

$$f^t \tilde{L} f = \sum_{i=1}^{n} \frac{f_i^2}{\Delta_{ii}} + \lambda \sum_{j=1}^{n} \frac{W_{ij}}{\sqrt{\Delta_{ii} \Delta_{jj}}} (f_i - f_j)^2$$

The matrix $\tilde{L}$ is definite positive and its associated kernel is defined as $K_{norm} = \tilde{L}^{-1}$. Note that, our regularized normalized Laplacian kernel is not defined as the inverse of the regularized normalized Laplacian $I + \lambda \Delta^{-\frac{1}{2}} l \Delta^{-\frac{1}{2}}$. This new formulation is consistent with the regularization constraint which should be added to equation 1 and provides significant advantages in the context of incoming data (Section 2.1).

2.1   Incoming Data

Let us first consider a kernel defined from the un normalized Laplacian. Given our learning set $D = \{G_1, \ldots, G_n\}$, the test of a new graph G within a regression or classification scheme requires to update the un normalized Laplacian l with this new graph and to compute the updated kernel defined as the inverse of the regularized and un normalized Laplacian $K = (I + \lambda l)^{-1}$. This direct method has a complexity equal to $\mathcal{O}((n+1)^3)$, where n is the size of our data set. Such a method is thus computationally costly, especially for large datasets. In this section, we propose a method to reduce the complexity of this operation.

Given the regularized and un normalized Laplacian $\tilde{l}_n = (I_n + \lambda(\Delta_n - W_n))$ defined on the dataset D, its updated version $\tilde{l}_{n+1}$ defined on $D \cup \{G\}$ may be expressed as follows:

$$\tilde{l}_{n+1} = \begin{pmatrix} \tilde{l}_n - \delta_n & B \\ B^t & 1 - \sum_i B_i \end{pmatrix}$$

where $B = (-\lambda exp(\frac{-d(G,G_i)}{\sigma}))_{i=\{1,\ldots,n\}}$ is deduced from the weights between the new input graph G and each graph $(G_i)_{i=\{1,\ldots,n\}}$ of our dataset and δ_n is a diagonal matrix with $(\delta_n)_{i,i} = B_i$.

The minimal eigen value of $\tilde{l}_{n+1}$ is equal to 1 (Section 2). This matrix is thus invertible, and its inverse may be expressed using a block inversion scheme:

$$K_{un} = (\tilde{l}_{n+1})^{-1} = \begin{pmatrix} \Gamma & \Theta \\ \Lambda & \Phi \end{pmatrix} \text{ with } \begin{cases} \Gamma = E^{-1} + \Phi E^{-1} B B^t E^{-1} \\ \Theta = -E^{-1} B \Phi \\ \Lambda = -\Phi B^t E^{-1} \\ \Phi = (1 - \sum_i B_i - B^t E^{-1} B)^{-1} \end{cases} \quad (3)$$

where $E = \tilde{l}_n - \delta_n$. Note that Φ corresponds to a scalar.

The computation of our new kernel, using equation 3, relies on the computation of the inverse of the matrix $E = \tilde{l}_n + \delta_n$ which may be efficiently approximated using a development to the order K of $(I - \tilde{l}_n^{-1}\delta_n)^{-1}$:

$$(\tilde{l}_n - \delta_n)^{-1} = \tilde{l}_n^{-1}(I - \tilde{l}_n^{-1}\delta_n)^{-1} \approx \sum_{k=0}^{K} l_n^{-k-1}\delta_n^k \tag{4}$$

Such a sum converges since $\|\tilde{l}_n^{-1}\delta_n\|_2 < 1$, for $\lambda < 1$. Indeed:

$$\|\tilde{l}_n^{-1}\delta_n\|_2 \leq \|\tilde{l}_n^{-1}\|_2\|\delta_n\|_2 \leq \|\delta_n\|_2 \leq \lambda \max_{i=1,n} exp(\frac{-d(G, G_i)}{\sigma})$$

The last term of this equation is strictly lower than one for any λ lower than one. Moreover, basic matrix calculus show that the approximation error is lower than ϵ for any K greater than:

$$\frac{log(2\epsilon)}{log(\max_{i=1,n} exp(\frac{-d(G,G_i)}{\sigma}))}. \tag{5}$$

Equation 4 allows to approximate the inverse of $(\tilde{l}_n - \delta_n)$ by a sum of pre computed matrices l_n^{-k-1} multiplied by diagonal matrices. Using such pre calculus, the inverse of $(\tilde{l}_n - \delta_n)$ and hence the computation of our new kernel may be achieved in KN^2.

If we now consider the regularized normalized Laplacian (Section 2) $\tilde{L} = \Delta^{-\frac{1}{2}}\tilde{l}\Delta^{-\frac{1}{2}}$, its inverse is defined as: $\tilde{L}^{-1} = \Delta^{\frac{1}{2}}\tilde{l}^{-1}\Delta^{\frac{1}{2}}$ and we have:

$$K_{norm} = \Delta^{\frac{1}{2}}K_{un}\Delta^{\frac{1}{2}} \tag{6}$$

The update of the regularized and normalized Laplacian kernel may thus be deduced from the one of the regularized un normalized Laplacian kernel.

3   Treelet Kernel

Kernels based on edit distance rely on a direct comparison of each pair of graph. An alternative strategy consists to represent each graph by a bag of patterns and to deduce the similarity between two graphs from the similarity of their bags. This strategy may provide semi definite kernels hereby avoiding the necessity to regularize the whole gram matrix for each incoming data (Section 2.1). As mentioned in Section 1, most of kernels of this family are based on linear patterns (bags of paths, trails or walks). Shervashidze et al. [12] describe a method to enumerate for any input unlabelled graph, all its connected subgraphs composed of up to 5 nodes. This efficient method provides up to 2048 patterns composed of connected subgraphs (called graphlets) of size lower or equal to 5. We propose here to adapt this method to the enumeration of sub-trees of acyclic unlabeled graphs up to size 6. The resulting patterns are called treelets (Fig. 1).

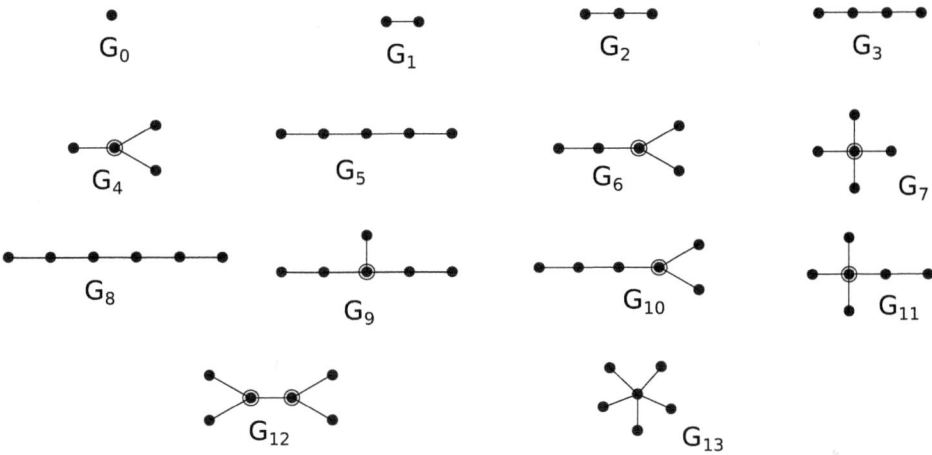

Fig. 1. Acyclic and unlabeled graphlets of maximum size equals to 6. Centers of 3-star and 4-star are surrounded.

3.1 Computing Embedded Distribution

Following [12], our enumeration of all treelets starts by an enumeration of all paths with a length lower or equal to 6. A recursive depth first search with a max depth equals to 6 from each node of the graph is thus performed. Note that using such an enumeration each path is retrieved from its two extremities and is thus counted twice. In order to prevent this problem, each path composed of at least two nodes is counted $\frac{1}{2}$ times. With this first step, the distribution of treelets G_0, G_1, G_2, G_3, G_5 and G_8 is computed (Fig. 1).

To compute the distribution of the remaining treelets, our method is based on the detection of nodes of degree 3 and 4. These nodes are respectively called R_{3-star} and R_{4-star} and are the center of the *3-star* and *4-star* treelets. Note that a 4-star treelet (G_7) contains four 3-star treelets (Fig. 2). This first degree analysis allows to compute the distribution of treelets G_4 and G_7. Treelets G_6, G_9, G_{10} and G_{12} are enumerated from the neighbourhood of 3-star treelets. For example, treelet G_6 requires a 3-star with at least one neighbour of R_{3-star} with a degree greater or equal to 2. Treelet G_{11} is the only sub tree derived from a 4-star. Properties characterizing treelets with a 3 or 4 star are summarized

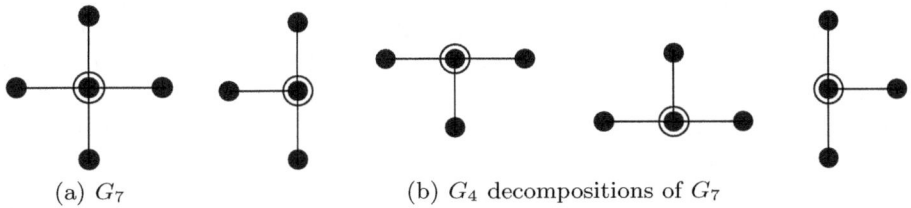

(a) G_7 (b) G_4 decompositions of G_7

Fig. 2. G_7 contains 4 G_4

Table 1. Conditions characterizing treelets derived from 3-star and 4-star. $N(v)$ and $d(v)$ denote respectively the set of neighbours and the degree of vertex v.

Treelet	Source treelet	Condition
G_6	3-star	$|\{v; v \in N(R_{3-star}); d(v) \geq 2\}| \geq 1$
G_9	3-star	$|\{v; v \in N(R_{3-star}); d(v) \geq 2\}| \geq 2$
G_{10}	3-star	$\exists v_0 \in N(R_{3-star}); d(v_0) \geq 2$ and $|\{v; v \in N(v_0) - \{R_{3-star}\}; d(v) \geq 2\}| \geq 1$
G_{11}	4-star	$|\{v; v \in N(R_{4-star}); d(v) \geq 2\}| \geq 1$
G_{12}	3-star	$|\{v; v \in N(R_{3-star}); d(v) \geq 3\}| \geq 1$

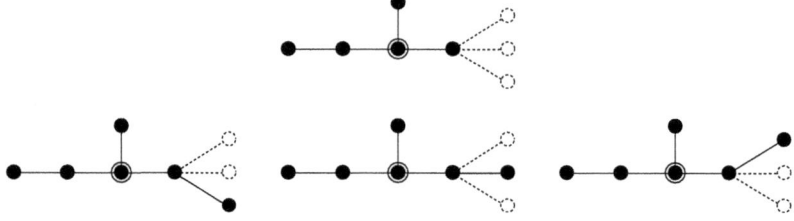

Fig. 3. Three permutations of G_9 sharing the same core

in Table 1. Note that treelet G_{12} is symmetric since it contains two centers of 3-star. Such a treelet will thus be counted twice (once from each of its 3-star) and must be counted for $\frac{1}{2}$.

Note that conditions summarized in Table 1 define necessary conditions for the existence of a given treelet centered around a 3 or 4 star. However, such conditions does not guarantee the uniqueness of such a treelet. Fig. 3 shows such an example: the rightest node of G_9 has a degree equals to 4 within the input graph whereas a degree greater or equal to 2 is required to define treelet G_9. Three different treelet G_9 may thus be built from the same five nodes. This configuration thus induces to count G_9 three times from the graph represented in Fig. 3(a) One may easily check that no isomorphism exists between treelets depicted in Fig. 1. Moreover, has shown by Read [10], the number of different alkanes composed of up to 6 carbons is equal to 13. Within our context, an alkane may be considered as an unlabeled acyclic graph whose vertex degree is bounded by 4. Therefore, treelet G_{13} which is the only treelet with a vertex of degree greater than 4 does not corresponds to a feasible alkane. The remaining 13 treelets in Fig. 1 represents, up to isomorphisms, all the unlabeled acyclic graphs whose size is lower than 6 and whose vertex degree is bounded by 4. Adding G_{13} to this set provides all unlabeled acyclic graphs with a size lower than 6.When all treelets from a graph G have been enumerated, a vector representing treelet distribution is computed. Each component of this vector, denoted the *spectrum* of G, is equal to the frequency of a given treelet in G:

$$f_i(G) = |(G_i \subset G)| \tag{7}$$

Table 2. Comparison of addition methods

Method	Classification Accuracy
KMean [15]	80% (55/68)
KWMean [4]	88% (60/68)
Trivial Similarity Kernel from Edit Distance [7]	90% (61/68)
Normalized Standard Graph Laplacian Kernel (Eq. 6)	90% (61/68)
Normalized Fast Graph Laplacian Kernel (Eq. 6)	90% (61/68)
Random Walk Kernel [16]	82% (56/68)

3.2   Definition of Treelet Kernel

A first idea to define a kernel from treelets consists to perform the inner product of vectors encoding the spectrum of graphs. Unfortunately, the inner product doesn't highlight spectrum similarities. For example, two graphs with nearly equal spectrum but with a low number of occurrences for each treelet are considered as less similar than two graphs having a same high number of treelet G_0 (ie same size) but a distribution of others treelets highly dissimilar. We thus use RBF kernels in order to better highlight differences between two spectra:

$$k_{Treelet}(G, G') = \sum_{k=0}^{N} e^{-\frac{(f_k(G) - f_k(G'))^2}{\sigma}} \tag{8}$$

where σ is a tuning variable used to weight the differences between treelet distribution and N is the number of enumerated treelets. Our kernel may thus be considered as a basic RBF kernel between two vectors and is hence definite positive.

4   Experiments

Our first experiment evaluates the graph Laplacian kernel on a classification problem. This problem is defined on the monoamine oxidase dataset(MAO)[1] which is composed of 68 molecules divided into two classes: 38 molecules inhibits the monoamine oxidase (antidepressant drugs) and 30 does not. These molecules are composed of different types of atoms with simple bonds and are thus encoded as labeled graphs. Classification accuracy is measured for each method using a leave one out procedure with a two-class SVM. This classification scheme is made for each of the 68 molecules of the dataset.

Table 2 shows results obtained by graph Laplacian kernel using approximate graph edit distance [11] with node substitution and edge deletion costs set to 1 and edge substitution cost set to the sum of incident node substitution costs. Graph Laplacian kernel methods obtain a classification accuracy of 90% which corresponds to the highest score. Note that the other method obtaining 90% of

---

[1] All databases in this section are available on the TC15 Web page:
   http://www.greyc.ensicaen.fr/iapr-tc15/links.html#chemistry

Table 3. Boiling point prediction on alkane dataset

Method	Average error (C)	Standard deviation (C)	Correlation
Neural Network [2]	3.11453	3.69993	0.9827
KMean [15]	4.65536	6.20788	0.9918
Random Walk Kernel [16]	10.6084	16.2799	0.9057
Graph Laplacian Kernel	10.7948	16.4484	0.9412
Treelet Kernel	1.40663	1.91695	0.9992

classification accuracy is also based on the edit distance. This last kernel may however be non definite positive.

We may additionally notice that the use of our fast inversion method (Section 2.1) does not modify graph Laplacian kernel's classification accuracy (Table 2, lines 4 and 5). The number of iterations required by this fast inversion method is determined by equation 5. Our experiments performed on the MAO database show that a value of ϵ equal to 10^{-4} induces a maximum of 9 iterations hence allowing to update the gram matrix in $\mathcal{O}(9N^2)$ instead of $\mathcal{O}(N^3)$ using a standard matrix inversion method. The low value of N on this dataset ($N = 68$) does not induce an important gain on execution time since the average time to update a Gram matrix using method described in Section 2.1 is $0.273ms$ on the MAO database while this time is equal to $0.498ms$ using a direct inverse matrix computation. The ratio between both execution times is nevertheless about 1.8 hence showing a significant gain. Our treelet kernel is not tested against this database since this kernel is devoted to unlabeled graphs.

Our second experiment is based on a database of alkanes [2]. An alkane is an acyclic molecule solely composed of carbons and hydrogens. A common encoding consists to implicitly encode hydrogen atoms using the valency of carbon atoms. Such an encoding scheme allows to represent alkanes as acyclic unlabeled graphs. The alkane dataset described in [2] is composed of 150 molecules, associated to their respective boiling points. Using the same protocol than [2], we evaluate the boiling point of each alkane using several test sets composed of 10% of the database, the remaining 90% being used as training set.

Table 3 shows results obtained by different methods. Poor results obtained by graph Laplacian kernel can be explained by the lack of information when dealing with unlabeled graphs. Indeed, using such graphs, the heuristic used to approximate graph edit distance [11] maps the set of vertices of both graphs using uniquely the degree of vertices. Such a method thus consider several mappings as equivalent if several vertices with a same degree exist in both graphs. In this case, the sub optimal graph edit distance induces a poor graph discrimination. This lack of local information within unlabeled graphs also explains the poor results obtained by Kmean and random walk kernels. Indeed, these kernels are based on linear structures which are only discriminated by their lengths within unlabeled graphs. On the other hand, treelet kernel (with $\sigma = 0, 25$) outperforms previous results of [2] based on neural networks combined with chemical descriptors.

5   Conclusion

In this paper we proposed a graph Laplacian kernel based on a sub optimal graph edit distance combined with an efficient update of the kernel in order to predict relevant properties of incoming data. Our experiments show the efficiency of this kernel on databases composed of complex molecules with several hetero atoms. However, this kernel performs poorly on unlabeled graphs. We thus propose a new kernel based on treelet enumeration for unlabeled acyclic graphs. This kernel outperforms results obtained by state of the art methods on this dataset but remains restricted to unlabeled graphs. Our future work will be devoted to overcome this last limitation by extending treelet kernel to labeled graphs.

References

1. Brun, L., Conte, D., Foggia, P., Vento, M., Villemin, D.: Symbolic learning vs. graph kernels: An experimental comparison in a chemical application. In: Proc. of the 1st Int. Workshop on Querying Graph Structured Data (2010)
2. Cherqaoui, D., Villemin, D.: Use of a neural network to determine the boiling point of alkanes. Journal of the Chemical Society, Faraday Transactions (1994)
3. Chung, F.R.K.: Spectral graph theory. In: AMSP (1997)
4. Dupé, F.-X., Brun, L.: Tree covering within a graph kernel framework for shape classification. In: Foggia, P., Sansone, C., Vento, M. (eds.) ICIAP 2009. LNCS, vol. 5716, pp. 278–287. Springer, Heidelberg (2009)
5. Kashima, H., Tsuda, K., Inokuchi, A.: Kernels for graphs, ch. 7, pp. 155–170. MIT Press, Cambridge (2004)
6. Mahé, P., Vert, J.-P.: Graph kernels based on tree patterns for molecules. Machine Learning 75(1), 3–35 (2008)
7. Neuhaus, M., Bunke, H.: Bridging the gap between graph edit distance and kernel machines. World Scientific Pub. Co. Inc., Singapore (2007)
8. Poezevara, G., Cuissart, B., Crémilleux, B.: Discovering emerging graph patterns from chemicals. In: Rauch, J., Raś, Z.W., Berka, P., Elomaa, T. (eds.) ISMIS 2009. LNCS, vol. 5722, pp. 45–55. Springer, Heidelberg (2009)
9. Ramon, J., Gärtner, T.: Expressivity versus efficiency of graph kernels. In: 1st Int. Workshop on Mining Graphs, Trees and Sequences, pp. 65–74 (2003)
10. Read, R.: Some recent results in chemical enumeration. In: Alavi, Y., Lick, D., White, A. (eds.) Graph Theory and Applications. Lecture Notes in Mathematics, vol. 303, pp. 243–259. Springer, Heidelberg (1972)
11. Riesen, K., Bunke, H.: Approximate graph edit distance computation by means of bipartite graph matching. Image and Vision Computing 27(7), 950–959 (2009)
12. Shervashidze, N., Vishwanathan, S.V.N., Petri, T.H., Mehlhorn, K., Borgwardt, K.M.: Efficient graphlet kernels for large graph comparison. In: Proceedings of AIStats, pp. 488–495 (2009)
13. Smola, A.J., Kondor, R.: Kernels and regularization on graphs. In: Learning theory and Kernel machines: 16th Annual Conference on Learning Theory and 7th Kernel Workshop, p. 144. Springer, Heidelberg (2003)
14. Steinke, F., Schölkopf, B.: Kernels, regularization and differential equations. Pattern Recogn. 41, 3271–3286 (2008)
15. Suard, F., Rakotomamonjy, A., Bensrhair, A.: Kernel on bag of paths for measuring similarity of shapes. In: European Symposium on Artificial Neural Networks (2002)
16. Vishwanathan, S.V.N., Borgwardt, K.M., Kondor, I.R., Schraudolph, N.N.: Graph Kernels. Neural Networks (2008)

Generalized Learning Graph Quantization

Brijnesh J. Jain and Klaus Obermayer

Berlin Institute of Technology, Germany
jbj@cs.tu-berlin.de

Abstract. This contribution extends generalized LVQ, generalized relevance LVQ, and robust soft LVQ to the graph domain. The proposed approaches are based on the basic learning graph quantization (`lgq`) algorithm using the orbifold framework. Experiments on three data sets show that the proposed approaches outperform `lgq` and `lgq2.1`.

1 Introduction

Learning vector quantization (LVQ) as introduced by Kohonen [11] is a supervised learning algorithm for pattern classification. To classify patterns, LVQ applies the nearest neighbor rule using a condensed set of prototypes. Prototypes are learned by combining competitive learning with supervision. LVQ is easy to implement, runs efficiently, allows to control the complexity of the resulting classifier, naturally deals with multiclass problems, constructs an informative rather than a black-box model, and in many cases provides state of the art performance. Due to well-known shortcomings of LVQ and LVQ2.1 more sophisticated and powerful learning vector quantizers such as generalized LVQ [16], generalized relevance LVQ [4], and soft robust LVQ [17] have been devised.

LVQ and related methods have been originally devised for feature vectors equipped with the Euclidean metric. Extensions have been proposed, for example, for vectors with arbitrarily differentiable distance functions [5], for variable length and warped feature sequences [18], for strings [12], and for graphs [8].

For graphs, LVQ and LVQ2.1 have been extended to the corresponding learning graph quantization algorithms `lgq` and `lgq2.1` and comparable results to state-of-the-art methods have been reported [8]. These findings give rise to the question at issue, whether extensions of more powerful learning vector quantizers to the graph domain yield improved learning graph quantizers.

In this contribution, we extend generalized LVQ, generalized relevance LVQ, and robust soft LVQ to the domain of attributed graphs. The proposed approaches are based on the orbifold framework for graphs [6] and on `lgq` [8]. Experiments on three data sets of the IAM graph database [14] show that the proposed algorithms outperform `lgq` and `lgq2.1`.

2 Graph Orbifolds

This section introduces attributed graphs and represent them as point of some orbifold [1]. Most of this presentation including proofs of statements and claims is based on the structure space formalism proposed by [6].

X. Jiang, M. Ferrer, and A. Torsello (Eds.): GbRPR 2011, LNCS 6658, pp. 122–131, 2011.

Representation of Graphs. Let $\mathbb{E}$ be a d-dimensional Euclidean space. An *attributed graph* $X = (V, E, \alpha)$ consists of a set V of *vertices*, a set $E \subseteq V \times V$ of *edges*, and an *attribute function* $\alpha : V \times V \to \mathbb{E}$, such that $\alpha(i, j) \neq \mathbf{0}$ for each edge and $\alpha(i, j) = \mathbf{0}$ for each non-edge. Attributes $\alpha(i, i)$ of vertices i may take any value from $\mathbb{E}$.

For simplifying the mathematical treatment, we assume that all graphs are of order n, where n is chosen to be sufficiently large. Graphs of order less than n, say $m < n$, can be extended to order n by including isolated vertices with attribute zero. For practical issues, it is important to note that limiting the maximum order to some arbitrarily large number n and extending smaller graphs to graphs of order n are purely technical assumptions to simplify mathematics. For pattern recognition problems, these limitations should have no practical impact, because neither the bound n needs to be specified explicitly nor an extension of all graphs to an identical order needs to be performed. When applying the theory, all we actually require is that the graphs are finite.

A graph X is completely specified by its *matrix representation* $\boldsymbol{X} = (\boldsymbol{x}_{ij})$ with elements $\boldsymbol{x}_{ij} = \alpha(i, j)$ for all $1 \leq i, j \leq n$. Let $\mathcal{X} = \mathbb{E}^{n \times n}$ be the Euclidean space of all $(n \times n)$-matrices with elements from $\mathbb{E}$ and let Π^n be the set of all $(n \times n)$-permutation matrices. For each $\boldsymbol{P} \in \Pi^n$ we define a mapping

$$\gamma_{\boldsymbol{P}} : \mathcal{X} \to \mathcal{X}, \quad \boldsymbol{X} \mapsto \boldsymbol{P}^\mathsf{T} \boldsymbol{X} \boldsymbol{P}.$$

Then $\mathcal{G} = \{\gamma_{\boldsymbol{P}} : \boldsymbol{P} \in \Pi^n\}$ is a finite group acting on $\mathcal{X}$. For $\boldsymbol{X} \in \mathcal{X}$, the *orbit* of $\boldsymbol{X}$ is the set defined by $[\boldsymbol{X}] = \{\gamma(\boldsymbol{X}) : \gamma \in \mathcal{G}\}$. The quotient set

$$\mathcal{X}_\mathcal{G} = \{[\boldsymbol{X}] : \boldsymbol{X} \in \mathcal{X}\}$$

consisting of all orbits is a *graph orbifold*. Its *orbifold chart* is the surjective continuous mapping

$$\pi : \mathcal{X} \to \mathcal{X}_\mathcal{G}, \quad \boldsymbol{X} \mapsto [\boldsymbol{X}]$$

that projects each matrix representation $\boldsymbol{X}$ to its orbit $[\boldsymbol{X}]$.

Suppose that $\boldsymbol{X}$ is a matrix representation of some attributed graph X. Then the orbit $[\boldsymbol{X}]$ consists of all possible matrices that represent X. By identifying the attributed graphs X with the orbits $[\boldsymbol{X}]$, we can regard graphs from $\mathcal{G}_\mathcal{A}$ as point of the graph orbifold $\mathcal{X}_\mathcal{G}$. The orbifold chart $\pi : \mathcal{X} \to \mathcal{X}_\mathcal{G}$ projects matrices $\boldsymbol{X}$ to the graphs X they represent.

For notational convenience, we identify $\mathcal{X}$ with $\mathbb{E}^N$, where $N = n^2$ and consider vector- rather than matrix representations of graphs. We obtain a *vector representation* $\boldsymbol{x}$ of graph X by concatenating the columns of a matrix $\boldsymbol{X}$ representing X. We write $\boldsymbol{x} \in X$ if $\boldsymbol{x} \in \mathcal{X}$ projects to $X \in \mathcal{X}_\mathcal{G}$ via the orbifold chart $\pi(\boldsymbol{x}) = X$.

Intrinsic Metric. The *intrinsic metric* of a graph orbifold $\mathcal{X}_\mathcal{G}$ is of the form

$$d(X, X') = \min \left\{ \|\boldsymbol{x} - \boldsymbol{x}'\|^2 : \boldsymbol{x} \in X, \boldsymbol{x}' \in X' \right\},$$

where $\|\cdot\|$ is the Euclidean distance on $\mathcal{X}$. We call a pair $(\boldsymbol{x}, \boldsymbol{x}') \in X \times X'$ with $\|\boldsymbol{x} - \boldsymbol{x}'\|^2 = d(X, X')$ an *optimal alignment* of X and X'. By $\mathcal{A}(X, Y)$ we denote the set of all optimal alignments of X and Y.

Suppose that $\boldsymbol{x} \in X$ is an arbitrary vector representation. Since $\mathcal{G}$ is a group, we have

$$d_{\boldsymbol{x}}(Y) = \min \left\{ \|\boldsymbol{x} - \boldsymbol{y}\|^2 \ : \ \boldsymbol{y} \in Y \right\} = d(X, Y).$$

By symmetry, we have $d_{\boldsymbol{y}}(X) = d(Y, X)$. Hence, the graph distance $d(X, Y)$ can be determined by fixing an arbitrary vector representation $\boldsymbol{x} \in X$ and then finding a vector representation $\boldsymbol{y}_*$ from Y that minimizes $\|\boldsymbol{x} - \boldsymbol{y}\|^2$ over all vector representations $Y \in Y$ and vice versa.

Note that the intrinsic metric is not a artificial construction for analytical purposes but rather is based on a generalized concept of maximum common subgraph and therefore appears in different guises as a common choice of proximity measure for graphs [2,3,19].

Orbifold Functions. Suppose that $\mathcal{X}_{\mathcal{G}}$ is a graph orbifold with orbifold chart $\pi : \mathcal{X} \to \mathcal{X}_{\mathcal{G}}$. An *orbifold function* is a mapping of the form $f : \mathcal{X}_{\mathcal{G}} \to \mathbb{R}$. The *lift* of f is a function $\tilde{f} : \mathcal{X} \to \mathbb{R}$ satisfying $\tilde{f} = f \circ \pi$. The lift $\tilde{f}$ is invariant under group actions of $\mathcal{G}$, that is $\tilde{f}(\boldsymbol{x}) = \tilde{f}(\gamma(\boldsymbol{x}))$ for all $\gamma \in \mathcal{G}$.

An example of an orbifold function is the parametrized metric $d_{\boldsymbol{x}}$ with $\boldsymbol{x} \in X$. In what follows, we investigate local analytical properties of $d_{\boldsymbol{x}}$. The *lift* $\tilde{d}_{\boldsymbol{x}}$ of the function $d_{\boldsymbol{x}}$ is defined by

$$\tilde{d}_{\boldsymbol{x}} : \mathcal{X} \to \mathbb{R}, \quad \boldsymbol{y} \mapsto \min \left\{ \|\boldsymbol{x} - \boldsymbol{y}'\|^2 \ : \ \boldsymbol{y}' \in Y \right\}.$$

Certainly, the lift satisfies $\tilde{d}_{\boldsymbol{x}} = d_{\boldsymbol{x}} \circ \pi$ and is invariant under group actions of $\mathcal{G}$, that is $\tilde{d}_{\boldsymbol{x}}(\boldsymbol{y}) = \tilde{d}_{\boldsymbol{x}}(\gamma(\boldsymbol{y}))$ for all $\gamma \in \mathcal{G}$.

By lifting the distance function $d_{\boldsymbol{x}}$ to the Euclidean space $\mathcal{X}$, we are in the position to transfer analytical concepts such as differentiability and gradients to functions on graph orbifolds. We say, the function $d_{\boldsymbol{x}}$ is continuous (locally Lipschitz, differentiable, generalized differentiable) at point $Y \in \mathcal{X}_{\mathcal{G}}$ if its lift $\tilde{d}_{\boldsymbol{x}}$ is continuous (locally Lipschitz, differentiable, generalized differentiable in the sense of Norkin [13]) at some vector representation $\boldsymbol{y} \in Y$. This definition is independent of the choice of the vector representation that projects to Y.

As a minimizer of a set of continuously differentiable distance functions, the function $d_{\boldsymbol{x}}$ is generalized differentiable at any point Y. Though $d_{\boldsymbol{x}}$ is not differentiable, it is locally Lipschitz and therefore differentiable almost everywhere.

Gradients. Suppose that $d_{\boldsymbol{x}}$ is differentiable at Y. Then the lift $\tilde{d}_{\boldsymbol{x}}$ is differentiable at any vector representation that projects to Y. The gradient $\nabla \tilde{d}_{\boldsymbol{x}}(\boldsymbol{y})$ of $\tilde{d}_{\boldsymbol{x}}$ at $\boldsymbol{y}$ is of the form

$$\nabla \tilde{d}_{\boldsymbol{x}}(\boldsymbol{y}) = -2(\boldsymbol{x} - \boldsymbol{y}_*)$$

where $(\boldsymbol{x}, \boldsymbol{y}_*) \in \mathcal{A}(X, Y)$ is an optimal alignment. Since $d_{\boldsymbol{x}}$ is differentiable at Y, the optimal alignment $(\boldsymbol{x}, \boldsymbol{y}_*)$ is unique. From

$$\nabla d_{\boldsymbol{x}}(\gamma(\boldsymbol{y})) = \gamma \left(\nabla \tilde{d}_{\boldsymbol{x}}(\boldsymbol{y}) \right)$$

for all $\gamma \in \mathcal{G}$ follows that the gradients of $\tilde{d}_{\boldsymbol{x}}$ at $\boldsymbol{y}$ and $\gamma(\boldsymbol{y})$ are vector representations of the same graph. Hence, at differentiable points Y, the gradient of $d_{\boldsymbol{x}}(Y)$ at Y is defined by the projection

$$\nabla d_{\boldsymbol{x}}(Y) = \pi \left(\nabla \tilde{d}_{\boldsymbol{x}}(\boldsymbol{y}) \right)$$

of the gradient $\nabla \tilde{d}_{\boldsymbol{x}}(\boldsymbol{y})$ at vector representation $\boldsymbol{y} \in Y$. Thus, the gradient of $d_{\boldsymbol{x}}$ at Y is a well-defined graph pointing to the direction of steepest ascent.

Generalized Gradients. Now suppose that $d_{\boldsymbol{x}}$ is generalized differentiable at Y. Then the lift $\tilde{d}_{\boldsymbol{x}}$ is generalized differentiable at any vector representation that projects to Y. The subdifferential $\partial \tilde{d}_{\boldsymbol{x}}(\boldsymbol{y})$ of $\tilde{d}_{\boldsymbol{x}}$ at $\boldsymbol{y}$ is a convex set containing

$$-2(\boldsymbol{x} - \boldsymbol{y}_*) \in \partial \tilde{d}_{\boldsymbol{x}}(\boldsymbol{y})$$

as generalized gradient, where $(\boldsymbol{x}, \boldsymbol{y}_*) \in \mathcal{A}(X, Y)$ is an optimal alignment. From

$$\partial d_{\boldsymbol{x}}(\gamma(\boldsymbol{y})) = \gamma \left(\partial \tilde{d}_{\boldsymbol{x}}(\boldsymbol{y}) \right)$$

for all $\gamma \in \mathcal{G}$ follows that the subderivatives of $\tilde{d}_{\boldsymbol{x}}$ at $\boldsymbol{y}$ and $\gamma(\boldsymbol{y})$ project to the same subset of graphs. Hence, at generalized differentiable points Y, the subderivative of $d_{\boldsymbol{x}}(Y)$ at Y is defined by the projection

$$\partial d_{\boldsymbol{x}}(Y) = \pi \left(\partial \tilde{d}_{\boldsymbol{x}}(\boldsymbol{y}) \right)$$

of the subderivative $\nabla \tilde{d}_{\boldsymbol{x}}(\boldsymbol{y})$ at an arbitrary vector representation $\boldsymbol{y} \in Y$. Thus, the subderivative of $d_{\boldsymbol{x}}$ at Y is well-defined and coincides with the gradient at differentiable points, that is $\partial d_{\boldsymbol{x}}(Y) = \{\nabla d_{\boldsymbol{x}}(Y)\}$.

3   Learning Graph Quantization

Learning graph quantization (lgq) aims at constructing a classifier $c : \mathcal{X}_{\mathcal{G}} \to \mathcal{C}$ that maps graphs from $\mathcal{X}_{\mathcal{G}}$ to class labels from a finite set $\mathcal{C}$. The classifiers are parameterized by a set of k prototypes $Y_1, \ldots, Y_k \in \mathcal{X}_{\mathcal{G}}$ with class labels $c_1, \ldots, c_k \in \mathcal{C}$. We predict the class label $c(X)$ of a new graph $X \in \mathcal{X}_{\mathcal{G}}$ by assigning it to the class label of the closest prototype according to the nearest neighbor rule. The goal of learning is to find a set of k prototypes that best predicts the class labels of graphs from $\mathcal{X}_{\mathcal{G}}$.

In the following, we first review lgq and lgq2.1 as proposed in [8]. Then we extend GLVQ, GRLVQ, and RSLVQ to the domain of graph orbifolds.

3.1   LGQ

Suppose that $\mathcal{S} = \{(X_i, y_i)\}_{i=1}^{n} \subseteq \mathcal{X}_{\mathcal{G}} \times \mathcal{C}$ is a training set consisting of n input graphs $X_i \in \mathcal{X}_{\mathcal{G}}$ together with class labels $y_i \in \mathcal{C}$. The algorithm first chooses k prototypes $\mathcal{Y} = \{(Y_j, c_j)\}_{j=1}^{k}$ such that each class is represented by at least one prototype. Next, during adaption, the algorithm randomly choses an example $(X, y) \in \mathcal{S}$ from the training set and modifies the closest prototype Y_X in accordance with the current example. The input graph X attracts its closest prototype Y_X if the class labels y of X and c_X of Y_X agree. Otherwise, if the class labels differ, the input X repels the closest prototype Y_X. To determine the closest prototype, lgq applies the nearest neighbor rule

$$Y_X = \arg \min_{Y \in \mathcal{Y}} \{d(X, Y)\}.$$

To update the closest prototype Y_X, the algorithm first selects an optimal alignment $(\boldsymbol{x}, \boldsymbol{y_x}) \in \mathcal{A}(X, Y)$. Then it applies the standard LVQ update rule

$$\boldsymbol{y_x} \leftarrow \begin{cases} \boldsymbol{y_x} + \eta(\boldsymbol{x} - \boldsymbol{y_x}) & : \quad y = c_x \\ \boldsymbol{y_x} - \eta(\boldsymbol{x} - \boldsymbol{y_x}) & : \quad y \neq c_x \end{cases},$$

where η is a monotonically decreasing learning rate following the guidelines of stochastic optimization. The updated vector representation projects to the updated graph prototype. This process continues until the procedure satisfies a termination criterion.

3.2 LGQ2.1

In contrast to lgq, the lgq2.1 procedure updates the two closest prototypes Y_X^1 and Y_X^2 in accordance to the current training example $(X, y) \in \mathcal{S}$. The algorithm adapts the prototypes Y_X^1 and Y_X^2 if the following conditions hold:

1. Exactly one of both prototypes Y_X^1 and Y_X^2 has the same class label as X
2. The input graph X falls in a window around the decision border defined by

$$\frac{d\left(X, Y_X^2\right)}{d\left(X, Y_X^1\right)} > \frac{1 - w}{1 + w},$$

where w is the relative width of the window.

For each prototype lgq2.1 uses the same update rule as lgq.

3.3 Generalized LGQ

We use the following notations: Suppose that (X, y) is an arbitrary training example from $\mathcal{S}$. Let Y^+ be the closest prototype hat belongs to the same class as the current input X, and likewise let Y^- be the closest prototype that belongs to a different class from X. By c^+ and c^- we refer to the class labels of Y^+ and Y^-, respectively. As before, Y_X denotes the closest prototype of X and c_X denotes the class of Y_X.

Following [16], generalized learning graph quantization (glgq) aims at minimizing the cost function

$$E = \sum_{i=1}^{n} f(\mu(X_i)),$$

where $f : \mathbb{R} \to \mathbb{R}$ is a monotonically increasing function and $\mu(X)$ is a function, which is positive if the class labels of X and Y_X agree and negative otherwise. We assume that $L = f \circ \mu$ is generalized differentiable. Then we can minimize E using the incremental generalized gradient method

$$Y^+ \leftarrow Y^+ - \eta G^+ \tag{1}$$

$$Y^- \leftarrow Y^- + \eta G^-, \tag{2}$$

where $G^\pm \in \partial L$ is a generalized gradient of L at $Y^\pm$. As for feature vectors [16], we can show that $\mathtt{lgq}$ and $\mathtt{lgq2.1}$ are special cases of $\mathtt{glgq}$.

Motivated by the robust and powerful performance of GLVQ for feature vectors, we choose

$$f(\mu) = \frac{1}{1 + \exp(-\mu)}$$

and

$$\mu(X) = \frac{d^+ - d^-}{d^+ + d^-},$$

where $d^+ = d(X, Y^+)$ and $d^- = d(X, Y^-)$. Then for any optimal alignment $(\boldsymbol{x}, \boldsymbol{y}^\pm) \in \mathcal{A}(X, Y^\pm)$ the vector representations

$$\boldsymbol{g}^+ = \frac{f'(\mu(X)) \cdot d^-}{(d^+ + d^-)^2} (\boldsymbol{x} - \boldsymbol{y}^+) \tag{3}$$

$$\boldsymbol{g}^- = -\frac{f'(\mu(X)) \cdot d^+}{(d^+ + d^-)^2} (\boldsymbol{x} - \boldsymbol{y}^-) \tag{4}$$

project to generalized gradients $G^\pm \in \partial L (Y^\pm)$ of L at $Y^\pm$.

3.4   Generalized Relevance LGQ

Generalized relevance learning graph quantization ($\mathtt{grlgq}$) extends an idea proposed by [4] to graph orbifolds. Following [4], we replace the distance metric $d(X, Y)$ by a prototype-dependent scaled version

$$d_\Lambda(X, Y) = \min \left\{ \|\boldsymbol{x} - \boldsymbol{y}\|_{\boldsymbol{\lambda}}^2 : \boldsymbol{x} \in X \right\},$$

where $\Lambda \in \mathcal{X}_\mathcal{G}$ is an attributed graph, $\boldsymbol{y} \in Y$ as well as $\boldsymbol{\lambda} \in \Lambda$ are arbitrary but fixed vector representation, and

$$\|\boldsymbol{x} - \boldsymbol{y}\|_{\boldsymbol{\lambda}}^2 = \sum_{i=1}^N \lambda_i (x_i - y_i)^2$$

is the scaled version of the squared Euclidean distance. Then updating amounts in updating the prototypes according to eqns. (1) and (2) accompanied by updating the relevance graph according to the rule

$$\Lambda^+ \leftarrow \Lambda^+ - \eta_1 H^+$$

$$\Lambda^- \leftarrow \Lambda^- - \eta_1 H^-,$$

where $\Lambda^{\pm}$ is the relevance graph of $Y^{\pm}$ and $H^{\pm} \in \partial L(\Lambda^{\pm})$ is a generalized gradient of L at $\Lambda^{\pm}$. Let

$$\boldsymbol{a} \circ \boldsymbol{b} = (a_1 b_1, \dots, a_n b_n)$$

denote the Schur product of vectors $\boldsymbol{a}, \boldsymbol{b} \in \mathbb{R}^n$. Suppose that $(\boldsymbol{x}, \boldsymbol{y}^{\pm}) \in \mathcal{A}(X, Y^{\pm})$ is an optimal alignment. Then vector representations of the form

$$\boldsymbol{g}^+ = \frac{f'(\mu(X)) \cdot d^-}{(d^+ + d^-)^2} \cdot \boldsymbol{\lambda} \circ (\boldsymbol{x} - \boldsymbol{y}^+) \tag{5}$$

$$\boldsymbol{g}^- = -\frac{f'(\mu(X)) \cdot d^+}{(d^+ + d^-)^2} \cdot \boldsymbol{\lambda} \circ (\boldsymbol{x} - \boldsymbol{y}^-) \tag{6}$$

project to generalized gradients $G^{\pm} \in \partial L(Y^{\pm})$ of L at $Y^{\pm}$. Furthermore, any vector representation

$$\boldsymbol{h}^+ = f'(\mu) \frac{d^-}{(d^+ + d^-)^2} (\boldsymbol{x} - \boldsymbol{y}^+)^2$$

$$\boldsymbol{h}^- = f'(\mu) \frac{d^+}{(d^+ + d^-)^2} (\boldsymbol{x} - \boldsymbol{y}^-)^2,$$

projects to a generalized gradient $H^{\pm} \in \partial L(\Lambda^{\pm})$. Observe that the update rule (5) and (6) of `grlgq` differs from the update rule (3) and (4) of `glgq` by including the relevance factors.

3.5 Robust Soft LGQ

Robust soft learning graph quantization (`rslgq`) is motivated by RSLVQ [17], which in difference to the other `lgq` aims at describing the distribution of the data by a Gaussian mixture model. The approach is to maximize the ratio L_r of the probability, that an example $(X, y) \in \mathcal{S}$ is generated by components of the model corresponding to those prototypes with a class label equal to y, and the probability, that the whole model generates X.

To extend RSLVQ to graph orbifolds, we assume that $(\boldsymbol{x}_j, \boldsymbol{y}_j) \in \mathcal{A}(X, Y_j)$ are optimal alignments of a given input graph X and the prototypes Y_j for $j \in \{1, \dots, k\}$. The update rule for Y_j is then of the form

$$Y_j \leftarrow Y_j + \frac{\eta}{\sigma^2} G_j,$$

where

$$\boldsymbol{g}_j = \begin{cases} (P_{\boldsymbol{y}}(\boldsymbol{y}_j | \boldsymbol{x}_j) - P(\boldsymbol{y}_j | \boldsymbol{x}_j))(\boldsymbol{x}_j - \boldsymbol{y}_j), & : \ \boldsymbol{y} = c_j \\ -P(\boldsymbol{y}_j | \boldsymbol{x}_j))(\boldsymbol{x}_j - \boldsymbol{y}_j), & : \ \boldsymbol{y} \neq c_j \end{cases}$$

projects to a generalized gradient $G_j \in \partial \log(L_r(Y_j))$ and

$$P_{\boldsymbol{y}}(\boldsymbol{y}_j | \boldsymbol{x}_j) = \frac{\exp \left(-\dfrac{(\boldsymbol{x}_j - \boldsymbol{y}_j)^2}{2\sigma^2} \right)}{\displaystyle \sum_{i : c_i = \boldsymbol{y}} \exp \left(-\dfrac{(\boldsymbol{x}_i - \boldsymbol{y}_i)^2}{2\sigma^2} \right)},$$

and

$$P(\boldsymbol{y}_j|\boldsymbol{x}_j) = \frac{\exp\left(-\dfrac{(\boldsymbol{x}_j - \boldsymbol{y}_j)^2}{2\sigma^2}\right)}{\displaystyle\sum_{i=1}^{k} \exp\left(-\dfrac{(\boldsymbol{x}_i - \boldsymbol{y}_i)^2}{2\sigma^2}\right)}.$$

It is important to note that the probabilistic interpretation of RSLVQ is no longer valid for its counterpart in graph orbifolds. A first step to remove this shortcoming is presented in [10].

4 Experiments

We conducted first experiments to compare the performance of the different `lgq` algorithms.

4.1 Data

We selected the following data sets from the IAM graph database repository: letter, grec, and fingerprint. Each data set is divided into a training, validation, and a test set. Table 1 provides a summary of the main characteristics of the data sets. For further details we refer to [14].

4.2 Experimental Setup

Setting of lgq algorithms. Given a data set, each `lgq` algorithm was first initialized with a single prototype for each class. To initialize the prototypes we computed a Frechet sample mean of all class members from the training set by using the incremental sample mean algorithm proposed in [7]. Next, we performed a parameter selection for the `lgq` algorithms. For each parameter configuration, we learned the prototypes using the training set and tested the learned model on both, the training and validation set. We selected the parameters that gave the best classification accuracy on the training and validation set. Finally, we assessed the generalization performance by applying the learned model on the test set. For all `lgq` algorithms we tuned the learning rate η. For `lgq2.1`, `grlgq`, and `rslgq`, we additionally calibrated the window width w, the learning rate η_λ of the relevance factors, and the width σ of the Gaussian. respectively.

Graph Distance Calculations and Optimal Alignment. For graph distance calculations and finding optimal alignments, we applied the extended Bron Kerbosch algorithm [9] with clique selection and $10\,|V_Z|$ as the maximum number of recursive calls, where V_Z denotes the vertex set of the association graph under consideration.

Protocol. All `lgq` algorithms have been applied to the training set of each data set 3 times. To assess the generalization performance on the test sets, we have chosen the model that best predicts the class labels on the training and validation set. We compared the `lgq` algorithms with the similarity kernel in conjunction

Table 1. Summary of main characteristics of the data sets. The tiny numbers in parentheses show the size of the training, validation, and test set, respectively.

data set	#(classes)	avg(nodes)	max(nodes)	avg(edges)	max(edges)
letter (750, 750, 750)	15	4.7	8	3.1	6
grec (286, 286, 528)	22	11.5	24	11.9	29
fingerprint (500, 300, 2000)	4	8.3	26	14.1	48

with the SVM (sk+svm) and the family of Lipschitz embeddings in conjunction with SVM (ls+svm) proposed by [15]. As a reference, we used the knn method based on the intrinsic metric, where the parameter k has been learned using the training and validation set.

4.3   Results

Table 2 summarizes the results. Since sk+svm and le+svm refer to a family of related methods rather than a single method, Table 2 presents the best result on the test set over all methods of the sk+svm and le+svm family for each data set. In doing so, the comparison is optimistically biased towards sk+svm and le+svm.

The first observation to be made is that the novel extensions, glgq, grlgq, and rslgq outperform lgq and lgq2.1. These finding are in line with results of LVQ algorithms for feature vectors. A fair comparison of glgq, grlgq, and rslgq, however, is difficult since the performance of any of the lgq variants critically depends on the proper choice of the parameters. An extensive parameter selection is only manageable for lgq (η), glgq (η) and to a certain extent also for lgq2.1 (η, w). For grlgq (η, η_λ) and rslgq (η, σ), however, exploring the parameter space is comparatively too time consuming for two reasons: (i) for a given learning rate, grlgq and rslgq require more iterations during learning until convergence than the other three algorithms, and (ii) both, grlgq and rslgq, critically depend on two rather than one parameter as this is the case for lgq and glgq.

The second observation to be made is that all novel extensions of lgq are comparable to state-of-the-art solutions. All lgq variants, however, are computationally faster than knn, sk+svm and le+svm. The largest portion of the computational effort to classify an unseen graph X is attributable to calculating (or approximating) graph distances between X and a set of prototypes specified by the respective classifier. While the prototype set for knn consists of the whole training set, sk+svm and le+svm use about $40\% - 60\%$ of the training set as prototypes. In contrast, the number of prototypes of the lgq algorithms in this setting corresponds to the number of classes (15 letters, 22 grec, 4 fingerprint).

Table 2. Classification accuracy (in %) of the lgq algorithms

	knn	sk+svm	le+svm	lgq	lgq2.1	glgq	grlgq	rslgq
letter	82.0	79.1	92.5	81.5	85.7	88.4	86.5	87.3
grec	96.8	94.9	96.8	86.2	92.6	97.5	97.0	97.4
fingerprint	80.0	41.0	82.8	79.9	81.5	84.8	84.0	84.1

5   Conclusion

Extensions of GLVQ, GRLVQ, and RSLVQ to the domain of graphs outperform
lgq and lgq2.1, provide state-of-the-art solution even if using a single prototype
for each class, and are superior than knn, sk+svm, and le+svm with respect to
run time during classification. In a practical setting, we recommend to use glgq
because of its simplicity and excellent performance. Future work aims at applying
the lgq algorithms to other data sets and exploring their performance with more
than one prototype per class.

References

1. Borzellino, J.E.: Riemannian geometry of orbifolds, PhD thesis, University of California, Los Angelos (1992)
2. Cour, T., Srinivasan, P., Shi, J.: Balanced graph matching. In: NIPS (2006)
3. Gold, S., Rangarajan, A.: Graduated Assignment Algorithm for Graph Matching. IEEE Transactions on PAMI 18, 377–388 (1996)
4. Hammer, B., Villmann, T.: Generalized relevance learning vector quantization. Neural Network 15, 1059–1068 (2002)
5. Hammer, B., Strickert, M., Villmann, T.: Supervised neural gas with general similarity measure. Neural Processing Letters 21(1), 21–44 (2005)
6. Jain, B., Obermayer, K.: Structure Spaces. Journal of Machine Learning Research 10, 2667–2714 (2009)
7. Jain, B.J., Obermayer, K.: Algorithms for the Sample Mean of Graphs. In: Jiang, X., Petkov, N. (eds.) CAIP 2009. LNCS, vol. 5702, pp. 351–359. Springer, Heidelberg (2009)
8. Jain, B.J., Srinivasan, S.D., Tissen, A., Obermayer, K.: Learning graph quantization. In: Hancock, E.R., Wilson, R.C., Windeatt, T., Ulusoy, I., Escolano, F. (eds.) SSPR&SPR 2010. LNCS, vol. 6218, pp. 109–118. Springer, Heidelberg (2010)
9. Jain, B., Obermayer, K.: Extending Bron Kerbosch for Solving the Maximum Weight Clique Problem. arXiv:1101.1266v1 (2011)
10. Jain, B., Obermayer, K.: Maximum Likelihood for Gaussians on Graphs. In: GbR (2011)
11. Kohonen, T.: Self-organizing maps. Springer, Heidelberg (1997)
12. Kohonen, T., Somervuo, P.: Self-organizing maps of symbol strings. Neurocomputing 21(1-3), 19–30 (1998)
13. Norkin, V.I.: Stochastic generalized-differentiable functions in the problem of nonconvex nonsmooth stochastic optimization. Cybernetics 22(6), 804–809 (1986)
14. Riesen, K., Bunke, H.: IAM graph database repository for graph based pattern recognition and machine learning. In: da Vitoria Lobo, N., Kasparis, T., Roli, F., Kwok, J.T., Georgiopoulos, M., Anagnostopoulos, G.C., Loog, M. (eds.) S+SSPR 2008. LNCS, vol. 5342, pp. 287–297. Springer, Heidelberg (2008)
15. Riesen, K., Bunke, H.: Graph Classification by Means of Lipschitz Embedding. IEEE Transactions on Systems, Man, and Cybernetics 39(6), 1472–1483 (2009)
16. Sato, A., Yamada, K.: Generalized learning vector quantization. In: NIPS (1996)
17. Seo, S., Obermayer, K.: Soft learning vector quantization. Neural Computation 15(7), 1589–1604 (2003)
18. Sumervuo, P., Kohonen, T.: Self-organizing maps and learning vector quantization for feature sequences. Neural Processing Letters 10(2), 151–159 (1999)
19. Umeyama, S.: An eigendecomposition approach to weighted graph matching problems. IEEE Transactions on PAMI 10(5), 695–703 (1988)

Parallel Graduated Assignment Algorithm for Multiple Graph Matching Based on a Common Labelling[*]

David Rodenas, Francesc Serratosa, and Albert Solé-Ribalta

Universitat Rovira i Virgili, Department d'Enginyeria Informàtica i Matemàtiques
43007 Tarragona, Spain
david.rodenas@gispert-rodenas.com,
{francesc.serratosa,albert.sole}@urv.cat

Abstract. This paper presents a new parallel algorithm to compute multiple graph-matching based on the Graduated Assignment. The aim of developing this parallel algorithm is to perform multiple graph matching in a current desktop computer, but, instead of executing the code in the generic processor, we execute a parallel code in the graphic processor unit. Our new algorithm is ready to take advantage of incoming desktop computers capabilities. While comparing the classical algorithm (executed in the main processor) respect our parallel algorithm (executed in the graphic processor unit), experiments show an important speed-up of the run time.

Keywords: Graph Common Labelling, Graduated Assignment, Parallel architecture, Low-cost computer, CUDA, Multiple Graph Matching.

1 Introduction

Classification is a task of pattern recognition that attempts to assign each input value to one of a given set of classes. Pattern recognition algorithms generally aim to provide a reasonable answer for all possible inputs and to match inputs with classes. Pattern recognition is studied in many fields such as psychology, cognitive science, computer science and so on. Depending on the application, inputs of the pattern recognition model or objects to be classified are described by different representations. The most usual representation is a set of real values but other common ones are strings, trees or graphs. Graph structures have more capacity to capture the knowledge of the model but their comparison or matching is also computationally more expensive. Sometimes in graph based pattern recognition applications, given a set of graphs, which all represent equivalent or related structures, it is required to find global consistent correspondences among all those graphs. These correspondences are called a Common Labelling (CL). Algorithms like [1] and [2] does pair matching and

---

[*] This research was partially supported by Consolider Ingenio 2010; project CSD2007-00018 and by the CICYT project DPI 2010-17112.

X. Jiang, M. Ferrer, and A. Torsello (Eds.): GbRPR 2011, LNCS 6658, pp. 132–141, 2011.
© Springer-Verlag Berlin Heidelberg 2011

reconstructs a general correspondence, other algorithms like [3] uses Graduated Assignment [4] to generate the CL by matching all graph nodes to a virtual node set in a polynomial time.

Nowadays desktop computer architectures have evolved towards supercomputing architectures. These architectures generally provide multiple processors and complex memory hierarchy. A simple desktop computer may contain tens of small processors called cores, some of them present at main processor [5], but most of them are present as auxiliary coprocessors like graphical processors [6]. Most algorithms are designed to be executed on a single core and general-purpose processor. Consequently, they are not designed to take advantage of all available resources on current desktop computers. By not taking into account available resources, it appears a gap between effective algorithm performance and potential algorithm performance. This gap will increase at the same rate that cores count on desktop computers increases.

This paper presents a new research project that aims to adapt classical graph algorithms to a up-to-date desktop computer. Intensive computation tasks are computed in the Graphic Processor, such as the graduated assignment algorithm [3]. In this framework, to compute the common labelling algorithm in desktop computers can make use available existing resources. The bases of our work are commented in the section 2 (original algorithm) and section 3 (computer architecture and parallel programming model). The new parallel algorithm is explained in section 4. Section 5 shows the runtime of the sequential algorithm in comparison to the new parallel algorithm with two well known graph databases. Section 6 concludes the paper.

2 Multiple Graph Matching and Computer Architecture

In this section, we introduce a common-labelling algorithm and its behaviour. We also present a set of equations which are used to transform the algorithm to be executed in multi-core computers.

2.1 Attributed Graphs and Multiple Graph Matching

Graduated Assignment [3] is one of the algorithms considered to have a good run-time performance between most popular common labelling algorithms. This algorithm approximates a distance and a labelling between many graphs using a polynomial time method respect the order of the graphs. The result of the CL algorithm is a set of probability matrices $\{ P_h^1, P_h^2, ..., P_h^N \}$ that represents, for each matrix, the probability of matching a node of one of p graph to a virtual node. Since any p matrix P_h^p values are continuous, a discretisation process of the probability matrix [7] is applied to obtain the final labelling between graphs nodes. This process is out of the scope of this paper.

Given a set of graphs $\{G^1, G^2, ..., G^N\}$ (that have R vertices) and their respective adjacency matrices $\{A^1, A^2, ..., A^N\}$, the general outline of the Graduated Assignment CL is shown in algorithm 1 and 2.

Algorithm 1. General diagram of the Graduated Assignment Common Labelling

```
β = β₀
Initialise Pₕ
begin Do until β ≥ βf
  begin Do until Pₕ convergence
    Pf^pq= Pₕᵖ · (Pₕ^q)ᵀ
    Q= Approx_Q(Pf, Pₕ, C)
    Pₕᵖ[a, w₁]= exp(β ·Qᵖₐ,w₁)
    Pₕᵖ= Stochastic(Pₕᵖ)
  end
  β= β · βr
end
Mᵖq = Discretise Pₕᵖ
```

Algorithm 2. *Approx_Q* function description

```
for 2 ≤ p ≤ N
  Qᵖ= 0
  for 1 ≤ q ≤ N ∧ p≠q
    for 1 ≤ a, i ≤ R
      v₁= 0
      for 1 ≤ b, j ≤ R ∧ b≠a ∧ j≠i
        v₁= v₁ + Pf^p,q[b, j] · C^pq_aibj
      end
      for 1 ≤ w₁ ≤ R
        Qᵖₐ,w₁= Qᵖₐ,w₁ + v₁·Pₕᵖ[i, w₁]
      end
    end
  end
end
```

C^{pq}_{aibj} represents the compatibility of labelling edge *(a,b)* of graph G^p to edge *(i,j)* of graph G^q and their respective ending nodes. In order to optimize C^{pq}_{aibj} computation it is defined as:

$$C^{pq}_{aibj}=\frac{1}{1+C^{pq}_{ai}+C^{pq}_{bj}+dist(A^p_{ab}, A^q_{ij})} \tag{1}$$

C^{pq}_{ai} is the precomputed distance between vertex *a* from graph *p* and vertex *i* from graph *q*, *dist* function determines the distance defined by the existence of graph *p ab* edges and graph *q ij* edges.

Function *Stochastic* obtains a double stochastic matrix [3] using the Sinkhorn method [8] as follows,

begin Do until convergence

$$\bigvee_{a=1}^{R} \bigvee_{i=1}^{R} P^p_h[a,i]=\frac{P^p_h[a,i]}{\sum_{k=1}^{R} P^p_h[a,k]} \tag{2}$$

$$\bigvee_{a=1}^{R} \bigvee_{i=1}^{R} P^p_{h'}[a,i]=\frac{P^p_{h'}[a,i]}{\sum_{k=1}^{R} P^p_{h'}[k,i]} \tag{3}$$

end

Sinkhorn method has been parallelised for many high-performance architectures, such as vector machines [9] and connection machines [10].

3   Computer Architecture and Programming Model

In this section, we introduce a desktop computer architecture and we relate it to a parallel programming model. We also present a set of directives to specify parallel algorithms.

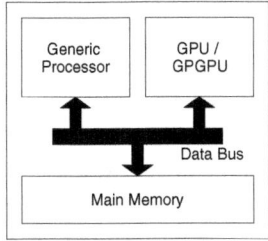

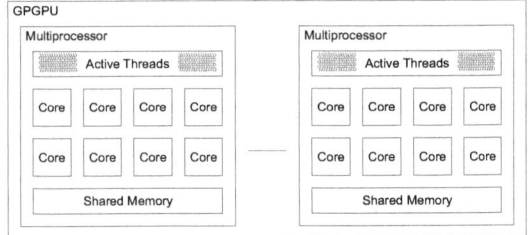

Fig. 1. General view of a desktop computer architecture

Fig. 2. Generic GPGPU Architecture

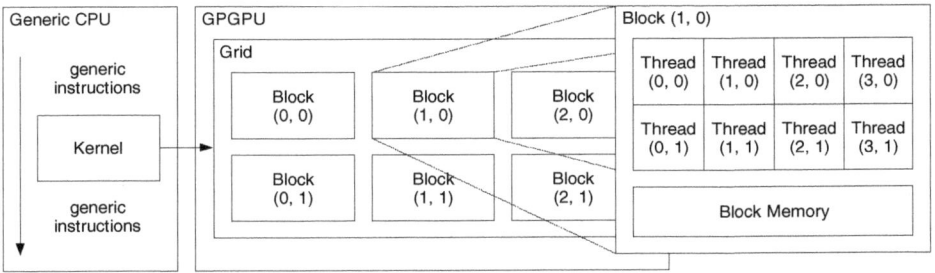

Fig. 3. CUDA Logical Execution Space

3.1 Desktop Computer Architecture

The current desktop computers are composed by 2 processors: A generic multi-core processor (composed by few cores) and a Graphics Processing Unit (GPU, composed by tens of small cores). Both processors have access to main memory (Figure 1).

Current GPUs are General Purpose GPUs (GPGPU) dedicated to intensive computations, mainly addressed to graphic tasks. They are able to execute simple functions usually called *kernels*. GPGPUs are massively multi-threaded architectures. They are composed by several multiprocessors (figure 2), each of which has multiple cores and a shared memory. *Cores* are the processing units that compute thread instructions. Shared memory has multiple banks, they can serve data simultaneously to multiple threads. This shared memory has small size but very low latency.

3.2 Parallel Programming Model

Our parallel programming model is CUDA [11]. This programming framework allows to mix sequential C code, executed in the generic processor, with kernels, executed in the GPGPU. When the sequential code reaches a kernel, it configures a logical grid of blocks (figure 3) and launches its execution on the GPGPU. A kernel code is executed concurrently by all threads of the grid of blocks. Each block is physically mapped into a GPGPU multiprocessor. The threads of a block are executed in the cores of one multiprocessor and the block memory is mapped into the shared memory of the multiprocessor (figure 2 and 3).

We define the following four directives to express parallel tasks:

Parallel Map $_{(x,y)=(X1,Y1)}^{(X2,Y2)}$ **Into Blocks** {*function*}: The same function is executed by all blocks *(x,y)* in the grid of blocks between block *(X1,Y1)* and block *(X2,Y2)*.

Parallel Map $_{(x,y)=(X1,Y1)}^{(X2,Y2)}$ **Into Threads** {*function*}: The same function is executed by all threads *(x,y)* of the current block between thread *(X1,Y1)* and thread *(X2,Y2)*.

Parallel Reduction *function*: A function is performed by a specific block.

Parallel Fetch $_{(x,y)=(X1,Y1)}^{(X2,Y2)}$ $U_{x,y}$: Given a matrix U stored in the main memory, an specific thread of a specific block reads a sub-matrix of elements of U between position *(X1,Y1)* and position *(X2,Y2)* and writes it to the shared memory of the block. Moreover, this directive acts as a barrier for all threads of the same block.

Fetch $U_{x,y}$: Given a matrix U stored in the main memory, a specific thread of an specific block reads the element $U_{x,y}$ and writes it to the shared memory of the block.

4   A Parallel Solution for the Graph Matching Problem

This section shows a parallel solution of the six main functions of the Graduated Assignment Common Labelling (algorithm 1 and 2): P_f *computation*, *Approx_Q*, *Exponentiation*, *Stochastic* and *Convergence computation*.

4.1   Parallel P_f Computation

P_f *computation* is very close to traditional matrix multiplication, but it is simpler because all accesses are ordered as consecutive row accesses. In the classical algorithm it computes the following equation:

$$\overset{N}{\underset{p=2}{\forall}} \overset{N}{\underset{q=1}{\forall}} \overset{R}{\underset{a=1}{\forall}} \overset{R}{\underset{i=1}{\forall}} P_f^{pq}[a,i] = \sum_{k=1}^{R} P_h^p[a,k] \cdot P_h^q[i,k] \quad \text{where} \quad p \neq q \tag{4}$$

but, with the aim of implementing a parallel algorithm, loops q and a are reordered:

$$\overset{N}{\underset{p=2}{\forall}} \overset{R}{\underset{a=1}{\forall}} \overset{N}{\underset{q=1}{\forall}} \overset{R}{\underset{i=1}{\forall}} P_f^{pq}[a,i] = \sum_{k=1}^{R} P_h^p[a,k] \cdot P_h^q[i,k] \quad \text{where} \quad p \neq q \tag{5}$$

We assume that R elements can be stored into block memory. The new algorithm that computes (5) parallelises loops p and a through blocks, thus, there is only one k row of $P_h^p[a,k]$ for each block. We fetch this k row of $P_h^p[a,k]$ into block memory and by this way, all threads of the same block are using the same data. The parallel algorithm is the following:

Algorithm 3. Parallel algorithm for P_f *Computation*

```
Parallel Map  (N,R)        Into Blocks {
             (p,a)=(2,1)
  Parallel Fetch    (R)   Pₕᵖ[a,k]
                 (k)=(1)
    Parallel Map   (R)     Into Threads {   Pᶠᵖᑫ[a,i] = Σ Pₕᵖ[a,k]·Pᶠᑫ[i,k]  }
                 (i)=(1)                       k=1
}
```

4.2 Parallel Computation of Approximate Q Matrix

Approx_Q function (presented at algorithm 2) can be rewritten as the following:

$$Q^p = 0; \ \bigvee_{p=2}^{N} \bigvee_{q=1}^{N} \bigvee_{a=1}^{R} \bigvee_{i=1}^{R} \left(v_1 = \left(\sum_{b=1}^{R} \sum_{j=1}^{R} P_f^{pq}[b,j] \cdot C_{aibj}^{pq} \right); \ \bigvee_{w_1=1}^{R} Q_{a,w_1}^p = Q_{a,w_1}^p + v_1 \cdot P_h^p[i,w_1] \right) \ \text{where} \ \begin{matrix} p \neq q \\ a \neq b \\ i \neq j \end{matrix} \quad (6)$$

If we convert v_1 into a hyper-matrix, we can split (6) as two expressions:

$$\bigvee_{p=2}^{N} \bigvee_{q=1}^{N} \bigvee_{a=1}^{R} \bigvee_{i=1}^{R} v_1^{pq}[a,i] = \sum_{b=1}^{R} \sum_{j=1}^{R} P_f^{pq}[b,j] \cdot C_{aibj}^{pq} \ \text{where} \ \begin{matrix} p \neq q \\ a \neq b \\ i \neq j \end{matrix} \quad (7)$$

$$Q^p = 0; \ \bigvee_{p=2}^{N} \bigvee_{q=1}^{N} \bigvee_{a=1}^{R} \bigvee_{i=1}^{R} \bigvee_{w_1=1}^{R} Q_{a,w_1}^p = Q_{a,w_1}^p + v_1^{pq}[a,i] \cdot P_h^p[i,w_1] \ \text{where} \ p \neq q \quad (8)$$

Loops q and i in (8) can be reordered to convert Q computation into summations:

$$\bigvee_{p=2}^{N} \bigvee_{a=1}^{R} \bigvee_{wI=1}^{R} Q_{a,w_1}^p = \sum_{q=1}^{N} \sum_{i=1}^{R} v_1^{pq}[a,i] \cdot P_h^q[i,w_1] \ \text{where} \ p \neq q \quad (9)$$

(9) is discussed in section 4.3. (7) is the most complex task of the whole algorithm. We assume that multiple matrices of B x B elements can be stored into block memory. We apply loop tiling technique [12] in order to expose sub-matrices small enough to fit on block memory:

$$\bigvee_{p=2}^{N} \bigvee_{q=1}^{N} \bigvee_{c=0}^{\frac{R}{B}} \bigvee_{k=0}^{\frac{R}{B}} \bigvee_{d=1}^{B} \bigvee_{l=1}^{B} v_1^{pq}[a,i] = \sum_{e=0}^{\frac{R}{B}} \sum_{u=0}^{\frac{R}{B}} \sum_{f=1}^{B} \sum_{v=1}^{B} P_f^{pq}[b,j] \cdot C_{aibj}^{pq} \ \text{where} \ \begin{matrix} a=c \cdot B + d \\ b=e \cdot B + f \\ i=k \cdot B + l \\ j=u \cdot B + v \end{matrix} \begin{matrix} p \neq q \\ a \neq b \\ i \neq j \end{matrix} \quad (10)$$

If we apply (1) to (10) we obtain the following expression:

$$\bigvee_{p=2}^{N} \bigvee_{q=1}^{N} \bigvee_{c=0}^{\frac{R}{B}} \bigvee_{k=0}^{\frac{R}{B}} \bigvee_{d=1}^{B} \bigvee_{l=1}^{B} v_1^{pq}[a,i] = \sum_{e=0}^{\frac{R}{B}} \sum_{u=0}^{\frac{R}{B}} \sum_{f=1}^{B} \sum_{v=1}^{B} \frac{P_f^{pq}[b,j]}{1 + C_{ai}^{pq} + C_{bj}^{pq} + dist(A_{ab}^p, A_{ij}^q)} \ \text{where} \ \begin{matrix} a=c \cdot B + d \\ b=e \cdot B + f \\ i=k \cdot B + l \\ j=u \cdot B + v \end{matrix} \begin{matrix} p \neq q \\ a \neq b \\ i \neq j \end{matrix} \quad (11)$$

We parallelise p, q, c and k through blocks. In order to avoid parallel reductions, we parallelise only d and l through threads. As a result each thread computes a unique $v_1^{pq}[a,i]$ element. Loop tiling is also applied to summations, b and j iterations are decomposed. The objective is that all operations inside f, v summations share the same data (sub-matrices f, v, d, l from $P_f^{pq}[b,j], C_{ai}^{pq}, C_{bj}^{pq}, A_{ab}^p$ and A_{ij}^q) between all d, l threads of the same block. For each e, u iteration, a new set of sub-matrices of data is fetched and all threads are synchronized in order to use the same data. As an optimization, data fetch is performed as soon as possible; for example: C_{ai}^{pq} indexes are constant for each thread for all summations, consequently it is fetched before any summation and reused for all summations of all threads of the same block. The parallelised algorithm is the following:

Algorithm 4. Parallel algorithm for V_1 *Computation* (11, 1, and 7)

```
Parallel Map (N,N,R/B,R/B) (p,q,c,k)=(2,1,0,0) Into Blocks {
  Parallel Map (B,B) (d,l)=(1,1) Into Threads {
    a = c·B + d; i = k·B + 1
    V_1^pq[a,i]=0
    Fetch C_ai^pq
    for 0 ≤ e ≤ R/B
      Parallel Fetch (B) (f)=(0) A_ab^p   where  b=e·B+f
      for 0 ≤ u ≤ R/B
        Parallel Fetch (B) (v)=(0) A_ij^q   where  j=u·B+v
        Parallel Fetch (B,B) (f,v)=(0,0) C_bj^pq   where  b=e·B+f ∧ j=u·B+v
        Parallel Fetch (B,B) (f,v)=(0,0) P_f^pq[b,j]   where  b=e·B+f ∧ j=u·B+v
        for 0 ≤ f,v ≤ B
          b = e·B + f; j = u·B + v
          v_1^pq[a,i]=v_1^pq[a,i]+ P_f^pq[b,j] / (1+C_ai^pq+C_bj^pq+dist(A_ab^p, A_ij^q))
        end
      end
    end
  }
}
```

4.3 Parallel Computation of P_h Matrix

We have grouped all P_h computation on the same parallel kernel: *Q Computation* (9), *Exponentiation*, *Stochastic* (2 and 3) and *Convergence test* (see algorithm 1).

Exponentiation, and *Convergence test* algorithm expressions are:

$$\forall_{p=2}^{N} \forall_{a=1}^{R} \forall_{i=1}^{R} P_f^p[a,i] = \exp\left(\beta \cdot Q_{a,wi}^p\right) \tag{12}$$

$$converged = \left(\sum_{p=2}^{N} \sum_{a=1}^{R} \sum_{i=1}^{R} P_f^p[a,i] - P_{f^0}^p[a,i]\right) < EPSILON \tag{13}$$

Convergence test is split in 2 parts in order to parallel compute local convergences:

$$\forall_{p=2}^{N} error^p = \sum_{a=1}^{R} \sum_{i=1}^{R} P_f^p[a,i] - P_{f^0}^p[a,i] \; ; \; converged = \left(\sum_{p=2}^{N} error^p\right) < EPSILON \tag{14}$$

All grouped computations presents an initial loop for p, and two more loops for a and i (or w_1) indexes. In order to minimize copies from main memory to block memory we assume that each block of threads will process a single and unique p value. As a result, all threads of the same block will keep the same P_h^p and no communication is required of P_h^p temporal values to other blocks. Block threads are parallelised through a and i (or w_1) indexes. The new algorithm is:

Algorithm 5. Parallel algorithm for P_h *Computation* (9, 12, 2, 3, 14)

```
Parallel Map (N)(p)=(2) Into Blocks {
  Parallel Map (R,R)(a,w1)=(1,1) Into Threads { P_h^p[a,i] = Σ_{q=1}^{N} Σ_{i=1}^{R} v_1^{pq}[a,i] · P_{h0}^q[i,w1] | p≠q }
  Parallel Map (R,R)(a,i)=(1,1) Into Threads { P_f^p[a,i] = exp(β · Q^p_{a,wi}) }
  begin Do until P_h^p convergence
     Parallel Map (R)(a)=(1) Into Threads {
       sum^P = Parallel Reduction Σ_{i=1}^{R} P_h^p[a,i]
       Parallel Map (R)(i)=(1,1) Into Threads { P_h^p[a,i] = P_h^p[a,i] / sum^p }
     }
     Parallel Map (R)(i)=(1) Into Threads {
       sum^P = Parallel Reduction Σ_{a=1}^{R} P_h^p[a,i]
       Parallel Map (R)(a)=(1,1) Into Threads { P_h^p[a,i] = P_h^p[a,i] / sum^p }
     }
  end
  error^P = Parallel Reduction Σ_{a=1}^{R} Σ_{i=1}^{R} P_f^p[a,i] − P_{f0}^p[a,i]
}
converged = ( Σ_{p=2}^{N} error^p ) < EPSILON
```

5 Experimental Evaluation

We have implemented the sequential algorithm [3] and the proposed parallel algorithm. Both algorithms have been tested over a GPGPU parallel architecture and over a generic Intel architecture. Table 1 shows the architecture characteristics in which the serial algorithm (first row) and the parallel algorithm (second row) are executed. Both are executed on the same desktop computer but the serial algorithm is executed on the main processor and the parallel algorithm is executed on the GPGPU.

We have used two databases in which nodes are defined over a two-dimensional domain that represents its plane position (x,y). Edges have binary attribute that represents the existence of a line between two terminal points. The first dataset is a subset of high noise level of the Letter dataset created the University of Bern [13]. This data set is composed of 15 classes and 150 graphs per class representing the Roman

Table 1. Architectures used in the comparative and their characteristics

Alg.	Computer	Proc. / GPGPU	GHz	Power	Cores	Threads	Bandwith
Serial [3]	ViewSonic VT132	Intel Atom 330	1.6	8W	2	4[1]	5 GB/s
New Parallel	ViewSonic VT132	NVIDIA 9400M	1.1	10W	16	1536	5 GB/s

[1] There are 4 threads available but algorithm [3] uses only one thread and one core.

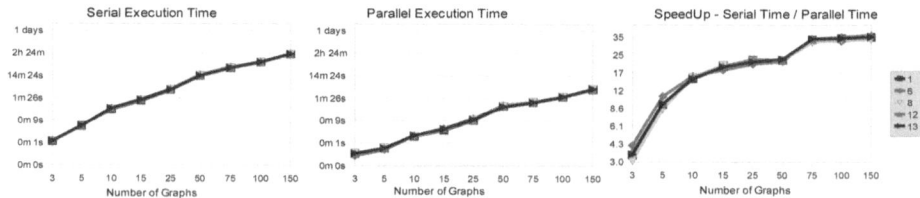

Fig. 4. Letter run time of Serial and Parallel and speedup respect to the number of graphs for each selected class. Vertical axis are in log. scale.

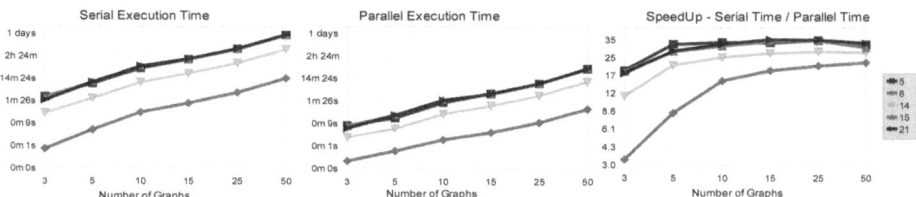

Fig. 5. GREC run time of Serial and Parallel and speedup respect to the number of graphs for each selected class. Vertical axis are in log. scale.

alphabet i.e. A, E, F, .., X, Y, and Z. The second dataset, called GREC dataset, created at the Universitat Autònoma de Barcelona [13], is composed of 22 classes and 50 graphs per class representing symbols from architectural and electronic drawings.

We have selected 5 classes of each dataset to compare execution speed. For each class we have randomly selected a number of graphs for $N \in$ [5, 10, 15, 25, 50, 75, 100, 150] for Letter dataset, and $N \in$ [5, 10, 15, 25, 50] for GREC dataset. Letter dataset classes selected are {1, 6, 8, 12, 13} each one with a mean number of nodes of {5.3, 5.3, 5.3, 5.4, 4.4}. GREC dataset classes selected are {5, 8, 14, 15, 21} each one with a mean of {19.4, 8.6, 12.7, 20.7, 17.14}.

Figure 4 shows the mean run time for each one of the five Letter dataset classes for serial and parallel algorithm experiments for a given different number of the graphs. Figure 5 shows the mean run time for each one of the five GREC dataset classes for serial and parallel algorithm experiments for a given different number of the graphs. The obtained distance is not shown since the sequential and parallel algorithm obtains exactly the same result. It can be observed a clear improvement on the run time when the parallel algorithm is used.

6   Conclusions and Future Work

We have presented a parallel algorithm which can take advantage of present computational resources on current desktop computers. Results show a significant speed-up of the run time of multiple graph-matching algorithms. The aim is to demonstrate that it is possible to perform some modifications to the classical algorithm in order to take advantage of existing resources and use low-cost computers to perform pattern recognition processes based on graphs. Future efforts will focus on parallelise other graph-matching algorithms, apply new solutions to other architectures (like common multi-core and vector operations) and identify a subset of common operations to simplify future algorithm adaptations.

References

[1] Bonev, B., Escolano, F., Lozano, M.A., Suau, P., Cazorla, M.A., Aguilar, W.: Constellations and the unsupervised learning of graphs. In: Escolano, F., Vento, M. (eds.) GbRPR 2007. LNCS, vol. 4538, pp. 340–350. Springer, Heidelberg (2007)

[2] Solé-Ribalta, A., Serratosa, F.: On the Computation of the Common Labelling of a set of Attributed Graphs. In: 14th Iberoamerican Congress On Pattern Recognition, pp. 137–144 (2009)

[3] Solé-Ribalta, A., Serratosa, F.: Graduated Assignment Algorithm for Finding the Common Labelling of a Set of Graphs. In: Hancock, E.R., Wilson, R.C., Windeatt, T., Ulusoy, I., Escolano, F. (eds.) SSPR&SPR 2010. LNCS, vol. 6218, pp. 180–190. Springer, Heidelberg (2010)

[4] Gold, S., Rangarajan, A.: A Graduated Assignment Algorithm for Graph Matching. IEEE TPAMI 18(4), 377–388 (1996)

[5] Gochman, S., Mendelson, A., Naveh, A., Rotem, E.: Introduction to Intel Core Duo Processor Architecture. Intel Technology Journal, 10(2), May 15 (2006)

[6] Owens, J.: Streaming architectures and technology trends. GPU Gems 2, 457–470 (2005)

[7] Kuhn, H.W.: The Hungarian method for the assignment problem Export. Naval Research Logistics Quarterly 2(1-2), 83–97 (1955)

[8] Sinkhorn, R.: A Relationship Between Arbitrary Positive Matrices and Doubly Stochastic Matrices. The Annals of Mathematical Statistics 35(2), 876–879 (1964)

[9] Zenios, S.A., Iu, S.-L.: Vector and parallel computing for matrix balancing. Annals of Operations Research 22, 161–180 (1990)

[10] Zenios, S.A.: Matrix balancing on a massively parallel Connection Machine. ORSA Journal on Computing 2, 112–125 (1990)

[11] NVIDIA CUDA http://developer.nvidia.com/object/cuda.html

[12] Xue, J.: Loop Tiling for Parallelism. Kluwer Academic Publishers, Dordrecht (2000)

[13] Riesen, K., Bunke, H.: IAM graph database repository for graph based pattern recognition and machine learning. In: da Vitoria Lobo, N., Kasparis, T., Roli, F., Kwok, J.T., Georgiopoulos, M., Anagnostopoulos, G.C., Loog, M. (eds.) S+SSPR 2008. LNCS, vol. 5342, pp. 287–297. Springer, Heidelberg (2008)

Smooth Simultaneous Structural Graph Matching and Point-Set Registration

Gerard Sanromà[1], René Alquézar[2], and Francesc Serratosa[1]

[1] Departament d'Enginyeria Informàtica i Matemàtiques,
Universitat Rovira i Virgili, Tarragona (Spain)
{gerard.sanroma,francesc.serratosa}@urv.cat
[2] Institut de Robtica i Informtica Industrial, CSIC-UPC, Barcelona (Spain)
ralquezar@iri.upc.edu

Abstract. We present a graph matching method that encompasses both a model of structural consistency and a model of geometrical deformations. Our method poses the graph matching problem as one of mixture modelling which is solved using the EM algorithm. The solution is then approximated as a succession of assignment problems which are solved, in a smooth way, using Softassign. Our method allows us to detect outliers in both graphs involved in the matching. Unlike the outlier rejectors such as RANSAC and Graph Transformation Matching, our method is able to refine an initial the tentative correspondence-set in a more flexible way than simply removing spurious correspondences. In the experiments, our method shows a good ratio between effectiveness and computational time compared with other methods inside and outside the graphs' field.

Keywords: correspondence problem, expectation-maximization, softassign, affine registration.

1 Introduction

The correspondence problem arises in many computer vision applications.

Tentative correspondences can be computed on the basis of the local image contents around some interest points [8][14]. However, a refinement process is often needed in order to remove erroneous correspondences in the tentative-set.

This is the case of RANSAC [6] and Graph Transformation Matching [2], which remove outlying correspondences by enforcing some kind of global consistency. The main drawback of these methods is that their success strongly depends on the reliability of the tentative-set. Since they are unable either to generate new correspondences or to modify the existing ones, a tentative-set with few successes may lead to sparse estimates. This is illustrated in figure 1.

Attributed Graph Matching techniques are another approach to refine the tentative correspondences which do not suffer from the aforementioned problem of the outlier rejectors.

Hancock *et al.* [13][4][11] present graph matching approaches that jointly solve the correspondence and alignment problems. The advantages of posing the graph

X. Jiang, M. Ferrer, and A. Torsello (Eds.): GbRPR 2011, LNCS 6658, pp. 142–151, 2011.

(a) Tentative correspondences computed by a matching by correlation method. The red dots are unmatched points. There are several misplaced correspondences.

(b) Only a few inliers are found by RANSAC. This may not be suitable in the cases when more dense correspondence-sets are needed.

Fig. 1. Matching results for two sample images from the class Resid (from ref. [1]) with superposed Harris corners [8]. Green lines represent the correspondences found by (a) a correlation method and (b) RANSAC applied to the correlation results.

matching as a joint correspondence and alignment problem, are twofold. On one hand, structural information may contribute to disambiguate the recovery of the alignment. On the other hand, geometrical information may aid to clarify the recovery of the correspondences in the case of structural corruption.

In [4][18][17], Hancock *et al.* propose a principled way of detecting outliers that consists in measuring the net effects of a node deletion in a reconfigured graph. This is a one-direction model, i.e., data-graph constraints are evaluated on the model-graph side. This implies that outliers can only be detected in the data-graph side, a practical limitation in computer vision where outliers can be found indistinguishably in both sides.

Gold and Rangarajan present *Graduated Assignment* [7], an optimization technique aimed at graph matching. They use Softassign [16][15][10] to handle continuous correspondences and to provide two-way constraints satisfaction.

We propose a method to solve the graph matching problem as one of mixture modelling [4][12]. Our mixture model evaluates the geometrical arrangement of the nodes as well as their structural relations. We use the EM algorithm to approximate the solution, in a principled way, as a succession of assignment problems which are solved using Softassign. This allows us to gradually move from continuous to discrete correspondences while being able to detect outliers in both graphs in a smooth way. We provide computational time results suggesting that our method can be used at specific moments during a real-time operation (e.g., when the tentative-sets are insufficient). Figure 2 shows that our approach arrives at a correct dense correspondence-state, while still leaving a few unmatched outliers in both images.

2   A Mixture Model

Consider two graph representations $G = (V, D, X)$ and $H = (W, E, Y)$, extracted from two images (e.g., figure 2).

The node-sets $V = \{v_a, \forall a \in \mathcal{I}\}$ and $W = \{w_\alpha, \forall \alpha \in \mathcal{J}\}$ contain the symbolic representations of the nodes, where $\mathcal{I} = 1...|V|$ and $\mathcal{J} = 1...|W|$ are their index-sets.

Fig. 2. The green lines are the result of applying our method using, as starting point, the tentative-set of figure 1(a). Nodes are placed in the locations of the Harris corners. Blue lines represent the edges generated by means of Delaunay triangulations on the Harris corners. Our method still detects a few outliers in both graphs.

The vector-sets $X = \{\boldsymbol{x}_a = (x_a^{ab}, x_a^{or}), \forall a \in \mathcal{I}\}$ and $Y = \{\boldsymbol{y}_\alpha = (y_\alpha^{ab}, y_\alpha^{or}), \forall \alpha \in \mathcal{J}\}$, contain the column vectors of the two-dimensional coordinates (abscissa and ordinate) of each node.

The adjacency matrices D and E contain the edge-sets, representing some kind of structural relation between pairs of nodes (e.g., connectivity or spatial proximity). Hence, $D_{ab} = \begin{cases} 1 \text{ if } v_a \text{ and } v_b \text{ are linked by an edge} \\ 0 \text{ otherwise} \end{cases}$ (the same applies for $E_{\alpha\beta}$).

We use continuous correspondence indicators S so, we denote as $s_{a\alpha} \in S$, the probability of node $v_a \in V$ being in correspondence with node $w_\alpha \in W$.

It is satisfied that

$$\sum_{\alpha \in \mathcal{J}} s_{a\alpha} \leq 1 , \ \forall a \in \mathcal{I} \tag{1}$$

where, $1 - \sum_\alpha s_{a\alpha}$ is the probability of node v_a being an outlier.

Our aim is to recover the correspondence indicators S and the registration parameters Φ that maximize the incomplete likelihood of the observed graph, $P(G|S, \Phi)$. The standard procedure to build likelihood functions for mixture distributions consists in factorizing over the observed data (i.e., observed graph nodes) and summing over the hidden variables (i.e., their corresponding reference nodes). Hence,

$$P(G|S, \Phi) = \prod_{a \in \mathcal{I}} \sum_{\alpha \in \mathcal{J}} P(v_a, w_\alpha|S, \Phi) \tag{2}$$

where $P(v_a, w_\alpha|S, \Phi)$ is the conditional likelihood of correspondence between nodes $v_a \in V$ and $w_\alpha \in W$.

Following a similar development than Luo and Hancock [12] we factorize, using the Bayes rules, the conditional likelihood in the right hand side of equation (2) into terms of individual correspondence indicators, in the following way.

$$P(v_a, w_\alpha|S, \Phi) = K_{a\alpha} \prod_{b \in \mathcal{I}} \prod_{\beta \in \mathcal{J}} P(v_a, w_\alpha|s_{b\beta}, \Phi) \tag{3}$$

where $K_{a\alpha} = [1/P(v_a|w_\alpha, \Phi)]^{|\mathcal{I}| \times |\mathcal{J}| - 1}$. If we assume that the observed node v_a is conditionally dependant on the reference node w_α and the registration parameters Φ only in the presence of the correspondence matches S, then $P(v_a|w_\alpha, \Phi) = P(v_a)$. Further assuming equiprobable priors $P(v_a)$, we can safely discard these quantities in the maximization of equation (2), since they do not depend neither on S or Φ.

We propose a measure for the conditional likelihood of equation (3) that uses the same model for the structural errors as in [12], augmented with a geometric compatibility measure.

On one hand, given two corresponding pairs of points $(v_a, v_b) \rightarrow (w_\alpha, w_\beta)$, we consider that there will be lack of edge-support (i.e., $D_{ab} = 0 \vee E_{\alpha\beta} = 0$) with a constant probability P_e. On the other hand, we consider that it is an affine-invariant density measurement on the point position errors $P(\boldsymbol{x}_b, \boldsymbol{y}_\beta|\Phi)$ (for brevity $P_{b\beta}$), that it is appropriate for weighting the conditional likelihood in the case of correspondence between nodes v_b and w_β. In the case of no correspondence we assign a constant probability ρ that controls the outlier process.

Accordingly, our expression for the conditional likelihood is

$$P(v_a, w_\alpha|s_{b\beta}, \Phi) = \left[(1 - P_e)P_{b\beta}\right]^{D_{ab}E_{\alpha\beta}s_{b\beta}} \left[P_e P_{b\beta}\right]^{(1-D_{ab}E_{\alpha\beta})s_{b\beta}} \left[P_e\rho\right]^{(1-s_{b\beta})} \tag{4}$$

Substituting equation (4) into equation (3), the final expression for the correspondence likelihood between v_a and w_α, expressed in the exponential form, is

$$P(v_a, w_\alpha|S, \Phi) = \exp\left[\sum_{b \in \mathcal{I}} \sum_{\beta \in \mathcal{J}} s_{b\beta} D_{ab} E_{\alpha\beta} \ln\left(\frac{1-P_e}{P_e}\right) + s_{b\beta}\ln\left(\frac{P_{b\beta}}{\rho}\right) + \ln\rho\right] \tag{5}$$

3 Expectation Maximization

The EM algorithm has been previously used by other authors to solve the Graph Matching problem [4] [12]. We seek the affine registration parameters Φ and the correspondence indicators S, that maximize the expected log-likelihood of our mixture distribution. Dempster *et al.* [5] showed that this could be posed as an iterative estimation of a weighted sum of log-likelihoods.

Accordingly, we seek the parameters $\hat{S}, \hat{\Phi}$ that maximize the following objective function

$$\Lambda\left(\hat{S}, \hat{\Phi}|S^{(n)}, \Phi^{(n)}\right) = \sum_{a \in \mathcal{I}} \sum_{\alpha \in \mathcal{J}} P\left(w_\alpha|v_a, S^{(n)}, \Phi^{(n)}\right) \ln P\left(v_a, w_\alpha|\hat{S}, \hat{\Phi}\right) \tag{6}$$

where $P(w_\alpha|v_a, S^{(n)}, \Phi^{(n)})$ are the posterior probabilities of the missing data given the most recent available parameters $S^{(n)}, \Phi^{(n)}$.

The basic idea is to alternate between Expectation and Maximization steps until convergence is reached. The expectation step involves computing the a posteriori probabilities of the missing data using the most recent available parameters. In the maximization phase, the parameters are updated in order to maximize the expected log-likelihood of the incomplete data.

3.1 Expectation

In the expectation step, the posteriori probabilities of the missing data (i.e., the reference graph measurements w_α) are computed using the current parameter estimates $S^{(n)}, \Phi^{(n)}$.

The posterior probabilities can be expressed in terms of conditional likelihoods, using the Bayes rule, in the following way

$$P\left(w_\alpha | v_a, S^{(n)}, \Phi^{(n)}\right) = \frac{P\left(v_a, w_\alpha | S^{(n)}, \Phi^{(n)}\right)}{\sum_{\alpha'} P\left(v_a, w_{\alpha'} | S^{(n)}, \Phi^{(n)}\right)} \equiv R_{a\alpha}^{(n)} \tag{7}$$

We substitute the conditional likelihoods of the above equation by the expression of equation (5).

3.2 Maximum Likelihood Affine Registration Parameters

ML affine registration parameters and correspondence indicators are recovered in separate steps.

We are interested in the registration parameters $\Phi^{(n+1)}$ that lead to the maximum likelihood of equation (6). We use the expressions in equations (7) and (5) for the posterior probability and conditional likelihood terms, respectively. Discarding the terms that are constant w.r.t. the registration parameters we obtain the following expression

$$\Phi^{(n+1)} = \arg\max_{\hat\Phi} \left\{ \sum_{a\in\mathcal{I}} \sum_{\alpha\in\mathcal{J}} R_{a\alpha}^{(n)} \sum_{b\in\mathcal{I}} \sum_{\beta\in\mathcal{J}} s_{b\beta}^{(n)} \ln\left(\frac{\hat{P}_{b\beta}}{\rho}\right) \right\} \tag{8}$$

Rearranging and further removing other terms constant w.r.t. the registration parameters, we get

$$\Phi^{(n+1)} = \arg\max_{\hat\Phi} \left\{ \sum_{b\in\mathcal{I}} \sum_{\beta\in\mathcal{J}} s_{b\beta}^{(n)} \ln\left(\frac{\hat{P}_{b\beta}}{\rho}\right) \sum_{a\in\mathcal{I}} \sum_{\alpha\in\mathcal{J}} R_{a\alpha}^{(n)} \right\} = \arg\max_{\hat\Phi} \left\{ \sum_{b\in\mathcal{I}} \sum_{\beta\in\mathcal{J}} s_{b\beta}^{(n)} \ln\hat{P}_{b\beta} \right\} \tag{9}$$

We assume that the geometrical compatibilities $\hat{P}_{b\beta}$ follow a multivariate gaussian distribution of the point errors.

Substituting $\hat{P}_{b\beta}$ by its appropriate expression and, removing constant terms, we arrive to the minimization of the following objective function

$$\mathcal{F} = \sum_{b\in\mathcal{I}} \sum_{\beta\in\mathcal{J}} s_{b\beta}^{(n)} \left(\tilde{x}_b - \hat\Phi\tilde{y}_\beta\right)^{\mathrm{T}} \Sigma^{-1} \left(\tilde{x}_b - \hat\Phi\tilde{y}_\beta\right) \tag{10}$$

where $\tilde{x}_b$ and $\tilde{y}_\beta$ are the augmented vectors of homogeneous coordinates, $\Phi = \begin{bmatrix} \phi_{11} & \phi_{12} & \phi_{13} \\ \phi_{21} & \phi_{22} & \phi_{23} \\ 0 & 0 & 1 \end{bmatrix}$ is the matrix of affine registration parameters and, Σ is diagonal matrix of variances.

Affine registration parameters are computed by solving the set of linear equations $\delta\mathcal{F}/\delta\phi_{ij} = 0$.

3.3 Maximum Likelihood Correspondence Indicators

One of the key points in our work, is to approximate the solution of the graph matching problem by a succession of easier assignment problems. As it is done in Graduated Assignment [7], we use the *Softassign* [16][15][10] to solve the assignment problems in a continuous way.

According to the EM development, we compute the correspondence indicators $S^{(n+1)}$ that maximize equation (6). Substituting equations (7) and (5) into (6) and, discarding the constant term $\ln\rho$ of equation (5), we obtain

$$S^{(n+1)} = \arg\max_{\hat{S}} \left\{ \sum_{a\in\mathcal{I}}\sum_{\alpha\in\mathcal{J}} R_{a\alpha}^{(n)} \sum_{b\in\mathcal{I}}\sum_{\beta\in\mathcal{J}} \hat{s}_{b\beta} \left[D_{ab}E_{\alpha\beta}\ln\left(\frac{1-P_e}{P_e}\right) + \ln\left(\frac{P_{b\beta}^{(n)}}{\rho}\right) \right] \right\} \tag{11}$$

Rearranging we obtain the following assignment problem [7]

$$S^{(n+1)} = \arg\max_{\hat{S}} \left\{ \sum_{b\in\mathcal{I}}\sum_{\beta\in\mathcal{J}} \hat{s}_{b\beta} Q_{b\beta}^{(n)} \right\} \tag{12}$$

where

$$Q_{b\beta}^{(n)} = \sum_{a\in\mathcal{I}}\sum_{\alpha\in\mathcal{J}} R_{a\alpha}^{(n)} \left[D_{ab}E_{\alpha\beta}\ln\left(\frac{1-P_e}{P_e}\right) + \ln\left(\frac{P_{b\beta}}{\rho}\right) \right] \tag{13}$$

is the benefit coefficient for the assignment $v_b \to w_\beta$.

Softassign computes the correspondence indicators in two steps. First, the correspondence indicators are updated with the exponentials of the benefit coefficients

$$s_{b\beta} = \exp\left(\mu\, Q_{b\beta}\right) \tag{14}$$

where μ is a control parameter. Second, two-way constraints are imposed by alternatively normalizing across rows and columns the matrix of exponentiated benefits. This is known as the *Sinkhorn normalization* and, it is applied either until convergence of the normalized matrix or, a predefined number of times.

Note that, as the control parameter μ of equation (14) approaches to ∞, the correspondence indicators $s_{b\beta}$ tend to discrete values ($s_{b\beta} = \{0,1\}$) after the Sinkhorn normalization.

3.4 Outlier Rejection

Outliers can dramatically affect the performance of a matching and therefore, it is important to develop techniques aimed at minimizing their influence [3].

According to our purposes, a node $v_b \in \mathbf{v}$ (or $w_\beta \in \mathbf{w}$) will be considered an outlier to the extent that there is no node w_β, $\forall\beta\in\mathcal{J}$ (or v_b, $\forall b\in\mathcal{I}$) which presents a matching benefit $Q_{b\beta}^{(n)}$ above a given threshold.

Note that, ρ establishes the threshold at which the geometrical terms (i.e., $\ln\left(P_{b\beta}/\rho\right)$) contribute positively (i.e., $\rho < P_{b\beta}$) or negatively (i.e., $\rho > P_{b\beta}$) to the benefit measure.

Our strategy for controlling the outlier process is the following. We set the outlying threshold to zero and, create an augmented benefit matrix $\tilde{Q}^{(n)}$ by adding to $Q^{(n)}$ an extra row and column of zeros (the *slack* variables of the Softassign [7]). Then, we apply the Softassign (exponentiation and Sinkhorn normalization) to the augmented benefit matrix. Last, the slack variables are removed leading to the resulting matrix of correspondence parameters $S^{(n+1)}$.

Note that, as the control parameter μ of the Softassing increases, the rows and columns of $S^{(n+1)}$ associated to the outlier nodes, tend to zero. This fact reduces the influence of these nodes in the maximization phases of the next iteration that, in turn, lead to even lower benefits, and so on.

It is now turn to define the value of the outlying threshold ρ. Since ρ is to be compared with $P_{b\beta}$, it is convenient to define it in terms of a multivariate gaussian of a distance threshold. This is,

$$\rho = \frac{1}{2\pi |\Sigma|^{1/2}} \exp \left[-\frac{1}{2} d^{\mathrm{T}} \Sigma^{-1} d \right] \tag{15}$$

where, $\Sigma = \mathrm{diag} \left((\sigma^{ab})^2, (\sigma^{or})^2 \right)$ is a diagonal variance matrix and, $d = (d^{ab}, d^{or})$ is a column vector with the abscissa and ordinate thresholding distances.

Cancelling the gaussian constant terms in the numerator and denominator of the geometrical term and, expressing the thresholding distance proportionally to the standard deviations of the data (i.e., $d = (N\sigma^{ab}, N\sigma^{or})$), the expression of ρ to be compared with $P_{b\beta}$ becomes

$$\rho = \exp \left\{ -\frac{1}{2} \left[\left(\frac{N\sigma^{ab}}{\sigma^{ab}} \right)^2 + \left(\frac{N\sigma^{or}}{\sigma^{or}} \right)^2 \right] \right\} = \exp \left(-N^2 \right) \tag{16}$$

So, we define ρ as a function of the number N of standard deviations permitted in the registration errors, in order to consider a plausible correspondence.

4   Experiments and Results

In the first set of experiments, we have evaluated the effectiveness of our method in front of non-rigid deformations in the positions of the features (i.e., nodes). In each experiment, a pattern of 15 randomly generated points is matched against a deformed version of itself. Deformations are introduced by applying gaussian noise, independently, to each point. Graphs' edges are generated by Delaunay triangulations on the point-sets. Figure 3 shows the comparison of our method (denoted as *Smooth*) to the graph matching + alignment methods in refs. [4] (*Dual-Step*) and [13] (*Unified*). We have used the values $P_e = 0.3$ and $\rho = \exp \left(-1.9^2 \right)$ for our method. The parameters of the other methods have been accurately set to have a good performance. All the methods have been initialized by the correspondences obtained by simple nearest neighbour association.

The mean execution times of the MATLAB implementations of each method are: *Smooth* 0.66 sec., *Dual-Step* 14.08 sec. and, *Unified* 0.91 sec. The *Dual-step* method is run without the outlier detection scheme (otherwise it slows down).

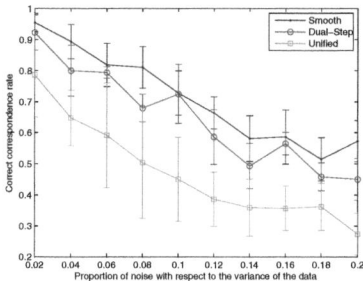

Fig. 3. Correct correspondence rate vs. noise level (represented as a proportion of the data variance). Each point is the mean of 25 experiments (5 random patterns by 5 random deformations of each pattern).

The last set of experiments evaluates the matching performance on real images under zoom and rotation from the database in [1]. Features are extracted with the Harris operator [8] and edges with Delaunay triangulations. Figure 4 show the results of applying our method to some images. We have used the values $P_e = 0.3$ and $\rho = \exp\left(-1.3^2\right)$ for our method. The value of ρ has been set so as to enable an actual detection of outliers.

We have compared some methods with explicit outlier detection mechanisms. These are the outlier rejectors *RANSAC* [6] and Graph Transformation Matching (*GTM*) [2] and, the graph matching method *Dual-Step* [4] (with the outlier detection scheme enabled). All the methods have been initialized with the matching by correlation (*Corr*) results.

From the resulting correspondences of each method, we have estimated the corresponding homographies with the DLT algorithm [9]. Since it is available the ground truth homography between each pair of images, we have measured the mean projection error (MPE) of the feature-points in the origin images. Table 1 shows the results.

Table 1. Mean Projection Error (MPE, in pixels) and execution times (in seconds) obtained by each method using a MATLAB implementation. In the case of accurate correlation results (e.g., Laptop), slight errors may be introduced due to the approximations done by each method in the model assumptions such as the affine one (in *Smooth* and *Dual-Step*) or a purely structural one (in *GTM*).

Methods	Resid MPE	Resid time	Boat MPE	Boat time	NewYork MPE	NewYork time	Laptop MPE	Laptop time	Eastpark MPE	Eastpark time
Corr	835.4	1.5	24.48	1.5	31.1	0.98	**0.28**	1.34	463.4	1.62
Smooth	1.5	13.8	**0.72**	18.2	**0.69**	3.6	0.29	8.18	**1.08**	19.7
Dual-Step	**1.33**	3615	1.68	3794	**0.69**	1429	0.3	2693	153.46	3027
RANSAC	20.23	0.2	1.6	0.1	17.11	0.12	**0.28**	0.11	350.13	0.42
GTM	24.15	0.02	0.8	0.1	3.2	0.02	0.34	0.02	359.9	0.04

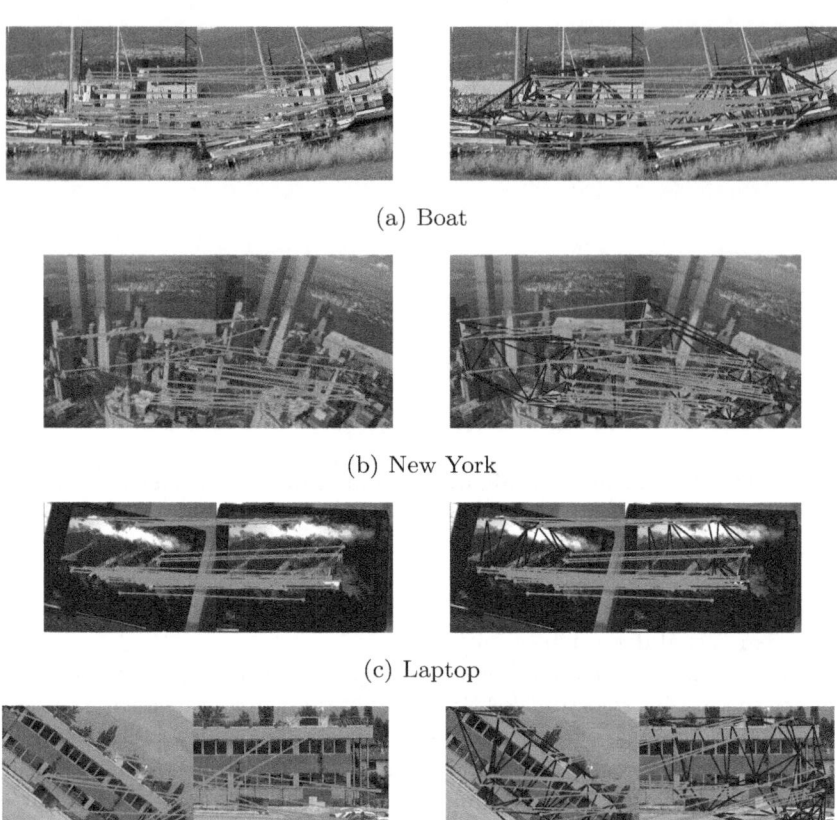

(a) Boat

(b) New York

(c) Laptop

(d) Eastpark

Fig. 4. Right column shows the results of our method using the correlation results (left column) as starting point

5 Conclusions

We have presented a method that uses the EM algorithm to approximate the graph matching problem as a succession of assignment problems which are then solved in a smooth way using Softassign. Our method refines an initial tentative correspondence-set in a more flexible way than the outlier rejectors such as RANSAC and Graph Transformation Matching that are only able to remove the spurious correspondences. Furthermore, it is capable of detecting outliers in both graphs. Results show that our method performs faster and better than other graph matching methods in the literature in the matching of synthetic graphs. Results in the matching of real images show that our method performs generally better than the others, within an admissible time. Methods with comparable efficiency than ours show computational times of two orders of magnitude higher.

Acknowledgements. This research is supported by Consolider Ingenio 2010 (CSD2007-00018), by the CICYT (DPI 2010-17112) and by project DPI-2010-18449.

References

1. Http://www.featurespace.org/
2. Aguilar, W., Frauel, Y., Escolano, F., Martinez-Perez, M.: A robust graph transformation matching for non-rigid registration. Image and Vision Computing 27, 897–910 (2009)
3. Black, M., Rangarajan, A.: On the unification of line processes, outlier rejection, and robust statistics with applications in early vision. International Journal of Computer Vision 19(1), 57–91 (1996)
4. Cross, A., Hancock, E.: Graph matching with a dual-step em algorithm. IEEE Trans. Pattern Analysis and Machine Intelligence 20(11), 1236–1253 (1998)
5. Dempster, A., Laird, N., Rubin, D.: Maximum likelihood from incomplete data via the em algorithm. Journal Royal Stat. Soc., Series B 39(1), 1–38 (1977)
6. Fischler, M.A., Bolles, R.C.: Random sample consensus: a paradigm for model fitting with applications to image analysis and automated cartography. Comunications of the ACM 24(6), 381–395 (1981)
7. Gold, S., Rangarajan, A.: A graduated assignment algorithm for graph matching. IEEE Trans. Pattern Analysis and Machine Intelligence 18(4) (April 1996)
8. Harris, C., Stephens, M.: A combined corner and edge detection. In: Proceedings of The Fourth Alvey Vision Conference, pp. 147–151 (1988)
9. Hartley, R., Zisserman, A.: Multiple View Geometry in Computer Vision. Cambridge University Press, Cambridge (2003)
10. Kosowsky, J., Yuille, A.: The invisible hand algorithm - solving the assignment problem with statistical physics. Neural Networks 7(3), 477–490 (1994)
11. Luo, B., Hancock, E.: Feature matching with procrustes alignment and graph editing. In: 7th Int. Conf. on Image Proc. and Apps., vol. 465, pp. 72–76 (1999)
12. Luo, B., Hancock, E.: Structural graph matching using the em algorithm and singular value decomposition. IEEE Transactions on Pattern Analysis and Machine Intelligence 23(10) (October 2001)
13. Luo, B., Hancock, E.: A unified framework for alignment and correspondence. Computer Vision and Image Understanding 92(1), 26–55 (2003)
14. Mikolajczyk, K., Schmid, C.: A performance evaluation of local descriptors. IEEE Trans. on Pattern Analysis and Machine Intelligence 27(10), 1615–1630 (2005)
15. Rangarajan, A., Gold, S., Mjolsness, E.: A novel optimizing network architecture with applications. Neural Computation 8(5), 1041–1060 (1996)
16. Sinkhorn, R.: Relationship between arbitrary positive matrices + doubly stochastic matrices. Annals of Mathematical Statistics 35(2), 876–879 (1964)
17. Wilson, R., Cross, A., Hancock, E.: Structural matching with active triangulations. Computer Vision and Image Understanding 72(1), 21–38 (1998)
18. Wilson, R., Hancock, E.: Structural matching by discrete relaxation. IEEE Trans. Pattern Analysis and Machine Intelligence 19(6), 634–648 (1997)

Automatic Learning of Edit Costs Based on Interactive and Adaptive Graph Recognition*

Francesc Serratosa, Albert Solé-Ribalta, and Xavier Cortés

Universitat Rovira i Virgili, Departament d'Enginyeria Informàtica i Matemàtiques, Spain
{francesc.serratosa,albert.sole}@urv.cat, xavi1984@gmail.com

Abstract. We propose a new method to automatically obtain edit costs for error-tolerant graph matching based on interactive and adaptive graph recognition. Values of edit costs for deleting and inserting nodes and vertices are crucial to obtain good results in the recognition ratio. Nevertheless, these parameters are difficult to be estimated and they are usually set by a naïve trial and error method. Moreover, we wish to seek these costs such that the system obtains the correct labelling between nodes of the input graph and nodes of the model graph. We consider the labelling imposed by a specialist is the correct one, for this reason, we need to present an interactive and adaptive graph recognition method in which there is a human interaction. Results show that when cost values are automatically found, the quality of labelling increases.

Keywords: Interactive Learning, Adaptive Learning, Graph Edit Distance.

1 Introduction

Graphs refer to a collection of nodes and a collection of edges that connect pairs of nodes. Attributed Graphs are graphs in which some attributes are added on nodes and edges to represent local information or characterisation. Attributed graphs have been widely used in several fields to represent objects composed by local parts and relations between these parts. More precisely, in Pattern Recognition and Computer Vision, attributed graphs have been used to represent structural objects that have to be identified or classified. These graphs can represent 2D or 3D objects, handwritten characters, proteins, fingerprints, and so on. Before using graphs, the pattern recognition process has to extract them from these objects. This is not a trivial task included in the Image Understanding field since the quality of graphs is crucial for the rest of the process. When attributed graphs have been extracted, a process to compare them is needed. This process is called graph matching. Usually, it obtains a similarity value and also a labelling between nodes and arcs of the involved graphs. This labelling between nodes and arcs represents the matching between the local parts that graphs represent.

When we characterise the role that attributed graphs and attributed graph matching do in pattern recognition, we realise that in most of the applications, labelling between

---

* This research was partially supported by Consolider Ingenio 2010 and by the CICYT project DPI 2010-17112.

X. Jiang, M. Ferrer, and A. Torsello (Eds.): GbRPR 2011, LNCS 6658, pp. 152–163, 2011.
© Springer-Verlag Berlin Heidelberg 2011

nodes is only partially considered. This is because it is considered in the first stages of pattern recognition process, in which it is desired to find a similarity between graphs, but when this similarity value is obtained, the knowledge of the labelling is not considered any more. Nevertheless, we consider that although the graph (that is, the object that represents) is properly classified or identified, the result of the comparison has not sense if the matching between their local parts is not correct.

In this paper, we present an interactive and adaptive graph recognition model with the aim of increasing the quality of the labelling between the graph to be identified and the reference graphs of the database. To that aim, we have extended the graph recognition model to consider the labelling between nodes proposed by a human specialist. This new knowledge is incorporated into the system and used to modify the weights that tune the similarity function between graphs. Usually, these weights are imposed before executing the learning process in a naïve way. We believe that if the quality of the labelling is increased also should do it the quality of the pattern recognition process. In [1], they considered the problem of partial matchings.

The rest of the paper is organised as follows, in sections 2 and 3, we first introduce concepts related to graph matching and then we define a new space of labellings. In sections 3, 4 and 5, we first present the classical pattern recognition and then we depict the new model of interactive and adaptive graph recognition method. In section 7, we compare the quality of the labellings obtained by the new models respect the classical pattern recognition. Section 8 drops some conclusions.

2   Error-Tolerant Graph Matching Based on Edit Operations

One of the most widely used methods for error-tolerant graph matching is the graph edit distance. The basic idea behind the graph edit distance is to define a dissimilarity measure between two graphs by the minimum amount of distortion required to transform one graph into the other [2,3]. To this end, a number of distortion or edit operations ε, consisting of insertion, deletion and substitution of both nodes and edges must be defined. Then, for every pair of graphs (G and G'), there exists a sequence of edit operations, or edit path $path(G, G') = (\varepsilon_1, ..., \varepsilon_k)$ (where each ε_i denotes an edit operation) that transforms one graph into the other. In general, several edit paths may exist between two given graphs. This set of edit paths is denoted by $\vartheta(G, G')$. To quantitatively evaluate which is the best edit path, edit cost functions are introduced. The basic idea is to assign a penalty cost C to each edit operation according to the amount of distortion that it introduces in the transformation. The edit distance between two graphs G and G', denoted by $dist_{K_n, K_e}(G, G')$, is defined as the minimum cost of edit path that transforms one graph into the other given parameters K_n (cost of node insertion or deletion) and K_e (cost of edge insertion or deletion) . Several edit paths may obtain the minimum cost. More formally, the edit distance is defined by,

$$dist_{K_n, K_e}(G, G') = \frac{min}{(\varepsilon_1, ..., \varepsilon_k) \in \vartheta(G, G')} \left\{ \sum_{i=1}^{k} C(\varepsilon_i) \right\} \qquad (1)$$

Usually, edit costs K_n and K_e are estimated in a naïve way or they are learned by trial and error method. The works presented in [4,5,6] are the only ones that aim to automatically estimate these costs. Nevertheless, their method minimises an energy

Table 1. Edit operations, their costs and the relation with the labelling function

Edit operation	Cost	Labelling f
Node deletion (a)	$K_n \in [0, \infty)$	$f^{-1}(a) = null\ element$
Node insertion (a')	$K_n \in [0, \infty)$	$f^{-1}(a') = null\ element$
Node substitution (a,a')	$distance(a, a') \in [0, \infty)$	$f^{-1}(a) = a'$
Edge deletion (b)	$K_e \in [0, \infty)$	$f^{-1}(b) = null\ element$
Edge insertion (b')	$K_e \in [0, \infty)$	$f^{-1}(b') = null\ element$
Edge substitution (b,b')	$distance(b, b') \in [0, \infty)$	$f^{-1}(b) = b'$

related to the recognition ratio in a pattern recognition framework without considering the goodness of the labellings of the involved graphs. Contrarily, in [7], they forced some specific cost. With these costs, they demonstrate the similarity between the edit-distance problem and the maximum common sub-graph problem.

In this paper, we aim to study the impact of K_n and K_e values to the quality of the labellings and present a method to automatically impose them. It seems logical to think that if the labelling between graphs is correct, the recognition ratio of the pattern recognition system has to be the best. First and second columns of Table 1 show the edit operations and edit costs we define throughout this paper. We impose the restriction that insertions and deletions of vertices or arcs have the same cost value. This is done to assure the symmetry property of (1).

Optimal [8] and approximate algorithms [9,10] for the graph edit distance computation have been presented so far, which are out of the scope of this paper. These algorithms obtain the distance value $dist(G, G')$ as well as a labelling f from vertices and arcs of the first graph to vertices and arcs of the second graph. Column 3 of Table 1 shows the labelling related to edit operations. Given any edit path, $path(G, G')$, a labelling $f(G, G')$ can be defined univocally. The cost of this labelling is,

$$Cost_{K_n, K_e}(G, G', f) = \sum_{i=1}^{k} C(\varepsilon_i)\ being\ f\ related\ to\ path(G, G') = (\varepsilon_1, \dots, \varepsilon_k) \qquad (2)$$

3   Labelling Space Based on K_n and K_e Values

Given an input graph G, a reference graph G' and a labelling f between them, we define the range of values of K_n and K_e that the labelling f has the minimum cost (2). That is, the values of K_n and K_e that f is the labelling that obtains the distance (1) between graphs G and G'. These values (K_n, K_e) hold condition E,

$$E: Cost_{K_n, K_e}(G, G', f) = \mathop{min}_{\forall f' \in \vartheta(G, G')} \{Cost_{K_n, K_e}(G, G', f')\} \qquad (3)$$

Let $R_{G, G', f}$ be a region defined by points (K_n, K_e) in which condition E is true,

$$R_{G, G', f} = \{(K_n, K_e)|E\} \qquad (4)$$

Table 2. Edit operations, their costs and the relation with the labelling function

Edit operation	Cost R2	Cost R3
Node del. (*a*)	$K_n = 0.3$	$K_n = 0.3$
Node ins. (*a'*)	$K_n = 0.3$	$K_n = 0.3$
Node subs. (*a,a'*)	distance(*a,a'*) = 2D Euclidean	distance(*a,a'*) = 2D Euclidean
Edge del. (*b*)	$K_e = 0.3$	$K_e = 0.5$
Edge ins. (*b'*)	$K_e = 0.3$	$K_e = 0.5$
Edge subs. (*b,b'*)	distance(*b,b'*) = 0	distance(*b,b'*) = 0

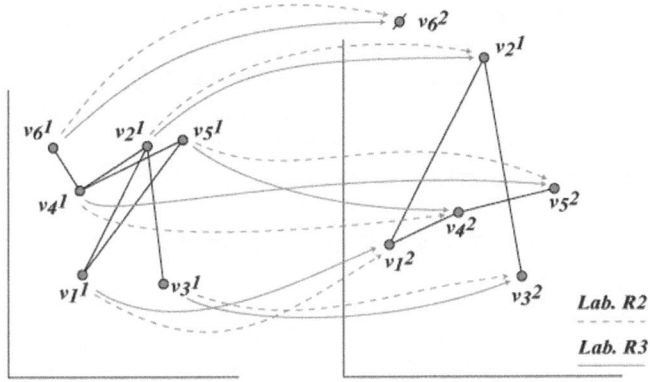

Fig. 1. Two graphs and two different labellings: R2 and R3

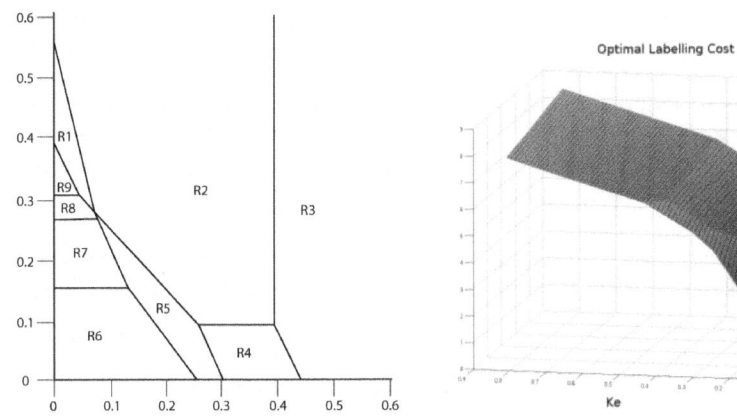

Fig. 2. (a) All possible labellings throughout the labelling space

Fig. 2. (b) Edit cost of the optimal labelling given Kn and Ke

Note that $Cost_{K_n,K_e}(G, G_{out}, f')$ takes different values when different values of K_n and K_e are used (2). For this reason, this cost, although it is the minimum one given a point (K_n, K_e) through all labellings f', is not constant throughout region $R_{G,G_{out},f}$.

Figure 1 shows two graphs and two possible labellings between them (called *R2* and *R3*). Table 2 shows the edit costs we have used.

Note that erasing an arc in R3 is more costly than in R2. For this reason R2 erases 3 arcs in the first graph and inserts 1 arc in the second. On the contrary, R3 does not erase any arc and only inserts 1 arc in the second graph. To conclude, R2 gives more importance to the position of the nodes and R3 gives more importance to the structural deformation.

Figure 2.a shows the labelling space between these two graphs respect K_n (vertical axis) and K_e (horizontal axis). There are only 9 different labellings which is a reduced number since the maximum number of labellings could arise 6! = 720. Finally, figure 2.b shows the minimum cost throughout the labelling space. We realise of four properties: 1. Regions $R_{G,G',f}$ are convex, 2. When costs K_n or K_e increase, it does the cost labelling. 3. Increase of labelling cost is linear throughout a region. 4. Regions located in bigger positions of K_n or K_e are less steeply.

4   Classical Graph Recognition Paradigm

Let G be an input graph and h a hypothesis or output of the graph recognition system, which the system derives from G. Let M be a model used by the graph recognition system to derive this output. M is obtained through a previous or batch training process from a given sequence of pairs composed by a graph and a class (G_i, C_i).

In some applications (for instance, character recognition), the hypothesis h is composed by only the class or cluster obtained by the system $h = \{C\}$. Nevertheless, in some applications, it is desired to know not only the class, but also the graph of the model with the minimum error G_{out} and the labelling f between them (for instance, scene identification). That is, $h = \{C, G_{out}, f\}$.

Figure 3 shows the three main modules of a classical graph recognition system. It is composed by a training process, a recognition processes and a model. The model is generated through a batch training process with a set of pairs (graph, class) and other parameters (such as K_n and K_e in (1)) and it represents the knowledge of the recognition process.

In classical pattern recognition [11] (and specifically, in graph recognition) the output of the system is computed through a function that aims to minimise the number

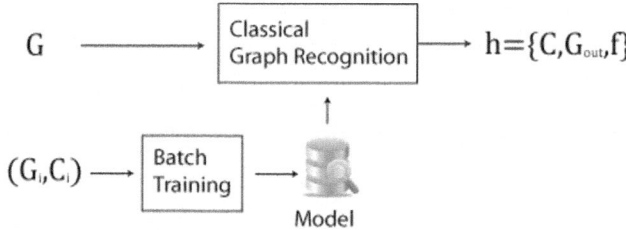

Fig. 3. Scheme of a Classical graph recognition system

of wrong hypothesis, that is, the number of misclassified objects. The best hypothesis is the one that minimises the misclassification of graphs, that is, it is the one which maximises the posterior probability of class C given G, $Pr(C|G)$ and given model M, $P_M(C|G)$.

$$h = \{C, G_{out}, f\} = \begin{matrix} max \\ \forall C \in M \end{matrix} P_M(C|G) \tag{5}$$

The maximum likelihood approach (5) is the most popular method to estimate the classification not only in the pattern recognition process but also in the training process to train the model M. However, in many cases, it is difficult to directly estimate $P_M(C|G)$ ($P(C|G)$ since now) and it is better to apply the Bayes rule in (5),

$$h = \begin{matrix} max \\ \forall C \in M \end{matrix} \frac{P(G|C)P(C)}{P(G)} = \begin{matrix} max \\ \forall C \in M \end{matrix} P(G|C)P(C) \tag{6}$$

The probability $P(C|G)$ is estimated independently for each pair (G, C) and from the available training pairs (G_i, C_i). The probability of each class $P(C)$ depends on the application and, sometimes, it is defined as constant for all classes. Finally, $P(G)$ has been dropped since does not depend on the hypothesis and so, it does not affect the maximisation criteria of (6). With these considerations, we have only to define the probability of having an input graph conditioned to a class. Several methods have been defined, for instance,

$$P(G|C) = \begin{matrix} max \\ \forall G' \in C \end{matrix} \{e^{-dist_{K_n,K_e}(G,G')}\} \vee P(G|C) = \begin{matrix} max \\ \forall G' \in C \end{matrix} \left\{ \frac{1}{1+dist_{K_n,K_e}(G,G')} \right\} \tag{7}$$

5 Interactive Graph Recognition Paradigm

Placing human interaction requires adaptation to the way we look at the graph recognition problem [12]. Few research has been done in this field [13] and none related to graphs. One of the main applications has been the semi-automatic transcription of handwritten texts [14, 15].

The paradigm we presented above obtains the minimum graph that minimises (6) independently of the obtained labelling f. Nevertheless, the user aims to obtain a correct labelling f independently of the model parameters imposed in the batch training. Moreover, although the class obtained by the system is considered the correct one by an expert, the hypothesis has to be considered non-correct if the labelling f between the input and output graph is far away from the one considered by an expert. For this reason, the feedback provided by the human is a new labelling g between G and G_{out} (that may be totally or partially equal to f) and then the hypothesis is estimated again.

Without varying the model M, human interaction offers an opportunity to improve the quality of the hypothesis h only using labelling g. Note that, this labelling g between the input graph G and the output graph G_{out}, is the most natural way for a human to interact with the system since it takes direct advantage of the intelligence and general knowledge of the expert. Moreover, it is one of the most complicated tasks to be processed in graph recognition. From an application point of view, f has to be considered as a first approximation of the labelling that helps the specialist to make a final (and possibly better) labelling g.

Since in our new framework, we have two labellings f and g between graphs G and G', it is interesting to have a measure of similarity between both labellings $S_{G,G'}(f,g) \in [0,1]$. We define this measure as follows,

$$S_{G,G'}(f,g) = \frac{1}{|G| + |G'|}\left(\sum_{\forall v \in G} T_1(v) + \sum_{\forall v' \in G'} T_2(v')\right) \in [0,1] \tag{8}$$

Where v and v' are nodes of graphs G and G', respectively. T_1 and T_2 are,

$$T_1(v) = \begin{cases} 1 & if \ f(v) = g(v) \\ 0 & otherwise \end{cases} \quad and \quad T_2(v') = \begin{cases} 1 & if \ f^{-1}(v') = g^{-1}(v') \\ 0 & otherwise \end{cases} \tag{9}$$

Figure 4 shows the new scheme, in which, it appears the human interaction through labelling g. The batch training generates a model similar to the classical scheme. The interactive recognition process generates a first hypothesis h through the model. Then, it generates the final hypothesis h' using the model and also the human interaction g. Note that, h' can be completely different from h. Not only the graph and labelling can be different but also the class. This is because, with the imposed labelling g, the cost (2) can be higher than the distance (1) between G and another graph of another class.

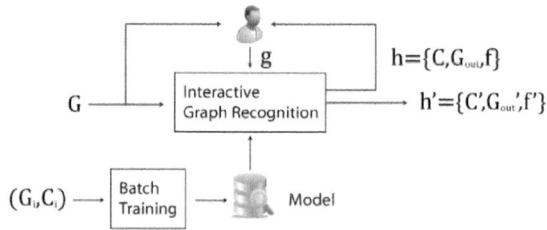

Fig. 4. Scheme of an Interactive graph recognition system

Now, interaction allows adding more conditions to (5) to obtain the new hypothesis,

$$h' = \{C', G'_{out}, f'\} = \overset{max}{\underset{\forall C \in M}{}} P_M(C|G,g) \tag{10}$$

Applying Bayes rule, we arrive at the following expression,

$$h' = \overset{max}{\underset{\forall C \in M}{}} \frac{P(G|C,g)P(C|g)P(g)}{P(G,g)} \tag{11}$$

In which probabilities $P(g)$ and $P(G,g)$ do not depend on C and they are dropped off,

$$h' = \overset{max}{\underset{\forall C \in M}{}} P(G|C,g)P(C|g) \tag{12}$$

Probability $P(C|g)$ represents the probability of a class conditioned to a labelling g between two graphs. We consider in this point that there is no information about the process to extract a graph from an object, commented in the introduction section (for instance, image or social net). Therefore, the probability of a class C has to be defined independent of having a specific labelling g between two graphs and we assume $P(C|g) = P(C)$.

The conditioned probability of the input graph G, $P(G|C)$, is defined depending on the data (7), but now, we add another conditional, which is the specialist labelling g,

$$P(G|C,g) = e^{-Cost_{K_n,K_e}(G,G',g)} \quad \text{or} \quad P(G|C,g) = \frac{1}{1+Cost_{K_n,K_e}(G,G',g)} \tag{13}$$

And the probability of class $P(C)$ is defined to be constant if there is no information. In some examples, it depends on the information extracted from the learning process.

6 Adaptive Graph Learning

In the previous section the model M has been assumed to be fixed. But now, the human interaction offers another opportunity to improve the system not only modifying the current hypothesis but modifying the model M when new information is available. We present a methodology to update parameters K_n and K_e of equation (1). These parameters are crucial for the quality of the recognition but they are very difficult to be estimated since they are data dependent. We propose a model that each time the specialist imposes a labelling g between an input graph G and an output graph G_{out}, these parameters are updated.

The adaptive-graph learning scheme is composed of three modules (figure 5). The training and recognition module are similar to the interactive scheme. The adaptive-learning module updates values of K_n and K_e in model M.

This scheme is implemented in an iterative algorithm that each time a new graph is introduced into the system to be identified, the following steps are carried out:

- Given input G, the graph recognition module outputs $h = \{C, G_{out}, f\}$
- Specialist analyses G and h and introduces labelling g into the system (g is the same than f if h is an ideal prediction)
- The interactive recognition module outputs h'. (C, G_{out} and f might be different to C', G'_{out} and f')
- The adaptive-learning module inputs h', g and G and modifies values of K_n and K_e using intersection of regions $R_{G,G'_{out},g}$.

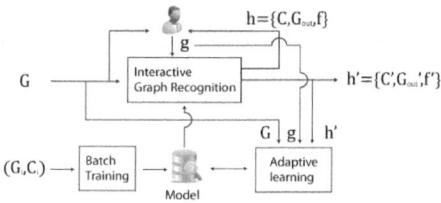

Fig. 5. Scheme of an Adaptive graph recognition system

7 Practical Evaluation

To evaluate and compare the quality of output labelling f using the Interactive Learning scheme (IL) with respect to the Classical Learning (CL) scheme, we propose to use the similarity $S(f,g)$ (8) between labelling f and the labelling imposed by the specialist g.

The evaluation measure we used is the Area under the Learning Curve (ALC). Our Learning curve plots in each point the addition of similarities obtained by processing the whole test set normalised by the cardinality of the test set. That is,

$$Learning\ Curve = \frac{1}{|TestSet|} \sum_{\forall G \in TestSet} S_{G,G_{out}}(f,g) \in [0,1] \tag{15}$$

Moreover, we obtain a global measure of the quality of the learning process when it is considered to be finished. It is independent of the cardinality of the test and reference set and also the cardinality of the graphs,

$$GlobalScore = \frac{ALC - A_{rand}}{A_{max} - A_{rand}} \tag{16}$$

Values A_{max} and A_{rand} are obtained through two baseline learning curves. The first one is the ideal learning curve that at first point goes up to the maximum value that equals 1. The second one is the "lazy" learning curve that follows a straight line. It is obtained by making random predictions. If $m = |test\ set|$ and n is the average cardinality of the graphs in the test set, the lazy learning curve can be approximated as $\frac{1}{(n!)^m}$.

We have used the Tarragona-graph database [16]. It is a database of graphs that represents handwritten letters and it has the peculiarity that nodes of graphs of the same class have been manually labelled. Nodes represent junctions or ending points of strokes and non-directional edges are strokes. Attributes on the nodes are the 2D position and edges do not have attributes. Figure 6 shows some examples of character A. Nodes have been manually labelled and so the ith node of each graph represents the ith ending-point or junction. Some nodes are missing in some characters.

There are 26 different characters and 11 elements for each character. From each character, we have used the 4 first elements as the test set and the other 7 elements as the reference set.

We have performed 4 experiments using CL and other 4 experiments using IL. The difference between these four experiments is the initial values of K_n and K_e. Note that, in CL, these values are constant through the learning process but in IL they are not constant. Figure 7 shows the average learning curves of the 8 experiments.

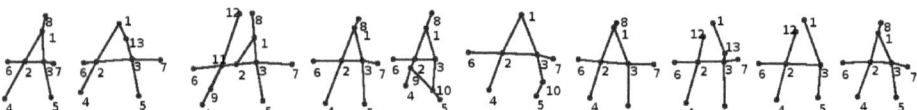

Fig. 6. Character A of Tarragona-graph database

We have carried out the following process to plot each of the 8 curves:

```
Model empty
For i = 1 to 7 do
    For C = 1 to 26 do
        Gᶜᵢ = Select_New_Graph_from_Reference_Set_belonging_to_Class_C
        Include Gᶜᵢ into de model
    If IL recompute Kₙ and Kₑ
    Sₐᵤₓ = 0
    For j = 1 to 4 do
        For C = 1 to 26 do
            Gᶜⱼ = Select_New_Graph_from_Test_Set_belonging_to_Class_C
            (Gₒᵤₜ,f) = 1-Nearest_Neighbour_belonging_to_Class_C (Gⱼ)
            g = Ideal_Labelling (Gⱼ,Gₒᵤₜ)
            Sₐᵤₓ = Sₐᵤₓ + Similarity (g,f)
    Plot (Sₐᵤₓ/(4x26),i)
```

Figure 8 shows the evolution of values of K_n and K_e when new graphs are added into the IL system. The table on the left shows the initial values. Each of the 4 curves has 7 points. The model has 1 graph per class in the first point, 2 graphs per class in the second point and so on. Note that the four curves move to the same point, which has the be considered the optimal one, and it is independent of the initial values.

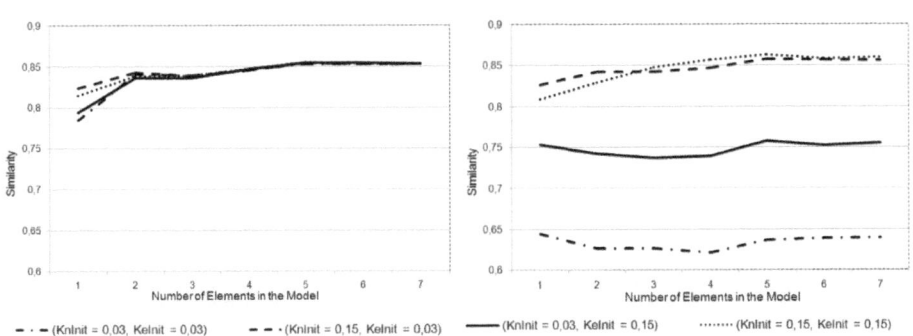

Fig. 7. Average learning curves respect the number of graphs in the model and given 4 different initial K_n and K_e values. Left plot: CL and right plot: IL.

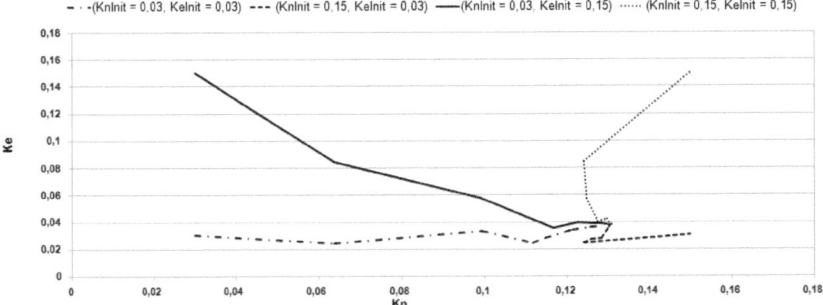

Fig. 8. Evolution of K_n and K_e. in the 4 IL experiments

Table 3. Global score of CL and IL depending on the initial K_n and K_e values

K_n	0.03	0.15	0.03	0.15
K_e	0.03	0.03	0.15	0.15
CL	0.632	0.848	0.747	0.848
IL	0.841	0.846	0.840	0.844

Finally, Table 3 shows the Global scores (16) of the learning curves shown in figure 7. Global scores of CL curves that the initial points are far from the final point in figure 8 have lower values. Conversely, there is a slight difference between CL and IL global scores when the initial points are close to the optimal point.

8   Conclusions and Future Work

We have presented an interactive and adaptive graph recognition method. It is based on the idea that if the quality of the labelling between graphs is increased, it also does the quality of the recognition or identification process. The feedback of the human specialist is introduced into the system as the ideal labelling between graphs and it is used to tone the weights of the edit costs in the similarity function between graphs. These weights are application dependent and difficult to be manually toned. Practical evaluation shows an example of obtaining K_n and K_e values. We conclude the final values are independent of the initial ones and also there is an increase of the quality of the labellings when these costs are automatically obtained.

References

1. Caetano, T.S., McAuley, J.J., Cheng, L., Le, Q.V., Smola, A.J.: Learning Graph Matching. IEEE Trans. Pattern Anal. Mach. Intell. 31(6), 1048–1058 (2009)
2. Sanfeliu, A., Fu, K.: A distance measure between attributed relational graphs for pattern recognition. IEEE Trans. on Sys., Man and Cybern. 13, 353–362 (1983)
3. Bunke, H., Allerman, G.: Inexact graph matching for structural pattern recognition. Pattern Recognition Letters 1(4), 245–253 (1983)
4. Neuhaus, M., Bunke, H.: A Probabilistic Approach to Learning Costs for Graph Edit Distance. ICPR (3), 389–393 (2004)
5. Neuhaus, M., Bunke, H.: Self-organizing maps for learning the edit costs in graph matching. IEEE Trans. on Sys., Man, and Cybernetics, Part B 35(3), 503–514 (2005)
6. Neuhaus, M., Bunke, H.: Automatic learning of cost functions for graph edit distance. Inf. Sci. 177(1), 239–247 (2007)
7. Bunke, H.: On a relation between graph edit distance and maximum common subgraph. Pattern Recognition Letters 18(8), 689–694 (1997)
8. Lladós, J., Martí, E., Villanueva, J.J.: Symbol Recognition by Error-Tolerant Subgraph Matching between Region Adjacency Graphs. IEEE Trans. Pattern Anal. Mach. Intell. 23(10), 1137–1143 (2001)
9. Gold, S., Rangarajan, A.: A Graduated Assignment Algorithm for Graph Matching. IEEE TPAMI 18(4), 377–388 (1996)
10. Christmas, W.J., Kittler, J., Petrou, M.: Structural matching in computer vision using probabilistic relaxation. IEEE TPAMI 17(8), 749–764 (1995)

11. Duda, R.O., Hart, P.E.: Pattern Classification and Scene Analysis. J. Wiley, Chichester (1973)
12. Canny, J.: The future of human-computer interaction. Queue,ACM 4(6), 24–32 (2006)
13. Vidal, E., Rodríguez, L., Casacuberta, F., García-Varea, I.: Interactive Pattern Recognition. In: MLMI, pp. 60–71 (2007)
14. Toselli, A.H., Romero, V., Pastor, M., Vidal, E.: Multimodal interactive transcription of text images. Pattern Recognition 43(5), 1814–1825 (2010)
15. Casacuberta, F., Civera, J., Cubel, E., Lagarda, A.L., Lapalme, G., Macklovitch, E., Vidal, E.: Human interaction for high-quality machine translation. Commun. ACM 52(10), 135–138 (2009)
16. http://deim.urv.cat/~francesc.serratosa/Tarragona_Graph_Data base

Exploration of the Labelling Space Given Graph Edit Distance Costs[*]

Albert Solé-Ribalta and Francesc Serratosa

Universitat Rovira i Virgili, Departament d'Enginyeria Informàtica i Matemàtiques, Spain
{albert.sole,francesc.serratosa}@urv.cat

Abstract. Graph Edit Distance is the most widely used measure of similarity between attributed graphs. Given a pair of graphs, it obtains a value of their similarity and also a path that transforms one graph into the other through edit operations. This path can be expressed as a labelling between nodes of both graphs. Important parameters of this measure are the costs of edit operations. In this article, we present new properties of the Graph Edit Distance and we show that its minimization lead to a few different labellings and so, most of the labellings in the labelling space cannot be obtained. Moreover, we present a method that using some of the new properties of the Graph Edit Distance speeds up the computation of all possible labellings.

Keywords: Graph Edit Distance, Graph Edit Costs, Graph Labelling Space.

1 Introduction

Usually graphs have been applied to pattern recognition, specifically in the context of classification. Many improvements have been made in this field, graph distance functions and graph matching algorithms are the main research lines. Some graph distance functions have been developed [1, 2, 3]. And there also exist some graph matching algorithms that minimize these distances [4, 5, 6]. One of the most commonly used distances is the Graph Edit Distance [1, 7]. Graph Edit Distance essentially depends on some constants, K_n and K_e, and an attribute distance function.

These constants and the distance function are application dependant and they are usually adjusted by an expert or learned using some not supervised process [4, 8]. Fig. 1 illustrates the problem when it is applied to classification. A Batch training process (an expert or some automatic procedure) learns K_n and K_e according to the decided model. Later, a classical Graph Recognition technique, such as KNN, SVM, Kernels and so on, is applied to perform classification tasks.

Another type of learning is the Adaptive Learning [9] (Fig. 2). In this type of learning an initial Batch Training process, that could be the same as the Classical Learning shown in Fig. 1, decides initial values for K_n and K_e constants. After this

---

[*] This research is supported by Consolider Ingenio 2010: project CSD2007-00018, by the CICYT project DPI 2007-61452 and by the Universitat Rovira i Virgili through a PhD research grant.

X. Jiang, M. Ferrer, and A. Torsello (Eds.): GbRPR 2011, LNCS 6658, pp. 164–174, 2011.
© Springer-Verlag Berlin Heidelberg 2011

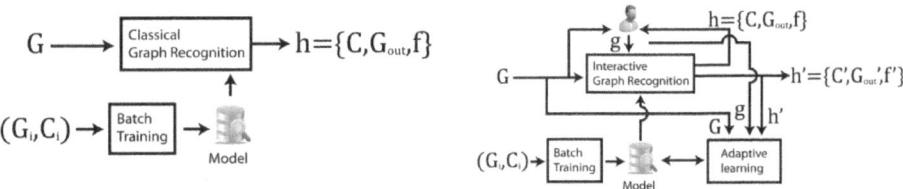

Fig. 1. Classical Graph Training   **Fig. 2.** Adaptive Graph Training

initial training, as graphs enter the system, the system output is compared with an expert decision and the Model is updated accordingly. See [10] for a detailed description of both learning types.

We aim to use Adaptive Learning methodology to learn K_n and K_e values using the support of an expert. Note that the cornerstone of the process is performed by the Adaptive Learning block which must decide given the most similar output G_{out} of the Model and the given graph G which K_n and K_e values will produce the most similar expert output g and update the system according to these parameters. The system faces some questions. How to compute the range of values of K_n and K_e that produce the desired labelling? All labellings that the user can provide are susceptible to be learned? Which type (optimal or suboptimal) of algorithms can be used to minimize the Graph Edit Distance cost? Taking into account these problems, the aim of the article is manifold. On the one hand, we aim to study how many labellings can be found changing the Graph Edit Distance constants K_e and K_n. In addition, we aim to determine how these labellings are distributed in the (K_n, K_v) space and derive an algorithm to compute them. On the other hand, we aim to give an orientation about how optimal is the Graduated Assignment [11] algorithm applied to the minimization of the Graph Edit Distance and give some orientation if it can be used for the type of learning we desire.

2   Basic Notions

Attributed Graph [12]: let Δ_v and Δ_e denote the domains of possible values for attributed vertices and arcs, respectively. An attributed graph AG over $(\Delta_v$ and $\Delta_e)$ is defined by a tuple $G = (\Sigma_v, \Sigma_e, \gamma_v, \gamma_e)$, where $\Sigma_v = \{v_k \mid k = 1, ..., R\}$ is the set of vertices (or nodes), $\Sigma_e \in \{e_{ij} \mid i, j \in \{1, ..., R\}, i \neq j\}$ is the set of arcs (or edges) and $\gamma_v: \Sigma_v \rightarrow \Delta_v$ and $\gamma_e: \Sigma_e \rightarrow \Delta_e$ assign attribute values to vertices and arcs respectively. In case it is required, any AG can be extended with null nodes [12]. A null node is a special node which has special attribute $\emptyset \in \Delta_v$. In the special case where edges do not have attributes, a binary attribute indicating the existence of the edge can be implicitly deduced:

$$\gamma_e(e_{ij}) = \begin{cases} 1 & if\ e_{ij} \in \Sigma_e \\ 0 & else \end{cases} \tag{1}$$

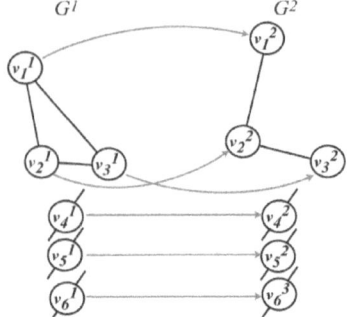

	v_1^2	v_2^2	v_3^2	v_4^2	v_5^2	v_6^2
v_1^1	1	0	0	0	0	0
v_2^1	0	1	0	0	0	0
v_3^1	0	0	1	0	0	0
v_4^1	0	0	0	1	0	0
v_5^1	0	0	0	0	1	0
v_6^1	0	0	0	0	0	1

Fig. 3.a. Example of two graphs[1] and a bijection between them

Fig. 3.b. Bijection of Fig. 3.a. in matrix representation

Bijection between graphs. Let $G^p = (\Sigma_v^p, \Sigma_e^p, \gamma_v^p, \gamma_e^p)$ and $G^q = (\Sigma_v^q, \Sigma_e^q, \gamma_v^q, \gamma_e^q)$ be two attributed graphs of R nodes. If the selected attributed graphs initially have different order or it is desired to allow some extra null-to-vertex labellings, G^p and G^q can be extended with any number of null nodes. Furthermore, let T be a set of all possible bijections between two vertex sets Σ_v. The bijection $f^{p,q}: \Sigma_v^p \to \Sigma_v^q, f^{p,q} \in T$, assigns each vertex of G^p to only one vertex of G^q. There is no need to define the arc bijection, denoted by $f_e^{p,q}$, since arcs are mapped accordingly to the bijection of their terminal nodes. In other words:

$$f^{p,q}(v_a^p) = v_i^q \wedge f^{p,q}(v_b^p) = v_j^q \Rightarrow f_e^{p,q}(e_{ab}^p) = e_{ij}^q \qquad (2)$$

Each bijection can be interpreted as a permutation matrix where rows represent nodes of the first graph and columns represent nodes of the second graph.

Fig. 3.a shows two graphs and a bijection between graph nodes G^1 and G^2. Graphs are composed by 6 nodes (three initial nodes and 3 added null nodes). Fig. 3.b shows the same bijection than Fig. 3.a in matrix form.

Graph Edit Distance between two graphs [1]. Given two graphs, G^1 and G^2 and a bijection between their nodes $f^{1,2}$, the Graph Edit Cost between them is defined as the cost of all edit operations that must be performed to transform G^1 to G^2. Three types of edit operations are defined for nodes and three for edges. These operations, for nodes, are node insertion, node deletion and node substitution. Usually, node insertions and node deletions penalized with a constant cost, that we name K_n. Node substitutions are usually computed as the Euclidean distance of node attributes. In this way the cost of a node assignation is given by:

$$C_{a,i}^{G^1,G^2} = \begin{cases} K_n & \text{if } \gamma_v(v_a^{G^1}) = \emptyset \oplus \gamma_v(v_i^{G^2}) = \emptyset & (Ins.\ and\ Del.) \\ 0 & \text{if } \gamma_v(v_a^{G^1}) = \emptyset \wedge \gamma_v(v_i^{G^2}) = \emptyset & (Subs.\ null\ node) \\ Euclid\left(\gamma_v(v_a^{G^1}), \gamma_v(v_i^{G^2})\right) & else & (Substitution) \end{cases} \qquad (3)$$

---

[1] Graphs G^1 and G^2 correspond to graphs one and three of class 7 of Letter (high distortion) dataset [16].

Edge assignation cost is defined equivalently and given by:

$$C^{G^1,G^2}_{e^{G^1}_{a,b},e^{G^2}_{i,j}} = \begin{cases} K_e & if \; \gamma_e(e^{G^1}_{ab}) = \emptyset \oplus \gamma_e(e^{G^2}_{ij}) = \emptyset & (Ins. \; and \; Del.) \\ 0 & if \; \gamma_e(e^{G^1}_{ab}) = \emptyset \wedge \gamma_e(e^{G^2}_{ij}) = \emptyset & (Subs. \; null \; node) \\ Euclid\left(\gamma_e(e^{G^1}_{ab}), \gamma_e(e^{G^2}_{ij})\right) & else & (Substitution) \end{cases} \quad (4)$$

Using costs defined in (3) and (4) Graph Edit Cost of two graphs is defined as:

$$EditCost(G^1, G^2, f^{G^1,G^2}, K_n, K_e) = \sum_{a=1}^{R} \sum_{b=1}^{R} C^{G^1,G^2}_{e^{G^1}_{ab},e^{G^2}_{ij}} + \sum_{a=1}^{R} C^{G^1,G^2}_{a,i} \quad (5)$$

Where $i = f^{G^1,G^2}(a)$ and $j = f^{G^1,G^2}(b)$

Finally the Graph Edit Distance is defined as the minimum cost under any bijection in T:

$$EditDistance(G^1, G^2, K_n, K_e) \\ = \underset{f^{G_1,G_2} \in T}{argmin} EditCost(G^1, G^2, f^{G^1,G^2}, K_n, K_e) \quad (6)$$

Using this definition, Graph Edit Distance depends specifically on K_n and K_e values.

3 Exploring the Labelling Space

It can be almost inferred from [13] and it is proven in [14] that given two graphs G^1, G^2 several values for K_n and K_e of Graph Edit Distance constants minimizes at the same bijection between G^1 and G^2. However, the final Graph Edit Cost will be different even the bijection that minimizes the Graph Edit Distance is the same. This is because Graph Edit Cost depends linearly on the edit distance constants. In this article, we are interested in explore how K_n and K_e values that minimize at the same bijection are distributed. To that aim, we define, given two graphs G^1 and G^2, function $\Omega^{G^1,G^2} : \mathbb{R}^2 \to T$, which assigns each pair of $\{K_n, K_e\}$ values to a graph bijection that minimizes the Graph Edit Distance. Moreover, we define the discrete labelling space as $\Omega^{G^1,G^2}_{disc} : \{K^{set}_n, K^{set}_e\} \to T$, where $K^{set}_n = \{K^{SetStart}_n, K^{SetStart}_n + inc_n, K^{SetStart}_n + 2 * inc_n, ..., K^{SetEnd}_n\}$. K^{set}_e is defined analogously.

Proposition 1. Given two graphs G^1 and G^2, any region in Ω^{G^1,G^2}, where $\{K_n, K_e\}$ are assigned to the same bijection $t' \in T$ form a convex polygon.

Demonstration: Given two graphs G^1 and G^2 and the labelling f^{G^1,G^2} used to compute the Graph Edit Distance, combining (3), (4) and (5), we can compute the Graph Edit Cost as:

$$EditCost(G_1, G_2, f^{G_1,G_2}) = \\ = K_n(\#nIns^f + \#nDel^f) + nSubs^f + K_e(\#eIns^f + \#eDel^f) \quad (7)$$

Note that for f^{G^1,G^2} to be optimal given concrete values for K_n and K_e. The following system of inequalities must hold:

$$K_n(\#nIns^f + \#nDel^f) + nSubs^f + K_e(\#eIns^f + \#eDel^f)$$
$$<$$
$$K_n(\#nIns^{f'} + \#nDel^{f'}) + nSubs^{f'} + K_e(\#eIns^{f'} + \#eDel^{f'})$$
$$\forall f' \in (T - f)$$

(8)

Each of the above inequalities divides the space into two parts by a linear equation. It is known that the intersection of any set of linear inequalities is a convex polygon [15]. Consequently, each optimal labelling just appears into a single convex polygon. Note that the above formulation allows to divide the labelling space into convex polygons, each of which corresponding to a single optimal labelling. Observe that it is possible that (8) produces an empty result, in this case the tested labelling is never optimal given any pair $\{K_n, K_e\}$.

Equations given by (8) can help to compute all optimal labelling. However, the number of equations that must be computed is factorial with respect to the order of the graphs. Next section provides an algorithm that using **Proposition 1** helps on computing the discrete labelling space $\Omega_{disc}^{G^1,G^2}$.

Fig. 4.b shows the labelling space for graphs in Fig. 4.a. Fig. 4.b represents Ω^{G^1,G^2} for values of K_n and K_e in the range [0,0.6]. Each region Rx represents a different labelling. Two of these labelling, $R2$ and $R3$, are represented in Fig. 4.a.

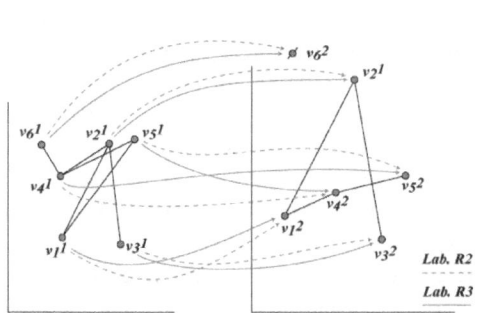
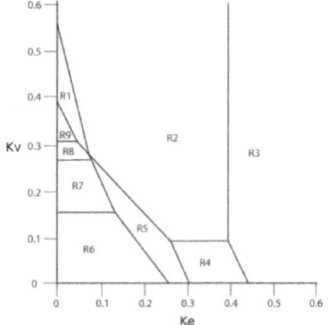

Fig. 4.a. Two optimal labellings taking different K_n and K_e values

Fig. 4.b. Labelling space given graph of Fig. 4.a

4   Computing the Discrete Labelling Space Using an Optimal Graph Matching Algorithm

With the aim of applying Adaptive Learning methodology to learn K_n and K_e constants of Graph Edit Distance, we present an algorithm which computes the discrete labelling space in an optimal way. The method is addressed to compute for each K_n

value the least possible number of optimal graph matching computations for all K_e values and induce the others using Proposition 1. To this aim, we define, given a pair of graphs and an optimal labelling f, computed with concrete values for K_n' and K_e', a range of values for K_e, lower than K_e', where the labelling f will lose its optimality. We will name this values K_e^{min} and K_e^{MAX}. $K_e^{min} < K_e^{MAX}$. K_e^{min} represents a K_e value where we certainly know that f will not be anymore optimal, and K_e^{MAX} indicates the K_e value from where it is possible that f loses its optimality. Once these ranges are computed, a dichotomic search is performed to locate the exact K_e value where labelling f loses optimality. Once the new labelling is located we continue the same process until K_e equals 0 or K_e^{min} and K_e^{MAX} indicates that the current labelling cannot lose its optimality. An outline of the algorithm is listed below.

```
Function ComputeLabelingSpaceGrid(G1, G2, maxKn, maxKe, interval)
    labelingSpaceGrid(*,*) = -1;
    for (currentKn=0 ; currentKn<=maxKn ; currentKn+=interval){
        currKe = maxKe;
        while (currKe > 0){
            currLab = optimalLabeling(G1,G2,Kn,currKe);
            labelingSpaceGrid(index(Kn),index(currKe)) = currLab;
            Kₑᵐⁱⁿ = comMinRange(G1,G2,Kn,currKe,currLab);
            KₑᴹᴬˣX = comMaxRange(G1,G2,Kn,currKe,currLab);
            if ((maxRange-minRange)>0){
                [currLab,currKe]=
                doDicotomicSearchOfBorder(G1,G2,currLab,minRange,maxRange);
                labelingSpaceGrid(index(Kn),index(currKe)) = currLab;
            }
        }
    }

    labelingSpaceGrid=fillNotComputedLabelings(labelingSpaceGrid);
End ComputeLabelingSpaceGrid returns labelingSpaceGrid
```

We know, using Proposition 1, that if two points of $\Omega_{disc}^{G^1,G^2}$ minimize at the same bijection, any bijection between them must be the same. Using this property of convex polygons `fillNotComputedLabelings` induces the rest of the grid labellings. The computation of values K_e^{min} and K_e^{MAX} is described in next sections.

4.1 Computing K_e^{min}

Taking into account that the algorithm outlined above traverses from high to low K_e values and under the assumption that contiguous labelling in $\Omega_{disc}^{G^1,G^2}$ should be somehow similar we propose to compute K_e^{min} using the following procedure. Given the initial optimal labelling f (computed using (K_n', K_e')) to compute K_e^{min} we first generate several offspring ($f_{1..LAST}$) of this initial labelling f by producing small modifications. We proceed by computing for each child f_i the $K_e^{f_i}$ value where this new bijection will gain the optimality. After computing all $K_e^{f_{1..LAST}}$ we take as K_e^{min} the maximum of $K_e^{f_{1..LAST}}$. Note that for K_e values lower K_e^{min}, f will lose its optimality, so we can consider that value a lower bound of K_e where f labelling will change in $\Omega_{disc}^{G^1,G^2}$ space. Example 1 shows the process we propose to obtain K_e^{min}.

	v_1^2	v_2^2	v_3^2	v_4^2	v_5^2	v_6^2
v_1^1	0	-Inf	-Inf	0.04	0.04	0.04
v_2^1	-Inf	0	-Inf	0.02	0.02	0.02
v_3^1	-Inf	-Inf	0	-0.10	-0.10	-0.10
v_4^1	0.04	0.02	-0.10	0	0	0
v_5^1	0.04	0.02	-0.10	0	0	0
v_6^1	0.04	0.02	-0.10	0	0	0

	v_1^2	v_2^2 v_4^2	v_4^2	v_5^2	v_6^2
v_1^1	1 0	0	0	0 1	0 0
v_2^1	0	1	0 0	0	0
v_3^1	0	0	1	0	0 0
v_4^1	0 1	0	0	1 0	0 0
v_5^1	0 0	0	0 0	1	0
v_6^1	0	0	0	0	0 1

Fig. 5.a. Data matrix for computing K_e^{min} **Fig. 5.b.** Bijection to compute value (v_1^1, v_4^2)

Example 1. Given the graphs in Fig. 3.a. and the labelling f in Fig. 3.b. which is optimal under the values $K_n = .29$ and $K_e = .6$, we compute offspring by performing a single node modification given the initial labelling f. A child of the initial labelling f is shown in Fig. 5.b, where assignation (v_1^1, v_1^2) has been changed to (v_1^1, v_4^2). Note that with this alteration the bijection we obtain is not correct. We must fix the bijection by assigning the doubled assigned node to the empty space that the imposed assignation leaves, in this concrete case assignation (v_4^1, v_4^2) goes to (v_4^1, v_1^2).

Given this new bijection f', child of the initial labelling, we compute K_e that makes the initial labelling lose its optimality in favour of f':

$$1.63 + K_e(2) < 1.48 + K_e(6) \rightarrow K_e < \frac{1.63 - 1.48}{6 - 2} = 0.04 \tag{9}$$

This value is placed in matrix of Fig. 5.a. at cell (v_1^1, v_4^2). The other values in matrix of Fig. 5.a are computed using the same procedure.

We take as K_e^{min} the maximum value of matrix in Fig. 5.a, in this case $K_e^{min} = 0.04$.

4.2 Computing K_e^{MAX}

Opposite to the definition of K_e^{min}, K_e^{MAX} corresponds to the highest value, lower than the current K_e, for which the current labelling can lose optimality. To optimally compute K_e^{MAX} we would need to know the labelling to which the current labelling would change to. Due to this would be an NP problem, we propose to use the best heuristic assignation possible to compute an upper bound of the K_e^{MAX} value. The first step to compute this value is to define an admissible heuristic for the minimum edit cost of node to node assignation. Considering that attributes of edges are computed using (1). The heuristic function we propose is:

$$heurNtoNAss\left(v_a^{G1}, v_i^{G2}\right) = C_{a,i}^{G_1,G_2} + K_e\left(\left|\deg\left(v_a^{G_1}\right) - \deg\left(v_i^{G_2}\right)\right|\right) \tag{10}$$

Using heuristic defined in (10) we are able to compute the cost of each node to node assignation considering K_e as an unknown. An example using graphs in Fig. 3.a is given in Fig. 6.

	v_1^2	v_2^2	v_3^2	v_4^2	v_5^2	v_6^2
v_1^1	0.73+Ke	2.76+Ke	3.20+Ke	0.29+ke	0.29+ke	0.29+ke
v_2^1	1.66	0.71	2.05	0.29+2ke	0.29+2ke	0.29+2ke
v_3^1	2.71+ke	1.28+ke	0.19+ke	0.29+2ke	0.29+ke	0.29+ke
v_4^1	0.29+2ke	0.29+2ke	0.29+2ke	0	0	0
v_5^1	0.29+2ke	0.29+2ke	0.29+2ke	0	0	0
v_6^1	0.29+2ke	0.29+2ke	0.29+2ke	0	0	0

Fig. 6. Heuristics given graphs in Fig. 3.a

Once all values are computed, we perform the best possible heuristic assignation. Recall that some values depend on K_e so depending on the values where we evaluate them we can have several minimums per row. The minimums for matrix in Fig. 6 are surrounded. In this case, row v_2^1 has two minimums. We propose to define K_e^{MAX} as the maximum K_e, lower than the current K_e, given all possible assignations. In the case of Fig. 6, two are the possible minimum assignations:

$$\{(v_1^1 \rightarrow v_4^2), (v_2^1 \rightarrow v_2^2), (v_3^1 \rightarrow v_3^2), (v_4^1 \rightarrow v_4^2), (v_5^1 \rightarrow v_4^2), (v_6^1 \rightarrow v_4^2)\}$$
$$\rightarrow 0.29 + K_e + 0.71 + 0.19 + K_e + 0 + 0 + 0$$
$$= 1.19 + 2K_e \rightarrow 1.63 + 2K_e > 1.19 + 2K_e \rightarrow K_e \qquad (11)$$
$$< Inf$$

$$\{(v_1^1 \rightarrow v_4^2), (v_2^1 \rightarrow v_4^2), (v_3^1 \rightarrow v_3^2), (v_4^1 \rightarrow v_4^2), (v_5^1 \rightarrow v_4^2), (v_6^1 \rightarrow v_4^2)\}$$
$$\rightarrow 0.29 + K_e + 0.29 + 2K_e + 0.19 + K_e + 0 + 0 + 0$$
$$= 0.77 + 4K_e \rightarrow 1.63 + 2K_e > 1.19 + 2K_e \qquad (12)$$

$$K_e < 0.42$$

In this case the maximum value is Inf, however the value is greater than the rent $K_e = .6$, so we take $K_e^{MAX} = 0.42$.

5 Graduated Assignment Algorithm for Computing the Grid

The Graduated Assignment is possibly the most commonly known algorithm for graph matching. We aim to evaluate how well can compute the discrete labelling space in comparison to an optimal algorithm. We think this comparison is necessary in order to use the graduated assignment algorithm for learning purposes.

In the original article [11], section 2.5 suggests the following function to optimize when graphs are attributed:

$$E_{awg}(M) = -\sum_{a=1}^{R} \sum_{i=1}^{R} \sum_{b=1}^{R} \sum_{j=1}^{R} M_{a,i} M_{b,j} Cp_{e_{ab}^{G1}, e_{ij}^{G2}}^{G_1, G_2} - \alpha \sum_{a=1}^{R} \sum_{i=1}^{R} M_{a,i} Cp_{a,i}^{G_1, G_2} \qquad (13)$$

Where $Cp_{e_{ab}^{G1}, e_{ij}^{G2}}^{G_1, G_2}$ indicates the compatibility of e_{ab}^{G1} to e_{ij}^{G2} and $Cp_{a,i}^{G_1, G_2}$ indicates the compatibility of node v_a^{G1} to v_i^{G2}. If we take a closer look at (13), we see that is very similar to the Graph Edit Cost (5). The principal difference is that the C in (5) represents a cost and Cp in (13) represents compatibility. To adapt the Graduated Assignment to optimize the Graph Edit Cost function we perform the following adaptation.

$$Cp_{e_{ab}^{G1}, e_{ij}^{G2}}^{G_1, G_2} = \left.1\middle/1 + C_{e_{ab}^{G1}, e_{ij}^{G2}}^{G_1, G_2}\right. \quad \text{and} \quad Cp_{a,i}^{G_1, G_2} = \left.1\middle/1 + C_{a,i}^{G_1, G_2}\right. \qquad (14)$$

6 Experiments

Three experiments are performed related to the aim of the article. The first experiment is addressed to evaluate how many optimal distance computations can be saved when using the algorithm presented in Section 4. The second experiment is addressed to analyze how different are the labelling found by the Optimal graph matching algorithm and the Graduated Assignment presented in Section 5. The last experiment is addressed to count the amount of different labellings that can be obtained varying the Graph Edit Distance costs. Experiments are performed using the Letter dataset (high distortion) created at the University of Bern [16]. It is composed of 15 classes and 150 graphs per class representing different letters of the Roman alphabet.

For the first experiment, we selected randomly 128 pairs of graph per class and we computed $\Omega_{disc}^{G^\circ,G^\circ}$ (Section 3) using two different manners. The first, considered the ground truth, computes $\Omega_{disc}^{G^\circ,G^\circ}$ by computing the optimal labelling for each pair of $\{K_n, K_e\}$ values. The algorithm used was the A*. The second manner computes $\Omega_{disc}^{G^\circ,G^\circ}$ using the methodology presented in Section 4. Again, the optimal labelling is found using the A* algorithm. Table 1, row 3, presents the mean number of distance computations performed when using algorithm of Section 4 with respect to compute all possible optimal labellings. We see that 90% of computations can be saved using the presented algorithm.

For the second experiment, we randomly chose 128 pairs of graphs per class. With each pair of graphs, we compute the discrete labelling space using the algorithm of Section 4 and the Graduated Assignment described in Section 5. Due to the Graduated Assignment algorithm is not optimal, we did not apply the algorithm of Section 4 due to we cannot ensure the results would be correct. To compute the discrete labelling space using the Graduated Assignment, we apply the algorithm to each pair of $\{K_n, K_e\}$ values. The last 4 rows of Table 1 show the results. Results are divided into 4 quadrants:

Analyzing the results, we see that most of errors are located in Q1 due to is the region where most of the different labellings are located, see as example Fig. 4.b. From these results, we conclude that the Graduated assignment has better performance with large $\{K_n, K_e\}$ values. The Graduated assignment algorithm tends to confuse regions

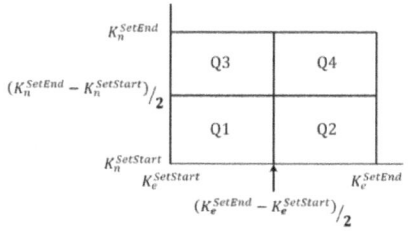

Fig. 7. Quadrant definition

Table 1. Results on Letter Dataset

	Mean of Experiments
#labelings A*	6,5
#labelings GA	10,1
% of done computation	10,11%
Q1	27,34%
Q2	23,23%
Q3	13,12%
Q4	17,77%

% of wrong labelings

in $\Omega_{disc}^{G^\circ,G^\circ}$ if these regions are small. However, when regions are of a considerable size, the Graduated assignment gives better results; this is clearly seen in Q3 and Q4.

Finally, for the last experiment we took again 128 randomly chosen pairs of graphs per class. Using the discrete labelling spaces computed by the optimal graph matching algorithm and the Graduated Assignment, we count and analyze the labellings found. Row one and two of Table 1 show the number of labellings found with both algorithms. We see that the Graduated Assignment generates more different labellings than the optimal algorithm. In addition, the number of labelling in both cases is very low. Graphs have a mean of 10 nodes, 5 real nodes and 5 null nodes which should give an amount of $\sum_{i=1}^{5} V_5^i * i!$ possible labellings which is much greater than the number of labellings we can obtain modifying $\{K_n, K_e\}$ values.

All experiments are performed using $K_n^{SetStart} = K_e^{SetStart} = 0$, $inc_n = inc_e = .01$ and $K_n^{SetEnd} = K_e^{SetEnd} = .8$.

7 Conclusions

The Graph Edit Distance is one of the most used Attributed Graph distance. Graph Edit Distance function clearly depends on two application dependant constants, node insertion/deletion cost K_n and edge insertion/deletion cost K_e. The article presents a study of the applicability of Adaptive Learning methodology to learn these constants. To this aim, results on the influence of Graph Edit Distance constants over labellings which minimizes the Graph Edit Distance are presented. Theoretical development confirms that for some different Graph Edit Distance constants, Graph Edit Distance can minimize at the same bijection. Moreover, it is mathematically proved that $\{K_n, K_e\}$ points that minimize at the same bijection form a convex polygon. In addition to the most theoretical concepts and propositions, the article presents an algorithm that using the convex polygon property computes all optimal bijections for a discrete set of $\{K_n, K_e\}$ values. The presented algorithm performs the task 90% faster than the trivial iterative process. Besides, with the objective of computing the labellings for a discrete set of $\{K_n, K_e\}$ values using a fast and suboptimal graph matching algorithm, the article presents a performance analysis of the Graduated Assignment algorithm adapted to minimize the Graph Edit Distance function. Results show that the algorithm commits more errors when $\{K_n, K_e\}$ values are small. Analyzing the regions that the Graduated Assignment produces, we see that these regions do not fulfil the convex polygon rule. With respect to the number of labellings that can be obtained modifying $\{K_n, K_e\}$ values, we see that it is very low. This small amount of possible labellings makes statistically difficult to find $\{K_n, K_e\}$ values that exactly matches expert decisions in a supervised learning system.

References

1. Sanfeliu, A., Fu, K.: A Distance measure between attributed relational graphs for pattern recognition. IEEE Transactions on Systems, Man and Cybernetics 13(3), 353–362 (1983)
2. Bunke, H., Shearer, K.: A graph distance metric based on the maximal common subgraph. Pattern Recognition Letters 19(3-4), 255–259 (1998)
3. Shapiro, L., Haralick, R.M.: A metric for comparing relational descriptions. IEEE Transactions on Pattern Analysis and Machine Intelligence 7, 90–94 (1985)
4. Riesen, K., Bunke, H.: Approximate graph edit distance computation by means of bipartite graph matching. Image Vision Comput. 27(7), 950–959 (2009)
5. Neuhaus, M., Bunke, H.: Bridging the Gap Between Graph Edit Distance and Kernel Machines. Ser. in Machine Perception and Art. Intelligence, ISBN 9812708170 (2007)
6. Conte, D., Guidobaldi, C., Sansone, C.: A Comparison of Three Maximum Common Subgraph Algorithms on a Large Database of Labeled Graphs. In: Hancock, E., Vento, M. (eds.) GbRPR 2003. LNCS, vol. 2726, pp. 130–141. Springer, Heidelberg (2003)

7. Gao, X., Xiao, B., Tao, D., Li, X.: A survey of graph edit distance. Pattern Analysis & Applications 13(1), 113–129 (2010)
8. Neuhaus, M., Bunke, H.: Automatic learning of costs functions for graph edit distance. Information Sciences 177(1), 239–247 (2007)
9. Toselli, A.H., Romero, V., Pastor, M., Vidal, E.: Multimodal interactive transcription of text images. Pattern Recognition 43(5), 1814–1825 (2010)
10. Serratosa, F., Solé-Ribalta, A., Cortés, X.: Automatic Learning of Edit Costs based on Interactive & Adaptive Graph Recognition. In: Jiang, X., Ferrer, M., Torsello, A. (eds.) GbRPR 2011. LNCS, vol. 6658, pp. 152–163. Springer, Heidelberg (2011)
11. Gold, S., Rangarajan, A.: A Graduated Assignment Algorithm for Graph Matching. IEEE Transactions on Pattern Analysis and Machine Intelligence 18(4), 377–388 (1996)
12. Wong, A.K.C., et al.: Entropy and distance of random graphs with application to structural pattern recognition. IEEE T. on Pat. An. and Mach. Intell. 7, 599–609 (1985)
13. Rice, S.V., Bunke, H., Nartker, T.A.: Classes of cost functions for string edit distance. Algorithmica 18(2), 271–280 (1997)
14. Bunke, H.: Error Correcting Graph Matching: On the Influence of the Underlying Cost Function. IEEE TPAMI 21(9), 917–922 (1999)
15. Grünbaum, B.: Convex Polytopes, 2nd edn. ISBN 0387404090 (2003)
16. Riesen, K., Bunke, H.: IAM graph database repository for graph based pattern recognition and machine learning. In: da Vitoria Lobo, N., Kasparis, T., Roli, F., Kwok, J.T., Georgiopoulos, M., Anagnostopoulos, G.C., Loog, M. (eds.) S+SSPR 2008. LNCS, vol. 5342, pp. 287–297. Springer, Heidelberg (2008)

Graph Matching Based on Dot Product Representation of Graphs

Jin Tang, Bo Jiang, and Bin Luo

School of Computer Science and Technology,
Anhui University, Hefei 230039, Anhui, China
zeyiabc@163.com, ahhftang@gmail.com

Abstract. This paper proposes an efficient algorithm for inexact graph matching. Our main contribution is that we render the graph matching process to a way of recovery missing data based on dot product representation of graph (DPRG). We commence by building an association graph using the nodes in graphs with high matching probabilities, and treat the correspondences between unmatched nodes as missing data in association graph. Then, we recover correspondence matches using dot product representation of graphs with missing data. Promising experimental results on both synthetic and real-world data show the effectiveness of our graph matching method.

Keywords: Graph matching; Random dot product graph; Association graph.

1 Introduction

Graph matching is the process of finding correspondences between the nodes and the edges of two graphs that satisfies some constrains. It is pivotal important in high-level vision. Unfortunately, in many applications, the observed graphs are subject to deformations due to noise, intrinsic variability of the patterns. So graph matching is invariably approached by inexact means [1], [2].

There is a considerable literature on the problem of inexact graph matching. Broadly speaking, the first category method is based on tree search techniques [1], [3], [4], [5], [6]. Relaxation and optimization have also been used in graph matching [7], [8], [9], [10]. Recently, one of the most popular approaches to the graph matching problem has been to use spectral method [11], [12], [13], [14], [15], [16]. For instance, Umeyama [14] has shown an eigendecomposition to recover the permutation matrix. Scott and Longuet-Higgins [12] used spectral method for correspondence analysis and align point-sets by performing singular value decomposition on a point association weight matrix between different images. Shapiro and Brady [13] developed an extension of the Scott and Longuet-Higgins method, in which point sets are matched by comparing the eigenvectors of the Gaussian point-proximity matrix. However, these methods often required graphs of equal size. In order to overcome the problems that match graphs with different numbers of the nodes and weighted graphs with weighted errors, one way is developed by Terry Caelli and Serhiy Kosinov [16]. They extend these methods by clustering algorithm. Additionally, Bai and Hancock [17] achieve graph matching by using spectral embedding and semidefinite programming. They firstly embed the nodes

X. Jiang, M. Ferrer, and A. Torsello (Eds.): GbRPR 2011, LNCS 6658, pp. 175–184, 2011.
© Springer-Verlag Berlin Heidelberg 2011

of a graph in Euclidean space and then use semidefinite programming to find point correspondences. Another way which is proposed by Luo and Hancock is to cast the problem of recovering correspondences in a statistical setting [18]. In this case the nodes of the input graph play the role of observed data while the nodes of the model graph act as hidden random variables. The matching is then found by EM algorithm. For computer vision area such as shape matching, image registration, there are some other special techniques to inexact matching [23], [24].

From Luo's [18] EM graph matching algorithm, we draw the following observation: we can render correspondences between two graphs to missing information and use statistic and matrix method to recovery the missing information. Based on this observation, we develop dot product representation of graph (DPRG) and cast matching of graphs into a missing data restoration based on DPRG.

Recently, DPRG [19], [20], [21], [22] is successfully used for graph embedding, which in turn facilitates clustering and image segmentation process. Given a (weighted) graph, DPRG produces a mapping of vertices to vectors in R^d; vertex i maps to vector $\mathbf{x}_i$. The desired property is that dot product of $\mathbf{x}_i$ and $\mathbf{x}_j$ should approximate the weight of edge ij. Furthermore, by introducing an iterative algorithm, Scheinerman [21] extended DPRG to deal with the graphs with missing value, i.e., the weights for some edges in the graph are unknown. In fact, this method put forward a way of recovery missing value for the graph with missing data. We follow this idea and expand it to graph matching. By treating the correspondences between the two graphs to be matched as the missing data, we proposed an iterative algorithm which recovers the correspondences in a similar manner with Scheinerman's.

2   Dot Product Representation of Graphs (DPRG)

2.1   Dot Product Representation of Graphs

We briefly recall the main idea of random dot product graphs [19] [20] [21] [22]. Given a set of vectors $\{\mathbf{x}_i\}_{i=1}^n$ where $\mathbf{x}_i \in R^d$ is a d-dimensional vector. For $i \neq j$, we assume that $\mathbf{x}_i \cdot \mathbf{x}_j \in [0,1]$ (we can normalize $\mathbf{x}_i, \mathbf{x}_j$ to make $\mathbf{x}_i \cdot \mathbf{x}_j \in [0,1]$), then a graph $G = (V, E)$ is generated at random. The vertex set V of this graph is $\{1, 2 \dots n\}$. For $i \neq j$, the probability of the edge connecting nodes v_i and v_j is set to $\mathbf{x}_i \cdot \mathbf{x}_j$. Let G_n denotes the set of all simple graphs with vertex set $\{1, 2 \dots n\}$, the probability measure $P_X(G)$ on G_n is defined as follow

$$P_X(G) = \left(\prod_{i<j, ij \in E} \mathbf{x}_i \cdot \mathbf{x}_j \right) \times \left(\prod_{i<j, ij \notin E} (1 - \mathbf{x}_i \cdot \mathbf{x}_j) \right). \tag{1}$$

Rather than studied on generating graphs at random from a set of vectors. Scheinerman [21] focused on the issue of finding the vectors $X = [\mathbf{x}_1, \mathbf{x}_2 \dots \mathbf{x}_n]$ that "best" model a given graph G in advance. The optimal solution of the vectors X is the one which maximizes the likelihood, i.e. $\arg\max P_X(G)$.

Let $A = (a_{ij})_{n \times n}$ be the adjacency matrix where a_{ij} is the edge weight with nonnegative value between nodes v_i and v_j. Then, the vectors X is sought to satisfy $\mathbf{x}_i \cdot \mathbf{x}_j \approx a_{ij}$ where $i \neq j$, i.e.

$$\min_X f_A(X) = \left\| X^T X - A - I \circ (X^T X) \right\|^2 \qquad (2)$$

where I is the $n \times n$ identity matrix, $\circ$ is the Haddamard product and $\|\cdot\|$ is the Frobenius norm. An iterant algorithm [21] is designed to find the "best" vectors X.

2.2 Dot Product Representation of Graphs with Missing Data

The problem Eq (2) takes information about the weight of all edges in the graph. However, it is possible that for some pairs of vertices, we simply do not have information as to whether there is an edge jointing those vertices and, if so, what its weight is. In the following, we introduce the idea that models these missing data graphs based on DPRG. The problem can be reformulated as follow

$$\min_X f_{A,M}(X) = \left\| X^T X - A - M \circ (X^T X) \right\|^2 \qquad (3)$$

where M is a labeling matrix with element valued 1 if the relationship between the two nodes v_i, v_j (weight of edge ij) is unknown or can not be determined, 0 otherwise. We call M as missing data label matrix (MDLM).

It should be noted that we can recover the missing value from DPRG (solution X). Let $X=[\mathbf{x}_1,\mathbf{x}_2,\ldots,\mathbf{x}_n]$ be the final solution of the Eq (3), the missing value for edge kl can be directly obtained from $\mathbf{x}_k \cdot \mathbf{x}_l$. As pointed by Scheinerman [21], the iterative algorithm (algorithm 1 in the following) can be used for DPRG with missing data by gradually eliminating the effect of the unknown or missing entries. Although the algorithm has been proved to converge to local optimum, it can give a unique solution to Eq. (3), because the initialization is definite.

Algorithm 1. Dot product representation of graph with missing data

Input:
— A nonnegative, symmetric, $n \times n$ matrix A
— A positive integer d, Missing data label matrix M
Output:
— A $d \times n$ matrix X
Step1: Let D be a zero matrix in order $n \times n$
Step2: Calculate the spectrum decomposition of matrix $A_M + D$. U is a $d \times n$ matrix builds on the first d eigenvectors. e is a diagonal matrix generating by the first d eigenvalues, where all negative eigenvalues are set to zero.
Step3: Calculate $X = U\sqrt{e}$ and $D = M \circ (X^T X)$
Step4: Repeat step 2 and step 3 until D converges.

3 Graph Matching Based on DPRG

In this section, we turn to formulate the graph matching problem in terms of finding the missing correspondences between unmatched nodes in the two graphs. We start from building an association graph using the nodes that have been matched already in both graphs, and treat the corresponding matches between unmatched nodes as missing data in association graph. Then the association graph is modeled using DPRG and the missing correspondences are then recovered by the iterative algorithm described in section 2.2.

3.1 Initialization

The initial correspondences can be the results from traditional matching algorithms, such as Umeyama [14], Bai method [17], Spectral embedding and so on. Our Spectral embedding matching algorithm can be described as following. We commence by solving Eq. (2) to embed the nodes of graphs in Euclidean space and then complete graph matching by aligning the points using Hungarian algorithm. The initial matching pairs are important for the following matching process. However, it is not very accurate and several of the corresponding pairs are likely to be incorrect. To make the matching more reliable, we introduce a consistency check process which is designed to remove these erroneous correspondences and obtain positive correspondences from initial result. Its principle is to enforce coherent spatial relationships of corresponding nodes between graphs. To do so, a compatibility matrix $C = (c_{ij})_{n \times n}$ is computed for each graph first. The element c_{ij} can simply be obtained from adjacency matrix A, i.e. $c_{ij} = a_{ij}$. Also, we can compute c_{ij} as the geodesic distance or Euclidean distance between two nodes of the graph in some special cases.

Algorithm 2. Consistency Check Algorithm

Input:
    — The initial correspondences I_{map}, Graph G_1 and G_2.
    — The number of positive correspondence nodes to be selected P_{num}.
Output:
    — Positive correspondence nodes set P_{set}, Positive correspondence mapping P_{map}.
Step1: Compute the compatibility matrices C_1, C_2 of graph G_1 and G_2 determined by
        correspondences I_{map}.
Step2: Initialization of P_{set} and P_{map}.
    1. Calculate the residual compatibility matrix $R=|C_1-C_2|$.
    2. Calculate the fitness of node i in G_1 as $f(i) = \dfrac{1}{d(i)}\sum_{j=1}^{|G_2|} R(i,j)$, where $d(i)$ is the
        degree of node i in G_1.
    3. Select the top P_{num} least fitness nodes as the initial positive correspondences
        P_{set} and hence the P_{map}.
Step3: Determine P_{set} and P_{map} iteratively.
    1. Re-calculate the fitness index of node i in G_1 as $f(i) = \dfrac{1}{d(i)}\sum_{j \in P_{set}} R(i,j)$, where
        P_{set} is the current positive correspondence node set.
    2. Select the top P_{num} least fitness nodes as the positive correspondences (P_{set}) and
        hence the P_{map} to update the ones in the last iteration.
    3. Go to 1, if P_{set} changed in the current iteration, Step 4 otherwise.
Step4: Output P_{set} and P_{map}.

3.2   Association Graph (AG)

With the positive correspondences (P_{map}) in hand, we turn our attention to build an association graph between the graphs to be matched. Let $P_{map}: I_1 \rightarrow I_2$ where I_1, I_2 are the label subsets of graph G_1 and G_2 respectively and $\overline{I_1}, \overline{I_2}$ be the complement of I_1, I_2. The adjacency matrices AM_1 and AM_2 of G_1 and G_2 are first normalized as follow

$$A_1 = 1 - \frac{AM_1}{\max_{i,j}(AM_1)}, A_2 = 1 - \frac{AM_2}{\max_{i,j}(AM_2)}$$

Then the normalized adjacency matrices are rearranged to form block matrices

$$A_1 = \begin{bmatrix} A_{c1} & B_1 \\ B_1^T & C_1 \end{bmatrix} \text{ and } A_2 = \begin{bmatrix} A_{c2} & B_2 \\ B_2^T & C_2 \end{bmatrix}$$

where $A_{c1}(i,j) = A_1(I_1(i), I_1(j)), i, j = 1 \cdots |I_1|$, $B_1(i,j) = A_1(I_1(i), \overline{I_1}(j)), i = 1 \cdots |I_1|, j = 1 \cdots |\overline{I_1}|$ and $C_1(i,j) = A_1(\overline{I_1}(i), \overline{I_1}(j)), i, j = 1 \cdots |\overline{I_1}|$, and A_{c2}, B_2, C_2 are similarly obtained. The adjacency matrix of our association graph (AG) is then taken the form

$$A_M = \begin{bmatrix} A_c & B_1 & B_2 \\ B_1^T & C_1 & O \\ B_2^T & O & C_2 \end{bmatrix} \tag{4}$$

where O is a zero matrix and $A_c = (A_{c1} + A_{c2})/2$.

The above association graph (AG) integrates both positive correspondences and two adjacency matrices at the same time. If we see AG as an incompletely observed network [19] in which the unobserved links consist of the corresponding relationship between nodes that are not yet matched, then we can use algorithm 1 to obtain the DPRG of AG. Moreover, if we assume i_0 belong to $\overline{I_1}$ and j_0 is its corresponding node in $\overline{I_2}$, then the vector $\mathbf{x}_{i0}$ will be similar to the vector $\mathbf{x}_{j0}$ ($\mathbf{x}_{i_0} \cdot \mathbf{x}_{j_0}$ will be larger) where $\mathbf{x}_{i_0}$ and $\mathbf{x}_{j_0}$ are from DPRG on AG. So we can see the missing value between two nodes in AG as the (similarity) matching probability for them in two graphs. In our paper, we predict these missing values using algorithm 1. In order to do so, we define missing data label matrix (MDLM) as follow.

$$MDLM = \begin{bmatrix} O & O & O \\ O & O & I \\ O & I & O \end{bmatrix} \tag{5}$$

where I is a unit matrix.

3.3   Missing Correspondences Recovery by DPRG

As discussed above, we see the missing values as the similarity between nodes and then can determine the correspondences based on Hungarian algorithm directly [23]. However, we don't determine the correspondences for all the unmatched nodes at the same time, because the unmatched nodes that have strong connections with nodes which have been already matched are most possibly determined robustly. So, we cast the recovery of correspondence matches for unmatched nodes in an iterative framework. Let P_{set} be the positive correspondence nodes set that contains the nodes which have been matched already in matching process and E_{set} be the unmatched node set in graph G_1. For node i_0 in E_{set}, we define the correlation strength RS between i_0 and P_{set} as $RS(i_0) = |E_0|/d(i_0)$ where $E_0 = \{e_{i_0 j} \mid A_1(i_0, j) \neq 0, j \in P_{set}\}$ and $d(i_0)$ is the node i_0 degree in graph G_1. In our iterative matching process, we only determine the correspondences for the nodes in E_{set} that have high correlation strength value every time. Let P_{map} denote the mapping that is relevant to the P_{set}, then our DPRG based graph matching can be described as follow.

Algorithm 3. Graph matching based on DPRG

Input:
— The initial correspondence I_{map}, Graph G_1 and G_2, threshold T.
— The number of positive correspondence nodes to be selected P_{num}.
Output:
— The correspondences E_{map} between graph G_1 and G_2.
Step1: Initialization. Calculate P_{map} using Algorithm 2.
Step2: Build an association graph (AG) model based on P_{map} and calculate the positive integer d by taking the integer value of $k \times |AG|$.
Step3: Find the correspondences or missing data using Algorithm 1.
Step4: Determine the correspondences for the nodes in the E_{set} that have high RS value using the missing data and Hungarian algorithm.
Step5: If the all nodes have their correspondences, go to step 6, otherwise update P_{map}, E_{set} and go to Step2.
Step6: Calculate fitness $f(i) = \dfrac{1}{d(i)} \sum_{j \in G_2} R(i, j)$ for every node in G_1. If $f(i) < T$, treat the node i as outlier node and delete it from E_{map}.

The last step guaranties that the algorithm can deal with graphs with some outliers and make the algorithm have the ability to handle with graphs with different sizes.

4   Experiments

We provide some experimental evaluation of the new matching technique. There are two aspects to this study. We commence with a sensitivity study on synthetic data. The aim is to evaluate how the method performs under controlled structural corruption. Then we evaluate the method on some real-world data. Both in synthetic data and real-world experiments we compare our method with some alternatives reported elsewhere in the literature. In our experiments, the parameter T chosen is $2 \times \text{mean}_i(f(i))$ where $f(i)$ is fitness value for node i described in step 6 in algorithm 3. The P_{num} is set to be integer value of $0.3 \times |G_1|$ and the k for calculating d is 0.3.

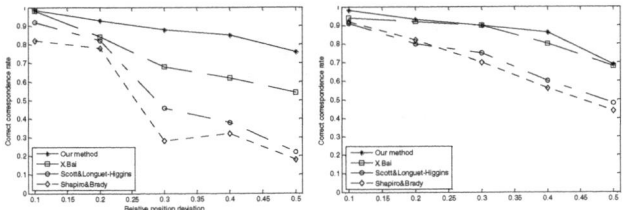

Fig. 1. Comparison results of the four methods on synthetic data

4.1 Synthetic Data Experiments

We commence with some synthetic data experiments. The aim is to evaluate how the new method works under controlled structural corruption and to compare it with some alternative methods. These alternatives are Bai [17], Shapiro and Brady [13] and Scott and Longuet Higgins [12] feature set matching algorithms. The last two methods use coordinate information for the feature points and do not incorporate the graph structural information. To construct our experiments, we have generated random points-sets and use the position of the points to generate a Delaunay graph. Firstly, we have fixed the number of points (50 in this experiment) and added Gaussian noise to the point positions then generated a Delaunay graph. The parameter of the noise process is the standard deviation of the positional jitter [18]. In our experiments, we express this parameter as a fraction of the average minimum distance between points. Fig. 1(left plot) shows the fraction of correct correspondences as a function of the noise standard deviation for different methods. We also evaluate the effect of structural noise. Here we have added a controlled fraction of additional nodes at random positions and recomputed the Delaunay triangulations. Fig. 1(right plot) shows the fraction of correct correspondences as a function of the fraction of added nodes. For every parameter, we perform our algorithm 30 times.

4.2 Real-World Data

In this section, we apply our matching method to three image sequences (YORK, CMU and MOVI). The sample images are shown in Fig. 2,5,7, and the detected corner features and their Delaunay triangulations are overlayed on the images.

We commence our real-world evaluation on YORK sequence. Fig. 3 shows some results obtained when we match the second image to some subsequent images in the sequence. The results are summarized in Fig. 4. We demonstrate our method with our spectral embedding which is the initial matching algorithm for our method. Fig. 4 shows that our method returns considerably better matches.

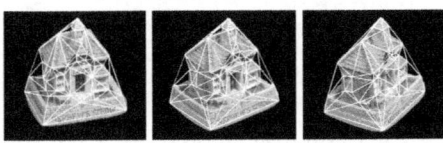

Fig. 2. Delaunay Graphs overlayed on the YORK house images

Fig. 3. Our algorithm for YORK house images

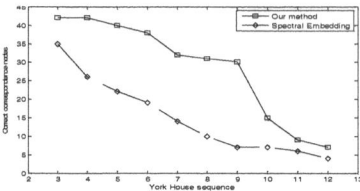

Fig. 4. Summary of experimental results for YORK House Sequence images

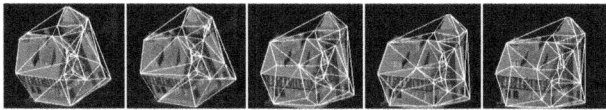

Fig. 5. Delaunay Graphs overlayed on the CMU house images

Fig. 6. Our algorithm for CMU house images

Table 1. Summary of two matching algorithms for the CMU House Sequence Images

Images	Points	methods	Correct correspondences	False correspondences	No correspondences
house1	30	EM	-	-	-
		DPRG	-	-	-
house2	32	EM	28	1	1
		DPRG	28	1	1
house3	31	EM	23	5	2
		DPRG	25	3	2
house4	30	EM	11	10	9
		DPRG	22	4	4
house5	30	EM	5	16	9
		DPRG	19	7	4

We then study our real-world evaluation of the method on CMU model house sequence. Fig. 6 shows the results obtained when we match the first image to each of the subsequent image in the sequence. We compare our method (DPRG) with EM method

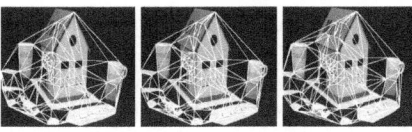

Fig. 7. Delaunay Graphs overlayed on the MOVI sample house images

Fig. 8. Our algorithm for MOVI house images

Table 2. Summary of Experimental results for the MOVI House Sequence Images

Images	Points	Correct correspondences	False correspondences	No correspondences
house1	140	-	-	-
house3	138	126	6	6
house5	142	114	19	7
house7	140	107	21	12

[18], and the results are summarized in table 1. From this result, our method performs better than EM method.

The final real-world example is furnished on MOVI sequence. Fig. 7 shows the some matching results in the sequence. The results are summarized in table 2.

5   Conclusions

An efficient graph matching algorithm is proposed in this paper. The main idea is to cast the graph matching process in a way of recovery missing data based on DPRG. We first build an association graph using the nodes that have been matched and treat the correspondences between unmatched nodes as missing data in association graph. Then, based on DPRG, we recover these correspondence matches. We have demonstrated the effectivity of the method on both synthetic and real-world data.

Acknowledgments. This research is supported in part by National Nature Science Foundation of China (61073116 &61003038), The Key Natural Science Project of Anhui Provincial Education Department (KJ2010A006 & KJ2009A145).

Reference

1. Sanfeliu, A., Fu, K.S.: A Distance Measure between Attributed Relational Graphs for Pattern Recognition. IEEE Trans. Syst. Man Cybernetics. 13(3), 353–362 (1983)
2. Shapiro, L.G., Haralick, R.M.: A Metric for Comparing Relational Descriptions. IEEE PAMI 7(1), 90–94 (1985)

3. Tsai, W.H., Fu, K.S.: Error-correcting isomorphisms of attributed relational graphs for pattern analysis. IEEE Trans. Syst. Man Cybernetics. 9, 757–768 (1979)
4. Eshera, M.A., Fu, K.S.: A similarity measure between attributed relational graphs for image analysis. In: Proc. 7th Int. Conf. Pattern Recognition, pp. 75–77 (1984)
5. Cordella, L.P., Foggia, P., Sansone, C., Vento, M.: An efficient algorithm for the inexact matching of ARG graphs using a contextual transformational model. In: Proc.13th Int. Conf. Pattern Recognition, pp. 180–184 (1996)
6. Llados, J., Marti, E., Villanueva, J.J.: Symbol recognition by error-tolerant sub-graph matching between region adjacency graphs. IEEE PAMI 23, 1137–1143 (2001)
7. Gold, S., Rangarajan, A.: A graduated assignment algorithm for graph matching. IEEE PAMI 18 (1996)
8. Christmas, W.J., Kittler, J., Petrou, M.: Structural matching in computer vision using probabilistic relaxation. IEEE PAMI 17, 749–764 (1995)
9. Myers, R., Wilson, R.C., Hancock, E.R.: Bayesian graph edit distance. IEEE PAMI 22, 628–635 (2000)
10. Huet, B., Hancock, E.R.: Shape recognition from large image libraries by inexact graph matching. Pattern Recognition Letter 20, 1259–1269 (1999)
11. Shokoufandeh, A., Dickinson, S., Siddiqi, K., Zucker, S.: Indexing using a spectral encoding of topological structure. In: CVPR, pp. 491–497 (1999)
12. Scott, G.L., Longuett-Higgins, H.C.: An algorithm for associating the features of two images. Proceedings of the Royal Society of London B 244, 21–26 (1991)
13. Shapiro, L.S., Brady, J.M.: Feature-based correspondence: an eigenvector approach. Image and Vision Computing 10, 283–288 (1992)
14. Umeyama, S.: An eigendecomposition approach to weighted graph matching problems. IEEE PAMI 10, 695–703 (1988)
15. Carcassoni, M., Hancock, E.R.: Weighted graph-matching using modal clusters. In: Proc. 3rd IAPR-TC15 Workshop Graph-Based Representations in Pattern Recognition, pp. 260–269 (2001)
16. Caelli, T., Kosinov, S.: An eigenspace projection clustering method for inexact graph matching. IEEE PAMI 26(4), 515–519 (2004)
17. Bai, X., Yu, H., Hancock, E.R.: Graph matching using spectral embedding and alignment. In: ICPR, vol. 3, pp. 23–26 (2004)
18. Luo, B., Hancock, E.R.: Structural graph matching using the EM algorithm and singular value decomposition. IEEE PAMI 23, 1120–1136 (2001)
19. David, J.M., Carey, E.P.: Predicting unobserved links in incompletely observed networks. Computational Statistics & Data Analysis 52, 1373–1386 (2008)
20. Young, S.J., Scheinerman, E.R.: Random Dot Product Graph Models for Social Networks. In: Bonato, A., Chung, F.R.K. (eds.) WAW 2007. LNCS, vol. 4863, pp. 138–149. Springer, Heidelberg (2007)
21. Scheinerman, E.R.: Kimberly Tucker. Modeling graphs using dot product representations. Computational Statistics 25(1), 1–16 (2010)
22. Zhang, D.M., Sun, D.D., Fu, M.S., Luo, B.: Extended dot product representations of graphs with application to radar image segmentation. Optical Engineering 49(11) (2010)
23. Bai, X., Latecki, L.J.: Path similarity skeleton graph matching. IEEE PAMI 30(7), 1282–1292 (2008)
24. Belongie, S., Puzhicha, J., Malik, J.: Shape Matching and Object Recognition Using Shape Contexts. IEEE PAMI 24(4), 509–522 (2002)

Indexing with Well-Founded Total Order for Faster Subgraph Isomorphism Detection

Markus Weber[1,2], Marcus Liwicki[1,2],
and Andreas Dengel[1,2]

[1] Knowledge Management Department,
German Research Center for Artificial Intelligence (DFKI) GmbH
Trippstadter Straße 122, 67663 Kaiserslautern, Germany
[2] Knowledge-Based Systems Group, Department of Computer Science,
University of Kaiserslautern, P.O. Box 3049, 67653 Kaiserslautern
firstname.lastname@dfki.de

Abstract. In this paper an extension of index-based subgraph matching is proposed. This extension significantly reduces the storage amount and indexing time for graphs where the nodes are labeled with a rather small amount of different classes. In order to reduce the number of possible permutations, a weight function for labeled graphs is introduced and a well-founded total order is defined on the weights of the labels. Inversions which violate the order are not allowed. A computational complexity analysis of the new preprocessing is given and its completeness is proven. Furthermore, in a number of practical experiments with randomly generated graphs the improvement of the new approach is shown. In experiments performed on random sample graphs, the number of permutations has been decreased to a fraction of 10^{-18} in average compared to the original approach by Messmer. This makes indexing of larger graphs feasible, allowing for fast detection of subgraphs.

Keywords: Graph isomorphism, Subgraph isomorphism, Tree search, Decision tree, Indexing.

1 Introduction

Graphs play a major role in structural pattern recognition. An important task in this field is to find similar structures (error-tolerant graph matching) or the same structure (exact graph matching). The focus of this paper is on the latter task, which is important if exactly the same structure or sub-structure needs to be retrieved.

Exact graph matching is needed when the user searches for specific constellations in molecules [11], in computer vision for the recognition of 3-D objects [8,14], shape matching in image analysis [6,2], or room-constellations in floor plans [13]. In most applications, the retrieval result should be available in real-time and the database of reference structures does not change too often. For those situations it is advisable to build an index of the reference structures in advance.

X. Jiang, M. Ferrer, and A. Torsello (Eds.): GbRPR 2011, LNCS 6658, pp. 185–194, 2011.
© Springer-Verlag Berlin Heidelberg 2011

Such a method has been proposed by Messmer et al. [9]. It builds an index using the permutated adjacency matrix of the graph. The real-time search is then based on a tree based. While the method has shown to be effective for reference set with small graphs, it is infeasible for graphs with more than 19 vertices.

In this paper we propose a method to overcome this problem. Assuming that the number of labels for the nodes is relatively small, we introduce a well-founded total order and apply this during index building. This optimization decreases the amount of possible permutations dramatically and allows building indexes of graphs with even more than 30 vertices.

The rest of this paper is organized as follows. First, Section 2 gives an overview over related work. Subsequently, Section 3 introduces definitions and notations which are used and Section 3.1 describes the new preprocessing step. Next, Section 4 will show that the number of computational steps will be significantly decreased. Finally, Section 5 concludes the work.

2   Related Work

In [7], Goa et al. give a survey of work done in the area of graph matching. The focus in the survey is the calculation of error-tolerant graph-matching; where calculating a graph edit distance (GED) is an important way. Mainly the GED algorithms described are categories into algorithms working on attributed or non-attributed graphs. Ullman's method [12] for subgraph matching is known as one of the fastest methods. The algorithm attains efficiency by inferentially eliminating successor nodes in the tree search.

Bunke [3,4] discussed several approaches in graph-matching. One way to cope with error-tolerant subgraph matching is using the maximum common subgraph as a similarity measure. Furthermore the application of graph edit costs which is an extension of the well-known string edit distances. A further group of suboptimal methods are approximate methods, they are based on neural networks, such Hopfield network, Kohonen map or Potts MFT neural net. Moreover methods like genetic algorithms, the usage of Eigenvalues, and linear programming are applied.

Graph matching is challenging in presence of large databases [1,4]. Consequently, methods for preprocessing or indexing are essential. Preprocessing can be performed by graph filtering or concept clustering. The main idea of the graph filtering is to use simple features to reduce to number of feasible candidates. Another concept clustering is used for grouping similar graphs. In principle, given a similarity (or dissimilarity) measure, such as GED [5], any clustering algorithm can be applied. Graph indexing can be performed by the use of decision trees.

Messmer and Bunke [9] proposed a decision tree approach for indexing the graphs. They are using the permutated adjacency matrix of a graph to build a decision tree. This technique is quite efficient during run time, as a decision tree is generated beforehand which contains all model graphs. However, the method has to determine all permutations of the adjacency matrices of the search graphs. Thus, as discussed in their experiments, the method is practically limited to graphs with a maximum of 19 vertices. The main contribution of this paper is to improve the method of Messmer and Bunke for special graphs by modifying the index building process.

3   Definitions and Notations

Basic definitions used throughout the paper are already defined in [9], such as a labeled graph $G = (V, E, L_v, L_e, \mu, v)$, adjacency matrix (M), and permutations on adjacency matrices $(A(G))$. Besides, definitions for orders on sets are needed.

Definition 1. *A **total order** is a binary relation $\leq$ over a set P which is transitive, anti-symmetric, and total, thus for all a, b and c in P, it holds that:*

- *if $a \leq b$ and $b \leq a$ then $a = b$ (anti-symmetry);*
- *if $a \leq b$ and $b \leq c$ then $a \leq c$ (transitivity);*
- *$a \leq b$ or $b \leq a$ (totality).*

Definition 2. *A partial or total order $\leq$ over a set X is **well-founded**, iff $(\forall\, Y \subseteq X\; :\; Y \neq \emptyset \rightarrow (\exists y \in Y\; :\; y$ minimal in Y in respect of $\leq))$.*

Additionally, a weight function is defined which assigns weight to a label of a graph.

Definition 3. *The **weight function** σ is defined as: $\sigma\; :\; L_v\; \rightarrow\; \mathbb{N}$.*

Using the weight function, a well-founded total order is defined on the labels of graph, for example $\sigma(L_1) < \sigma(L_2) < \sigma(L_3) < \sigma(L_4)$. Thus the labeled graph can be extended in its definition.

Definition 4. *A **labeled graph** consists of a 7-tuple, $G = (V, E, L_v, L_e, \mu, v, \sigma)$, where*

- *V is a set of vertices,*
- *$E \subseteq V \times V$ is a set of edges,*
- *L_v is a set of labels for the vertices,*
- *L_e is a set of labels for the edges,*
- *$\mu : V \rightarrow L_v$ is a function which assigns a label to the vertices,*
- *$v : E \rightarrow L_e$ is a function which assigns a label to the edges,*
- *$\sigma : L_v \rightarrow \mathbb{N}$ is a function which assigns a weight to the label of the vertices,*

and a binary relation $\leq$ which defines a well-founded total order on the weights of the labels:

$$\forall x, y \in L_v\; :\; \sigma(x) \leq \sigma(y)\; \vee\; \sigma(y) \leq \sigma(x)$$

3.1   Algorithm

The algorithm for subgraph matching is based on the algorithm proposed by Messmer and Bunke [9], which is a decision tree approach. Their basic assumption is that several graphs are known a priori and the query graph is just known during run time. Messmer's method computes all possible permutations of the adjacency matrices and transforms them into a decision tree. At run time, the adjacency matrix of the query graph is used to traverse the decision tree and find a subgraph which is identical.

Let $G = (V, E, L_v, L_e, \mu, v)$ be a graph from the graph database and M the corresponding $n \times n$ adjacency matrix and $A(G)$ the set of permuted matrices. Thus the total number of permutations is $|A(G)| = n!$, where n is the dimension of the permutation matrix, respectively the number of vertices.

Now, let $Q = (V, E, L_v, L_e, \mu, v)$ be a query graph and M' the corresponding $m \times m$ adjacency matrix, with $m \leq n$. So, if a matrix $M_P \in A(G)$ exists, such that $M' = S_{m,m}(M_P)$, the permutation matrix P which corresponds to M_P represents a subgraph isomorphism from Q to G, i.e

$$M' = S_{m,m}(M_P) = S_{m,m}(PMP^T).$$

Messmer proposed to arrange the set $A(G)$ in a decision tree, such that each matrix in $A(G)$ is classified by the decision tree. However, this approach has one major drawback. For building the decision tree, all permutations of the adjacency matrix have to be considered. Thus, for graphs with more than 19 vertices the number of possible permutations becomes intractable. In order to overcome this issue, the possibilities of permutations have to be reduced. One way is to define constraints for the permutations. Therefore a weight function σ (see Definition 3) is introduced which assigns a weight for each vertex according to its label. Thus each label has a unique weight and a well-founded total order (see Definition 1 and Definition 2) on the set of labels which reduces the number of allowed inversion for the adjacency matrix. Figure 1 illustrates an example for the modified matrices and the corresponding decision tree. Let us consider the following weights for the nodes:

$$L_v = \{L_1, L_2, L_3\}$$
$$\sigma(L_1) = 1,$$
$$\sigma(L_2) = 2,$$
$$\sigma(L_3) = 3.$$

Each inversion that violates the ordering is not allowed. Thus just the vertices which have the same label, respectively the same weights, have to be permuted and if the labels have a different weight, just the variations are required. Given the graph G, the following labels are assigned to the vertices,

$$V = \{v_1, v_2, v_3\}$$
$$\mu(v_1) = L_1,$$
$$\mu(v_2) = L_2,$$
$$\mu(v_3) = L_2.$$

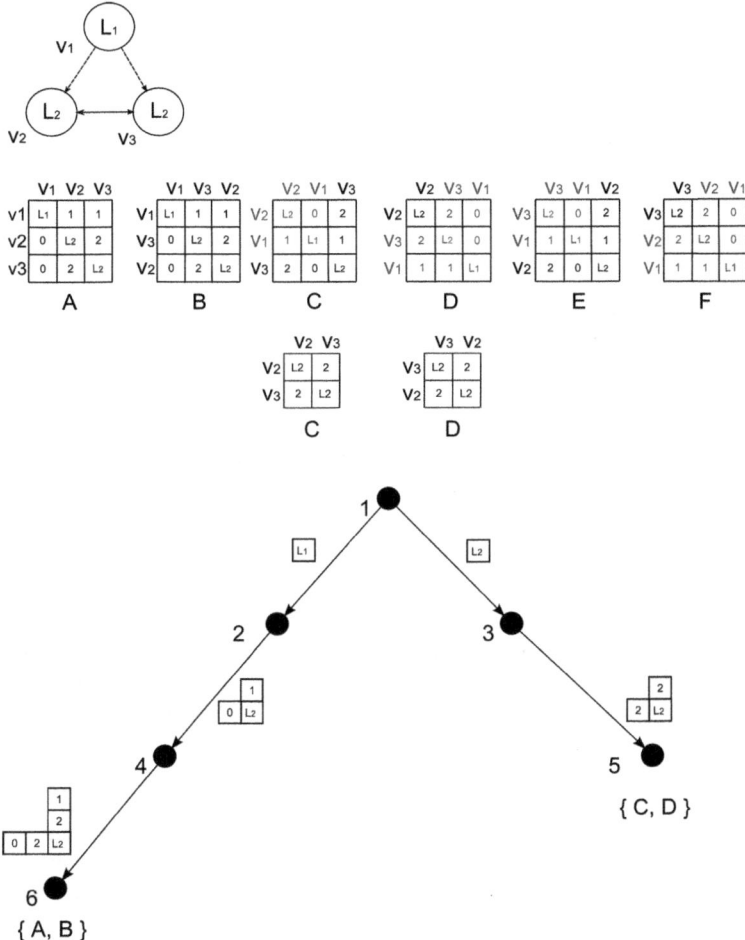

Fig. 1. Modified decision tree for adjacency matrices

Hence, the only valid permutations are:

1. $\sigma(\mu(v_1)) \leq \sigma(\mu(v_2)) \leq \sigma(\mu(v_3))$
2. $\sigma(\mu(v_1)) \leq \sigma(\mu(v_3)) \leq \sigma(\mu(v_2))$
3. $\sigma(\mu(v_2)) \leq \sigma(\mu(v_3))$
4. $\sigma(\mu(v_3)) \leq \sigma(\mu(v_2))$

Let $VA(G)$ be the set of all valid permutations. The decision tree is built according to the row-column elements of the adjacency matrices $M_P \in VA(G)$ and should cover all graphs from the database. So, let R be the set of semantics $R = \{G_1, G_2, ..., G_n\}$, where n is the total number of graphs in the repository, with their sets of corresponding adjacency matrices $VA(G_1), VA(G_2), ..., VA(G_n)$. Now, each set of adjacency matrices has to be added to the decision tree. An obvious advantage of the method is that the whole process can be done

Algorithm 1. BUILD_INDEX($G = (V, E, L_v, L_e, \mu, v, \sigma)$, Tree)

Require: Unsorted set V of vertices, μ labeling function, σ weight function.
1: sort(V, L_v, μ, σ)
Ensure: Vertices V are sorted according to the defined order.
2: Let O be an empty list.
3: **for all** $l_i \in L_V$ **do**
4:     Let interval $\{v_a, \ldots, v_b\}$ contain all v with $\mu(v) = l_i$
5:     $O_i \leftarrow VARIATIONS(\{v_a, \ldots, v_b\})$
6: **end for**
7: Let $AG \leftarrow O_1 \times \ldots \times O_{|Lv|}$.
8: **for all** m_i in AG **do**
9:     Add row column vector for sequence of m_i to Tree.
10: **end for**

a priori. The decision tree acts as an index for subgraphs. So, during run time the decision tree has been loaded into memory and by traversing the decision tree, the corresponding subgraph matrices are classified. For the query graph the adjacency matrix is determined following the constraints defined by ordering. Afterwards the adjacency matrix is split up into row-column vectors a_i. For each level i the corresponding row-column vector a_i is used to find the next node in the decision tree using an index structure. As query q_1 ends in a leaf of the decision tree, the labels of the leaf are the results, query q_2 stops in a node, thus the labels of all leafs beneath the node combine the result.

3.2   Proof of Completeness

For the proposed modified algorithm it has to be proven that the algorithm finds all solutions. The algorithm elaborated in the previous section reduces the number of valid permutations. So, it has to be shown that by leaving out permutations, no valid solution is lost.

Let $G = (V, E, L_v, L_e, \mu, v, \sigma)$ be a well-founded total ordered graph and let $A(G)$ be the set which contains all valid permutations of the graph's adjacency matrices. To be complete, the algorithm must find a solution if one exists; otherwise, it correctly reports that no solution is possible. Thus if every possible valid subgraph $S \subseteq G$, where the vertices of S fulfill the order, every corresponding adjacency matrix M has to be an element of the set $A(G)$, $M \in A(G)$.

For this reason to proof that the algorithm is complete it has to be shown that the algorithm generates all valid subgraphs $S \subseteq G$. Therefore the pseudo code of Algorithm 1 shows how the index is build. Algorithm 2 and Algorithm 3 are helping functions for calculating all variations of the set of vertices in an interval. The generation of the index starts with an unsorted set of vertices. By sorting the vertices with their associated labels using the well-founded total order, the set is ordered according to the weights of the labels.

Now, the algorithm iterates over all intervals of vertices $\{v_a, ..., v_b\}$ where the labels have the same weights, $\sigma(\mu(v_a)) == \sigma(\mu(v_b))$. For each interval $\{v_a, \ldots, v_b\}_i$ all variations with respect to the order have to be determined.

Algorithm 2. $PERMUTE(V, begin, end, R)$

Require: Sorted set V of vertices and $begin < end$, with V_{end-1} being last the element.
1: Adding sequence of vertices V to R.
2: **for** $i \leftarrow end - 2$ to $begin$ **do**
3:     **for** $j \leftarrow i + 1$ to $end - 1$ **do**
4:         Swapping position i and j in V.
5:         Call $PERMUTE(V, i + 1, end, R)$.
6:     **end for**
7:     Call $ROTATE(V, i + 1, end, R)$.
8: **end for**

Algorithm 3. $ROTATE(V, begin, end, R)$

1: Let $temp \leftarrow V_{end-1}$.
2: Shift elements in V in from position $begin$ to $end - 1$ one position right
3: Set $V_{begin} \leftarrow temp$.
4: Add sequence of vertices V to R.

These variations are computed in Algorithm 4, by determining all combination of the interval $\{v_a, \ldots, v_b\}_i$ including the empty set and calculating all permutations for these combinations. Algorithm 2 and Algorithm 3 realize the algorithm proposed by Rosen [10] which computes all permutations for a defined interval. It has been proven that Rosen's algorithm computes all permutations. In combinatorial mathematics, a k-variation of a finite set S is a subset of k distinct elements of S. For each chosen variation of k elements, where k is $L_{interval} = $ length of interval; $k = 1 \ldots L_{interval}$, again all permutations have to be considered. Now, assuming there would be a valid subgraph $Q = (V', E', L'_v, L'_e, \mu, v, \sigma)$, respectively the corresponding adjacency matrix A which depends on the alignment of the vertices. To be a valid subgraph, V' has to be a subset of V, $V' \subseteq V$. Furthermore the alignment of the vertices V' according to their labels has to fulfill the defined order, $\sigma(\mu(v_i)) \leq \sigma(\mu(v_{i+1}))$. For the alignment the intervals $\{v'_a, \ldots, v'_b\} \in V'$ where the weights of the labels have the same value $\sigma(\mu(v'_a))) == \sigma(\mu(v'_b))$ are important as they can vary. The Algorithm 4 determines all variations for intervals with the same weights for labels, thus the alignment $\{v'_a, \ldots, v'_b\}$ is considered. This holds for each interval, thus algorithm produces all valid permutations according to the well-founded total order. As the query graph Q also has to fulfill the order, its adjacency matrix A will be an element of $A(G)$, if Q is a valid subgraph of G. Thus, the solution will be found in the decision tree.

3.3 Complexity Analysis

The original algorithm by Messmer [9] as well as the proposed algorithm need an intensive preprocessing, the compilation of the decision tree. Messmer's method has to compute all permutations of the adjacency matrix of the graph, thus the

Algorithm 4. VARIATIONS($\{v_a, \ldots, v_b\}$)

Require: Sorted set $V = \{v_a, \ldots, v_b\}$ of vertices, $a \leq b$.
1: Let O be an empty list.
2: Determine all combinations C for $\{v_a, \ldots, v_b\}$ including the empty set.
3: **for all** c in C **do**
4:     Call $PERMUTE(c, 0, |c|, O)$.
5: **end for**
6: Return O.

compilation of the decision tree for a graph $G = (V, E, L_v, L_e, \mu, \upsilon, \sigma)$ has a run time complexity of $\mathcal{O}(|V|!)$.

Due to space limitations, we omit the detailed listing of all calculations. The final result for the complexity of our proposed approach is

$$\mathcal{O}(((n_{max} + 1)!)^{|L_v|}),$$

where n_{max} is the maximum number of vertices with the same weight. Thus for the worst case - where all vertices have the same label - $n_{max} = |V|$, $\mathcal{O}((|V|+1)!)$ which would be worse than the method proposed by Messmer and the best case - where all vertices have different labels ($n_{max} = 1$) is $\mathcal{O}(2^{|V|})$ To find the average case of the algorithm the distribution of the labels in the graph has to be considered. This distribution varies according to the represented data.

4   Evaluation

In order to examine run time efficiency of the modified subgraph matching experiments on randomly generated graphs were performed. The modified decision tree algorithm has been implemented in Java using a Java 6 virtual machine. The experiments ran on a Intel Core Duo P8700 (2.53 GHz) CPU with 4 GByte main memory. For the experiment 100 random graphs were generated with 15 to 30 vertices. It compares Messmers's algorithm with its required permutations to the modified algorithm. The permutations for the modified algorithm were determined according to the algorithm discussed in Section 3.1 and the formula in Section 3.3:

$$\prod_{i=1}^{|L_v|} \left(\sum_{j=1}^{n_i} \binom{n_i}{j} \cdot j! \right)$$

and as the original has to be calculate the permutations for all vertices ($|V|!$ permutations). In the second experiment the time to add a graph to the decision tree was measured and again the number of permutations of the adjacency matrix which were added to the decision tree. As the experiment was quite time-consuming on a desktop machine, only the performance for five smaller graphs was measured. The results of the experiment are listed in Tab. 2. The experiments show that the algorithm significantly reduces the number of permutations (see Tab. 1). Though, the time needed to compile the decision tree is

Table 1. Results of graph experiments (first 10 graphs)

Graph	Vertices #	Permutations (modified)	Permutations (original)	Same lables (max.)
1	17	3.26×10^6	3.55×10^{14}	5
2	21	3.59×10^9	5.10×10^{19}	8
3	17	1.08×10^7	3.55×10^{14}	5
4	20	2.50×10^8	2.43×10^{18}	6
5	24	1.64×10^{12}	6.20×10^{23}	10
6	17	1.63×10^6	3.55×10^{14}	3
7	21	2.04×10^7	5.10×10^{19}	3
8	30	1.39×10^{12}	2.65×10^{32}	5
9	22	8.01×10^8	1.12×10^{21}	6
⋮	⋮	⋮	⋮	⋮
100	23	1.00×10^9	2.58×10^{22}	6
∅	23.05	1.09×10^{13}	3.73×10^{31}	5.23

Table 2. Run time for compiling the decision tree for each graph

Graph	Vertices #	Run time (minutes)	Permutations #	Same lables (max.)
1	17	1.47	8.19×10^5	4
2	17	8.90	4.17×10^6	5
3	21	45.67	5.32×10^7	4
4	21	10.06	8.19×10^6	3
5	21	38.01	4.09×10^7	3

still quite long even for small problem instance, as shown in Tab. 2. However, as the method is designed for an off-line preprocessing and considered to run on a server machine, it is still reasonable for practical applications.

5   Conclusions and Future Work

In this paper an extension for the method of Messmer's subgraph matching has been proposed. The original method is very efficient to perform exact subgraph matching on a large database. However, it has a limitation for the maximum number of vertices. The modification discussed in this paper enables to increase this limit depending on how the vertices are labeled. As the number of permutations in the preprocessing step depends on the vertices with the same labels, an analysis of the data that will be represented in graph is necessary. If there are just a few vertices with the same label, e.g. less than five, even graphs with 30 vertices can be handled. It has been proven that the modification of the method does not affect its completeness.

Noteworthy, the proposed method can be applied in several areas, such as object recognition, matching of 2D or 3D chemical structures, and architectural floor plan retrieval. Future work will be to perform experiments on real graph data sets and research strategies for choosing appropriate weight functions. Furthermore, we plan to extend this method to provide a fast method for error-tolerant graph matching.

References

1. Bunke, H.: On a relation between graph edit distance and maximum common subgraph. Pattern Recognition Letters 18(8), 689–694 (1997)
2. Bunke, H., Messmer, B.: Efficient attributed graph matching and its application to image analysis, pp. 44–55 (1995)
3. Bunke, H.: Graph matching: Theoretical foundations, algorithms, and applications. In: Proc. Vision Interface (2000)
4. Bunke, H.: Recent developments in graph matching. International Conference on Pattern Recognition 2, 2117 (2000)
5. Bunke, H., Shearer, K.: A graph distance metric based on the maximal common subgraph. In: Pattern Recognition Letters, pp. 255–259 (1998)
6. Cheng, J., Huang, T.: Image registration by matching relational structures. Pattern Recognition 17(1), 149–159 (1984)
7. Gao, X., Xiao, B., Tao, D., Li, X.: A survey of graph edit distance. Pattern Analysis and Applications 13(1), 113–129 (2009)
8. Kim, W., Kak, A.: 3-d object recognition using bipartite matching embedded in discrete relaxation. IEEE Transactions on Pattern Analysis and Machine Intelligence 13, 224–251 (1991)
9. Messmer, B., Bunke, H.: A decision tree approach to graph and subgraph isomorphism detection. Pattern Recognition 32, 1979–1998 (1999)
10. Rosen, K.H.: Discrete mathematics and its applications, 2nd edn. McGraw-Hill, Inc., New York (1991)
11. Schomburg, I., Chang, A., Ebeling, C., Gremse, M., Heldt, C., Huhn, G., Schomburg, D.: BRENDA, the enzyme database: updates and major new developments. NUCLEIC ACIDS RESEARCH 32(SI) (January 2004)
12. Ullmann, J.: An algorithm for subgraph isomorphism. Journal of the ACM (JACM) 23(I), 31–42 (1976)
13. Weber, M., Langenhan, C., Roth-Berghofer, T., Liwicki, M., Dengel, A., Petzold, F.: a.SCatch: Semantic Structure for Architectural Floor Plan Retrieval. In: Bichindaritz, I., Montani, S. (eds.) ICCBR 2010. LNCS, vol. 6176, Springer, Heidelberg (2010)
14. Wong, E.K.: Model matching in robot vision by subgraph isomorphism. Pattern Recogn 25, 287–303 (1992)

Graph Transduction as a Non-cooperative Game

Aykut Erdem[1] and Marcello Pelillo[2]

[1] Hacettepe University, Beytepe, 06800, Ankara, Turkey
aykut.erdem@hacettepe.edu.tr
[2] "Ca' Foscari" University of Venice, Mestre, Venezia, 30172, Italy
pelillo@dsi.unive.it

Abstract. Graph transduction is a popular class of semi-supervised learning techniques, which aims to estimate a classification function defined over a graph of labeled and unlabeled data points. The general idea is to propagate the provided label information to unlabeled nodes in a consistent way. In contrast to the traditional view, in which the process of label propagation is defined as a graph Laplacian regularization, here we propose a radically different perspective that is based on game-theoretic notions. Within our framework, the transduction problem is formulated in terms of a non-cooperative multi-player game where any equilibrium of the proposed game corresponds to a consistent labeling of the data. An attractive feature of our formulation is that it is inherently a multi-class approach and imposes no constraint whatsoever on the structure of the pairwise similarity matrix, being able to naturally deal with asymmetric and negative similarities alike. We evaluated our approach on some real-world problems involving symmetric or asymmetric similarities and obtained competitive results against state-of-the-art algorithms.

1   Introduction

In the machine learning community, *semi-supervised learning* (SSL) has gained considerable popularity over the last decade [3,19] and within the existing paradigms, graph-based approaches to SSL, namely the *graph transduction* methods, constitute an important class. These approaches model the geometry of the data as a graph with nodes corresponding to the labeled and unlabeled points and edges being weighted by the similarity between points, and try to estimate the labels of unlabeled points by propagating the coarse information available at the labeled nodes to the unlabeled ones. Performing this propagation in a consistent way relies on a common a priori assumption, known as the *"cluster assumption"* [17,3], which states that (1) points which are close to each other are expected to have the same label, and (2) points in the same cluster (or on the same manifold) are expected to have the same label. Building on this assumption, traditional graph-based approaches formalize graph transduction as a regularized function estimation problem on an undirected graph [9,20,17].

In this paper, we present a novel framework for graph transduction, which is derived from a game-theoretic formulation of the competition between the multi-population of hypotheses of class membership. Specifically, we cast the problem of graph transduction as a *multi-player non-cooperative game* where the players are the data points that

X. Jiang, M. Ferrer, and A. Torsello (Eds.): GbRPR 2011, LNCS 6658, pp. 195–204, 2011.

play a classification game over and over until an equilibrium is reached in their respective strategies. In this game, the strategies played by the labeled points are already decided at the outset, as each of them knows which class it belongs to. On the other hand, the strategies available to unlabeled points are the whole set of hypotheses of being a member of one of the provided classes. The players compete with each other by selecting their own strategies, each choice obtains support from the compatible ones and competitive pressure from all the others. In the long run, the competition will reduce the population of strategies which assume the hypotheses that do not receive strong support from the rest, while it will allow populations with strong support to flourish. In this study, this evolutionary dynamics is modeled by a classic formalization of natural selection process used in the *evolutionary game theory* [16], commonly referred to as the *replicator dynamics*. It is worth-mentioning that our formulation is intrinsically a *multi-class* approach and does not impose any constraint on the value of the payoffs (similarities); in particular, *payoffs do not have to be nonnegative or symmetric*.

The remainder of this paper is structured as follows. In Section 2, we review basic notions from non-cooperative game theory. In Section 3, we formulate graph transduction in terms of a non-cooperative multi-player game. In Section 4, we present our experimental results on a number of real-world classification problems. Finally, in Section 5, we conclude the paper with a summary and directions for future work.

2   Non-cooperative Games and Nash Equilibria

Following the notations used in [16], a game with many players can be expressed in normal form as a triple $G = (\mathcal{I}, S, \pi)$, where $\mathcal{I} = \{1, \ldots, n\}$, with $n \geq 2$, is the set of *players*, $S = \times_{i \in \mathcal{I}} S_i$ is the *joint strategy space* defined as the Cartesian product of the individual pure strategy sets $S_i = \{1, \ldots, m_i\}$, and $\pi : S \to \mathbb{R}^n$ is the *combined payoff function* which assigns a real valued payoff $\pi_i(s) \in \mathbb{R}$ to each *pure strategy profile* $s \in S$ and player $i \in \mathcal{I}$.

A *mixed strategy* of player $i \in \mathcal{I}$ is a probability distribution over its pure strategy set S_i, which can be described as the vector $x_i = (x_{i1}, \ldots, x_{im_i})^T$ such that each component x_{ih} denotes the probability that the player chooses to play its hth pure strategy among all the available strategies. Mixed strategies for each player $i \in \mathcal{I}$ are constrained to lie in the *standard simplex* of the m_i-dimensional Euclidean space $\mathbb{R}^{m_i}$, $\Delta_i = \{x_i \in \mathbb{R}^{m_i} : \sum_{h=1}^{m_i} x_{ih} = 1, \text{ and } x_{ih} \geq 0 \text{ for all } h\}$. Accordingly, a *mixed strategy profile* $x = (x_1, \ldots, x_n)$ is defined as a vector of mixed strategies, each $x_i \in \Delta_i$ representing the mixed strategy assigned to player $i \in \mathcal{I}$, and each mixed strategy profile lives in the *mixed strategy space* of the game, given by the Cartesian product $\Theta = \times_{i \in \mathcal{I}} \Delta_i$.

For the sake of simplicity, let $z = (x_i, y_{-i}) \in \Theta$ denote the strategy profile where player i plays strategy $x_i \in \Delta_i$ whereas other players $j \in \mathcal{I} \setminus \{i\}$ play based on the strategy profile $y \in \Theta$, that is to say, $z_i = x_i$ and $z_j = y_j$ for all $j \neq i$. The expected value of the payoff that player i obtains can be determined by a weighted sum for any $i, j \in \mathcal{I}$ as

$$u_i(x) = \sum_{s \in S} x(s)\pi_i(s) = \sum_{k=1}^{m_j} u_i\left(e_j^k, x_{-j}\right) x_{jk} \qquad (1)$$

where $u_i \left(e_j^k, x_{-j} \right)$ denotes the payoff that player i receives when player j adopts its kth pure strategy, and $e_j^k \in \Delta_j$ stands for the *extreme mixed strategy* corresponding the vector of length m_j whose components are all zero except the kth one which is equal to one.

The *mixed best replies* for player i against a mixed strategy $y \in \Theta$, denoted by $\beta_i(y)$, is the set of mixed strategies which is constructed in such a way that no other mixed strategy other than the ones included in this set gives a higher payoff to player i against strategy y, defined as the set $\beta_i(y) = \{x_i \in \Delta_i : u_i(x_i, y_{-i}) \geq u_i(z_i, y_{-i}) \; \forall z_i \in \Delta_i\}$. Subsequently, the combined mixed best replies is defined as the Cartesian product of best replies of all the players $\beta(y) = \times_{i \in \mathcal{I}} \beta_i(y) \subset \Theta$.

Definition 1. *A mixed strategy* $x^* = (x_1^*, \ldots, x_n^*)$ *is said to be a Nash equilibrium if it is the best reply to itself,* $x^* \in \beta(x^*)$, *that is*

$$u_i(x_i^*, x_{-i}^*) \geq u_i(x_i, x_{-i}^*) \tag{2}$$

for all $i \in \mathcal{I}, x_i \in \Delta_i$, *and* $x_i \neq x_i^*$. *Furthermore, a Nash equilibrium* x^* *is called strict if each* x_i^* *is the unique best reply to* x^*, $\beta(x^*) = \{x^*\}$

Nash equilibrium constitutes the key concept of game theory. It is proven by Nash that any non-cooperative game with finite set of strategies has at least one mixed Nash equilibrium [11]. The algorithmic issue of computing a Nash equilibria for the proposed transduction game will be discussed later in Section 3.2.

3 Graph Transduction Game (GTG)

Consider the following *graph transduction game*. Assume each player $i \in \mathcal{I}$ participating in the game corresponds to a particular point in a data set $\mathcal{X} = \{\mathbf{x}_1, \ldots, \mathbf{x}_n\}$ and can choose a strategy among the set of strategies $S_i = \{1, \ldots, c\}$, each expressing a certain hypothesis about its membership to a class and $|S_i|$ being the total number of classes. Hence, the mixed strategy profile of each player $i \in \mathcal{I}$ lies in the c-dimensional simplex Δ_i. By problem definition, we can categorize the players of the game into two disjoint groups: those which already have the knowledge of their membership, which we call *determined players* and denote them with the symbol $\mathcal{I}_\mathcal{D}$, and those which don't have any idea about this in the beginning of the game, which are hence called *undetermined players* and correspondingly denoted with $\mathcal{I}_\mathcal{U}$.

The so-called determined players of the game can further be distinguished based on the strategies they follow without hesitation, coming from their membership information. In formal terms, $\mathcal{I}_\mathcal{D} = \{\mathcal{I}_{\mathcal{D}|1}, \ldots, \mathcal{I}_{\mathcal{D}|c}\}$, where each disjoint subset $\mathcal{I}_{\mathcal{D}|k}$ stands for the set of players always playing their k^{th} pure strategies. It thus follows from this statement that each player $i \in \mathcal{I}_{\mathcal{D}|k}$ plays its extreme mixed strategy $e_i^k \in \Delta_i$. In other words, x_i is constrained to belong to the minimal face of the simplex Δ_i spanned by $\{e_i^k\}$. In this regard, it can be argued that the determined players do not play the game to maximize their payoffs since they have already chosen their strategies. In fact, the transduction game can be easily reduced to a game with only undetermined players $\mathcal{I}_\mathcal{U}$ where the definite strategies of determined players $\mathcal{I}_\mathcal{D}$ act as bias over the choices of undetermined players.

It should be noted that any instance of the proposed transduction game will always have a Nash equilibrium in mixed strategies [11]. Recall that, for the players, such an equilibrium corresponds to a steady state such that each player plays a strategy that could yield the highest payoff when the strategies of the remaining players are kept fixed, and it provides us a globally consistent labeling of the data set. Once an equilibrium is reached, the label of a data point (player) i is simply given by the strategy with the highest probability in the equilibrium mixed strategy of player i as $y_i = \arg\max_{h \leq c} x_{ih}$.

3.1 Defining Payoff Functions

A crucial step in formulating transduction as a non-cooperative game is how the payoff function of the game is specified. Here, we make a simplification and assume that the payoffs associated to each player are additively separable, and this makes the proposed game a member of a special subclass of multi-player games, known as *polymatrix games* [8,7]. Formally speaking, for a pure strategy profile $s = (s_1, \ldots, s_n) \in S$, the payoff function of every player $i \in \mathcal{I}$ is in the form:

$$\pi_i(s) = \sum_{j=1}^{n} A_{ij}(s_i, s_j) \qquad (3)$$

where $A_{ij} \in \mathbb{R}^{c \times c}$ is the *partial payoff* matrix between players i and j. It follows that, in terms of a mixed strategy profile $x = (x_1, \ldots, x_n)$, the payoffs are computed as $u_i(e_i^h) = \sum_{j=1}^{n}(A_{ij}x_j)_h$ and $u_i(x) = \sum_{j=1}^{n} x_i^T A_{ij} x_j$.

In an instance of the transduction game, since each determined player is restricted to play a definite strategy of its own, all of these fixed choices can be reflected directly in the payoff function of a undetermined player $i \in \mathcal{I}_{\mathcal{U}}$ as follows:

$$u_i(e_i^h) = \sum_{j \in \mathcal{I}_{\mathcal{U}}}(A_{ij}x_j)_h + \sum_{k=1}^{c}\sum_{j \in \mathcal{I}_{\mathcal{D}|k}} A_{ij}(h, k) \qquad (4)$$

$$u_i(x) = \sum_{j \in \mathcal{I}_{\mathcal{U}}} x_i^T A_{ij} x_j + \sum_{k=1}^{c}\sum_{j \in \mathcal{I}_{\mathcal{D}|k}} x_i^T (A_{ij})_k \qquad (5)$$

Now, we are left with specifying the partial payoff matrices between each pair of players. Let the geometry of the data be modeled with a weighted graph $\mathcal{G} = (\mathcal{X}, \mathcal{E}, w)$ in which $\mathcal{X}$ is the set of nodes representing both labeled and unlabeled points, and $w : \mathcal{E} \to \mathbb{R}$ is a weight function assigning a similarity value to each edge $e \in \mathcal{E}$. Representing the graph with its weighted adjacency matrix $W = (w_{ij})$, we set the partial payoff matrix between two players i and j as $A_{ij} = I_c \times w_{ij}$ where I_c is the identity matrix of size c[1]. Note that when partial payoff matrices are represented in block form as $A = (A_{ij})$, the matrix A is given by the Kronecker product $A = I_c \otimes W$. Our experiments demonstrate that in specifying the payoffs, it is preferable to use the normalized

---

[1] The rationale for specifying partial payoffs in this way depends on the analysis of graph transduction on a unweighted undirected graph. Due to the page limit, the details will be reported in a longer version.

similarity data matrix $\widehat{W} = D^{-1/2}WD^{-1/2}$ where $D = (d_{ii})$ is the diagonal degree matrix of W with its elements given by $d_{ii} = \sum_j w_{ij}$.

3.2   Computing Nash Equilibria

In the recent years, there has been a growing interest in the computational aspects of Nash equilibria. The general problem of computing a Nash equilibrium is shown to belong to the complexity class PPAD-complete, a newly defined subclass of NP [4]. Nevertheless, there are many refinements and extensions of Nash equilibria which can be computed efficiently and moreover, the former result does not apply to certain classes of games. Here, we restrict ourselves to the well-established evolutionary approach [16], initiated by J. Maynard Smith [10]. This dynamic interpretation of the concept imagines that the game is played repeatedly, generation after generation, during which a selection process acts on the multi-population of strategies, thereby resulting in the evolution of the fittest strategies. The selection dynamics is commonly modeled by the following set of ordinary differential equations:

$$\dot{x}_{ih} = g_{ih}(x)x_{ih} \tag{6}$$

where a dot signifies derivative with respect to time, and $g(x) = (g_1(x), \ldots, g_n(x))$ is the growth rate function with open domain containing $\Theta = \times_{i \in \mathcal{I}} \Delta_i$, each component $g_i(x)$ being a vector-valued growth rate function for player i. Hence, g_{ih} specifies the growth rate at which player i's pure strategy h replicates. It is generally required that the function g be *regular* [16], *i.e.* (1) g is Lipschitz continuous and (2) $g_i(x) \cdot x_i = 0$ for all $x \in \Theta$ and players $i \in \mathcal{I}$. While the first condition guarantees that the system (6) has a unique solution through every initial state, the condition $g_i(x) \cdot x_i = 0$ ensures that the simplex Δ_i is invariant under (6).

The class of regular selection dynamics includes a wide subclass known as *payoff monotonic dynamics*, in which the ratio of strategies with a higher payoff increase at a higher rate. Formally, a regular selection dynamics (6) is said to be payoff monotonic if

$$u_i\left(e_i^h, x_{-i}\right) > u_i\left(e_i^k, x_{-i}\right) \Leftrightarrow g_{ih}(x) > g_{ik}(x) \tag{7}$$

for all $x \in \Theta$, $i \in \mathcal{I}$ and pure strategies $h, k \in S_i$.

A particular subclass of payoff monotonic dynamics, which is used to model the evolution of behavior by imitation processes, is given by

$$\dot{x}_{ih} = x_{ih}\left[\sum_{l \in S_i} x_{il}\left(\phi_i\left[u_i\left(e_i^h - e_i^l, x_{-i}\right)\right] - \phi_i\left[u_i\left(e_i^l - e_i^h, x_{-i}\right)\right]\right)\right] \tag{8}$$

where $\phi_i(u_i)$ is a strictly increasing function of u_i. When ϕ_i is taken as the identity function, *i.e.* $\phi_i(u_i) = u_i$, we obtain the multi-population version of the replicator dynamics:

$$\dot{x}_{ih} = x_{ih}\left(u_i(e_i^h, x_{-i}) - u_i(x)\right) \tag{9}$$

The following theorem states that the fixed points of (9) are Nash equilibria.

Theorem 1. *A point $x \in \Theta$ is the limit of a trajectory of (9) starting from the interior of Θ if and only if x is a Nash equilibrium. Further, if point $x \in \Theta$ is a strict Nash equilibrium then it is asymptotically stable, additionally implying that the trajectories starting from all nearby states converge to x.*

Proof. See [16].

In the experiments, we utilized the following discrete-time counterpart of (9), where we initialize the mixed strategies of each undetermined player to uniform probabilities, *i.e.* the barycenter of the simplex Δ_i.

$$x_{ih}(t+1) = x_{ih}(t) \frac{u_i(e_i^h)}{u_i(x(t))} \qquad (10)$$

The discrete-time replicator dynamics (10) has the same properties as the continuous version (See [12] for a detailed analysis). The computational complexity of finding a Nash equilibrium of a transduction game using (10) can be given by $\mathcal{O}(kcn^2)$, where n is the number of players (data points), c is the number of pure strategies (classes) and k is the number of iterations needed to converge. In theory, it is difficult to predict the number of required iterations, but experimentally, we noticed that it typically grows linearly on the number of data points[2]. We note that the complexity of popular graph transduction methods such as [20,17] is also close to $\mathcal{O}(n^3)$.

4   Experimental Results

Our experimental evaluation is divided into two groups based on the structure of similarities that arise in the problems. Basically, we test our approach on some real-world problems involving *symmetric* or *asymmetric* similarities. It is noteworthy to mention that the standard methods are restricted to work with *symmetric* and *non-negative* similarities but our game-theoretic interpretation imposes no constraint whatsoever, being able to naturally deal with *asymmetric* and *negative* similarities alike.

4.1   Experiments with Symmetric Similarities

We conducted experiments on three well-known data sets: *USPS*[3], *YaleB* [5] and *20-news*[4]. *USPS* contains images of hand-written digits 0-9 down-sampled to 16×16 pixels and it has 7291 training and 2007 test examples. As used in [17], we only selected the digits 1 to 4 from the training and test sets, which gave us a total of 3874 data points. *YaleB* is composed of face images of 10 subjects captured under varying poses and illumination conditions. As in [2], we down-sampled each image to 30×40

---

[2] We observed that the dynamics always converged to a fixed point in our experiments with symmetric and asymmetric similarities. It should be added that in the asymmetric case, the convergence is in fact not guaranteed since there is no Lyapunov function for the dynamics. Still, by Theorem 1, if the dynamics converges to a fixed point, it will definitely be a Nash equilibrium.

[3] http://www-stat.stanford.edu/~tibs/ElemStatLearn/

[4] http://people.csail.mit.edu/jrennie/20newsgroups/

pixels and considered a subset of 1755 images which corresponds to the individuals 2, 5 and 8. *20-news* is the text classification data set used in [17], which contains 3970 news-group articles selected from the 20-newsgroups data set, all belonging to the topic `rec` which is composed of the subjects `autos`, `motorcycles`, `sport.baseball` and `sport.hockey`. As described in [17], each article is represented in 8014-dimensional space based on the TFIDF representation scheme.

For *USPS* and *YaleB*, we treated each image pixel as a single feature, thus each example was represented in 256-, and 1200-dimensional space, respectively. We computed the similarity between two examples $\mathbf{x}_i$ and $\mathbf{x}_j$ using the Gaussian kernel as $w_{ij} = exp(-\frac{d(\mathbf{x}_i, \mathbf{x}_j)^2}{2\sigma^2})$ where $d(\mathbf{x}_i, \mathbf{x}_j)$ is the distance between $\mathbf{x}_i$ and $\mathbf{x}_j$ and σ is the kernel width parameter. Among several choices for the distance measure $d(\cdot)$, we evaluated the Euclidean distance $\|\mathbf{x}_i - \mathbf{x}_j\|$ for *USPS* and *YaleB*, and the cosine distance $d(\mathbf{x}_i, \mathbf{x}_j) = 1 - \frac{\langle \mathbf{x}_i, \mathbf{x}_j \rangle}{\|\mathbf{x}_i\|\|\mathbf{x}_j\|}$ for *20-news*.

In the experiments, we compared our approach, which we denote as GTG, against four well-known graph-based SSL algorithms, namely the Spectral Graph Transducer (SGT) [9][5], the *Gaussian fields and harmonic functions* based method (GFHF) [20][6], the *local and global consistency* method (LGC) [17][7] and Laplacian Regularized Least Squares (LapRLS) [1][8]. A crucial factor in the success of graph-based algorithms is the construction of the input graph as it represents the data manifold. To be fair in our evaluation, for all the methods, we used a fixed set of kernel widths and generated 9 different candidate 20-NN graphs by setting $w_{ij} = 0$ if x_j is not amongst the 20-nearest neighbors of x_i. In particular, the kernel width σ ranges over the set $linspace(0.1r, r, 5) \cup linspace(r, 10r, 5))$ with r being the average distance from each example to its 20th nearest neighbor and $linspace(a, b, n)$ denoting the set of n linearly spaced numbers between and including a and b.

In Fig. 1, we show the test errors of all methods averaged over 100 trials with different sizes of labeled data where we randomly select labeled samples so that each set contains at least one sample from each class. As it can be seen, LapRLS method gives the best results for the relatively small data set *YaleB*. However, for the other two, its performance is poor. In general, the proposed GTG algorithm is either the best or the second best algorithm; while its success is almost identical to that of the LGC method in *USPS* and *Yale-B*, it gives superior results for *20-news*.

4.2 Experiments with Asymmetric Similarities

We carried out experiments on three document data sets – *Cora*, *Citeseer* [14][9], and *We-bKB*[10]. *Cora* contains 2708 machine learning publications classified into seven classes, and there are 5429 citations between the publications. *Citeseer* consists of 3312

---

[5] We select the optimal value of the parameter c with the best mean performance from the set $\{400, 800, 1600, 3200, 6400, 12800\}$.

[6] In obtaining the hard labels, we employ the *class mass normalization* step suggested in [20].

[7] As in [17], we set the parameter α as 0.99.

[8] We select the optimal values of the extrinsic and intrinsic regularization parameters γ_A and γ_I from the set $\{10^{-6}, 10^{-4}, 10^{-2}, 1\}$ for the best mean performance.

[9] Both data sets are available at http://www.cs.umd.edu/projects/linqs/projects/lbc/

[10] Available at http://www.nec-labs.com/~zsh/files/link-fact-data.zip

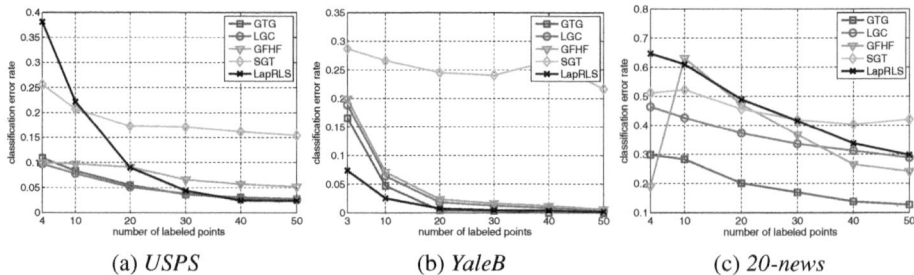

(a) USPS   (b) YaleB   (c) 20-news

Fig. 1. Performance comparisons on classification problems with *symmetric* similarities

scientific publications, each of which belongs to one of six classes, and there are a total of 4732 links. *WebKB* contains webpages collected from computer science departments of four universities (*Cornell*, *Texas*, *Washington* and *Wisconsin*), and each classified into seven categories. Following the setup in [18], here we concentrate on classifying student pages from the others. Each subset respectively contains 827, 814, 1166 and 1210 webpages and 1626, 1480, 2218 and 3200 links. In our experiments, as in [18], we only considered the citation structure, even though one can also assign some weights by utilizing the textual content of the documents. Specifically, we worked on the link matrix $W = (w_{ij})$, where $w_{ij} = 1$ if document i cites document j and $w_{ij} = 0$ otherwise.

Unlike our approach, the standard methods mentioned before, namely SGT, GFHF, LGC and LapRLS, are subject to symmetric similarities. Hence, in this context, they can be applied only after rendering the similarities symmetric but this could result in loss of relevant information in some cases. In our evaluation, we restrict ourselves to the graph-based methods which can directly deal with asymmetric similarities. Specifically, we compared our game-theoretic approach against our implementation of the method in [18], denoted here with LLUD. We note that LLUD is based on the notion of random walks on directed graphs and it reduces to LGC in the case of symmetric similarities. However, it assumes the input similarity graph to be strongly connected, so in [18] the authors consider the *teleporting random walk (trw)* transition matrix as input, which is given by $P^\eta = \eta P + (1 - \eta)P^u$ where $P = D^{-1}W$ and P^u is the uniform transition matrix. For asymmetric similarity data, we also define the payoffs in terms of this transition matrix and denote this version with GTGtrw. In the experiments, we fixed $\eta = 0.99$ for both LLUD and GTGtrw. To provide a baseline, we also report the results of our approach that works on the symmetrized similarity matrices, denoted with GTGsym. For that case, we used the transformation $\widetilde{W} = 0.5 \times (W + W^T)$ for *SCOP*, and the symmetrized link matrix $\widetilde{W} = (w_{ij})$ for the others, where $w_{ij} = 1$ if either document i cites document j or vice versa, and $w_{ij} = 0$ otherwise.

The test errors averaged over 100 trials are shown in Fig. 2. Notice that the performances of GTGtrw and LLUD are quite similar on the classification problems in *WebKB* data sets. On the other hand, GTGtrw is superior in the multi-class problems of *Cora* and *Citeseer*. We should add that symmetrization sometimes can provide good results. In *Cora* and *Citeseer*, GTGsym performs better than the other two methods.

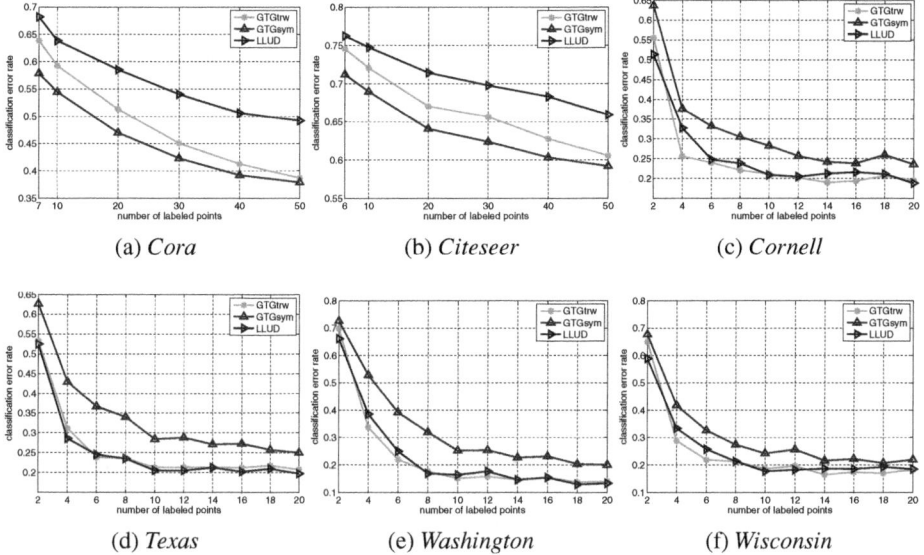

Fig. 2. Performance comparisons on classification problems with *asymmetric* similarities

5   Summary and Discussion

In this paper, we provided a game-theoretic interpretation to graph transduction. In the suggested approach, the problem of transduction is formulated in terms of a multi-player non-cooperative game where any equilibrium of the game coincides with the notion of a consistent labeling of the data. As compared to existing approaches, the main advantage of the proposed framework is that there is no restriction on the pair-wise relationships among data points; similarities and thus the payoffs can be negative or asymmetric. The experimental results show that the our approach is not only more general but also competitive with standard approaches. In the future, we plan to con-tinue exploring the generality of our approach when both similarity and dissimilarity relations exist in data [15,6]. Another possible future direction is to focus on improving the efficiency. In our current implementation, we use the standard replicator dynam-ics to reach an equilibrium but we can study other selection dynamics that are much faster [13].

Acknowledgments

We acknowledge financial support from the FET programme within EU FP7, under the SIMBAD project (contract 213250).

References

1. Belkin, M., Niyogi, P., Sindhwani, V.: Manifold regularization: A geometric framework for learning from labeled and unlabeled examples. J. Mach. Learn. Res. 7, 2399–2434 (2006)

2. Breitenbach, M., Grudic, G.Z.: Clustering through ranking on manifolds. In: ICML, pp. 73–80 (2005)
3. Chapelle, O., Schölkopf, B., Zien, A. (eds.): Semi-Supervised Learning. MIT Press, Cambridge (2006)
4. Daskalakis, C., Goldberg, P.W., Papadimitriou, C.H.: The complexity of computing a Nash equilibrium. Commun. ACM 52(2), 89–97 (2009)
5. Georghiades, A., Belhumeur, P., Kriegman, D.: From few to many: Illumination cone models for face recognition under variable lighting and pose. IEEE Trans. on PAMI 23(6), 643–660 (2001)
6. Goldberg, A., Zhu, X., Wright, S.: Dissimilarity in graph-based semi-supervised classification. In: AISTATS (2007)
7. Howson, J.T.: Equilibria of polymatrix games. Management Science 18(5), 312–318 (1972)
8. Janovskaya, E.B.: Equilibrium points in polymatrix games (in Russian) Litovskii Matematicheskii Sbornik 8, 381–384 (1968); (Math. Reviews 39 #3831)
9. Joachims, T.: Transductive learning via spectral graph partitioning. In: ICML, pp. 290–297 (2003)
10. Maynard Smith, J.: Evolution and the theory of games. Cambridge University Press, Cambridge (1982)
11. Nash, J.: Non-cooperative games. The Annals of Mathematics 54(2), 286–295 (1951)
12. Pelillo, M.: The dynamics of nonlinear relaxation labeling processes. J. Math. Imaging Vis. 7(4), 309–323 (1997)
13. Rota Bulò, S., Bomze, I.M.: Infection and immunization: a new class of evolutionary game dynamics. Games and Economic Behaviour (Special issue in honor of John F. Nash, jr.) 71, 193–211 (2011)
14. Sen, P., Namata, G.M., Bilgic, M., Getoor, L., Gallagher, B., Eliassi-Rad, T.: Collective classification in network data. AI Magazine 29(3), 93–106 (2008)
15. Tong, W., Jin, R.: Semi-supervised learning by mixed label propagation. In: AAAI, pp. 651–656 (2007)
16. Weibull, J.W.: Evolutionary Game Theory. MIT Press, Cambridge (1995)
17. Zhou, D., Bousquet, O., Lal, T.N., Weston, J., Schölkopf, B.: Learning with local and global consistency. In: NIPS, pp. 321–328 (2004)
18. Zhou, D., Huang, J., Schölkopf, B.: Learning from labeled and unlabeled data on a directed graph. In: ICML, pp. 1036–1043 (2005)
19. Zhu, X.: Semi-supervised learning literature survey. Tech. Rep. 1530, Computer Sciences, University of Wisconsin-Madison (2005)
20. Zhu, X., Ghahramani, Z., Lafferty, J.: Semi-supervised learning using Gaussian fields and harmonic functions. In: ICML, pp. 912–919 (2003)

A Graph-Based Approach to Feature Selection

Zhihong Zhang and Edwin R. Hancock

Department of Computer Science, University of York, UK

Abstract. In many data analysis tasks, one is often confronted with very high dimensional data. The feature selection problem is essentially a combinatorial optimization problem which is computationally expensive. To overcome this problem it is frequently assumed either that features independently influence the class variable or do so only involving pairwise feature interaction. To tackle this problem, we propose an algorithm consisting of three phases, namely, i) it first constructs a graph in which each node corresponds to each feature, and each edge has a weight corresponding to mutual information (MI) between features connected by that edge, ii) then perform dominant set clustering to select a highly coherent set of features, iii) further selects features based on a new measure called multidimensional interaction information (MII). The advantage of MII is that it can consider third or higher order feature interaction. By the help of dominant set clustering, which separates features into clusters in advance, thereby allows us to limit the search space for higher order interactions. Experimental results demonstrate the effectiveness of our feature selection method on a number of standard data-sets.

1 Introduction

High-dimensional data pose a significant challenge for pattern recognition. The most popular methods for reducing dimensionality are variance based subspace methods such as PCA. However, the extracted PCA feature vectors only capture sets of features with a significant combined variance, and this renders them relatively ineffective for classification tasks. Hence it is crucial to identify a smaller subset of features that are informative for classification and clustering. The idea underpinning feature selection is to select the features that are most relevant to classification while reducing redundancy. Mutual information provides a principled way of measuring the mutual dependence of two variables, and has been used by a number of researchers to develop information theoretic feature selection criteria. For example, Batti [1] has developed the Mutual Information-Based Feature Selection (MIFS) criterion, where the features are selected in a greedy manner. Given a set of existing selected features S, at each step it locates the feature x_i that maximize the relevance to the class $I(x_i; C)$. The selection is regulated by a proportional term $\beta I(x_i; S)$ that measures the overlap information between the candidate feature and existing features. The parameter β may significantly affect the features selected, and its control remains an open problem. Peng et al [7] on the other hand, use the so-called Maximum-Relevance

X. Jiang, M. Ferrer, and A. Torsello (Eds.): GbRPR 2011, LNCS 6658, pp. 205–214, 2011.

Minimum-Redundancy criterion (MRMR), which is equivalent to MIFS with $\beta = \frac{1}{n-1}$. Yang and Moody's [9] Joint Mutual Information (JMI) criterion is based on conditional MI and selects features by checking whether they bring additional information to an existing feature set. This method effectively rejects redundant features. Kwak and Choi [5] improve MIFS by developing MIFS-U under the assumption of a uniform distribution of information for input features. It calculates the MI based on a Parzen window, which is less computationally demanding and also provides better estimates.

However, there are two limitations for the above MI feature selection methods. Firstly, they assume that each individual relevant feature should be dependent with the target class. This means that if a single feature is considered to be relevant it should be correlated with the target class, otherwise the feature is irrelevant [2]. So only a small set of relevant features is selected, and larger feature combinations are not considered. The second weakness is that most of the methods simply consider pairwise feature dependencies, and do not check for third or higher order dependencies between the candidate features and the existing features. To overcome the above problem, we introduce the so called multidimensional interaction information (MII) $I(F; C) = I(f_1, \ldots, f_m; C)$ to select the optimal subset of features. The main reason for using $I(F; C)$ as feature selection criterion is that: because $I(F; C)$ is a measure of the reduction of uncertainty in class C due to the knowledge of feature vector $F = \{f_1, \ldots, f_m\}$, selecting features that maximize $I(F; C)$, from an information theoretic perspective, translates into selecting those features that contain the maximum information about class C.

Although an MII based on the second-order feature dependence assumption can be used to select features that both maximize class-separability and simultaneously minimize dependencies between feature pairs, there is no reason to assume that the final optimal feature subset formed by features that only exhibit pairwise interactions. In particular the approach neglects the fact that third or higher order dependencies feature combinations may determine the optimal feature subset.

The primary reason for using the approximation $\widehat{I}(F; C)$ for feature selection instead of directly using multidimensional interaction information $I(F; C)$ is that $I(F; C)$ requires estimation of the joint probability distribution for features using large training samples. To overcome this problem, in this paper, we propose a graph-based approach to feature selection. In this feature selection scheme, the original features are clustered into different dominant-sets based on dominant-set clustering and each dominant-set just includes a small set of features. Therefore, for each dominant set, we do not need to use the approximation $\widehat{I}(F; C)$. Instead we can directly use the multidimensional interaction information $I(F; C)$ criterion for feature selection. Using the Parzen window for probability distribution estimation, we then apply a greedy strategy to incrementally select the feature that maximizes the multidimensional mutual information between the already selected features and the output class set.

2   Dominant-Set Clustering Algorithm

Concept of Dominant Set: The dominant set[6], is a combinational concept in graph theory that generalizes the notion of a maximal complete subgraph from simple graphs to edge-weighted graphs. In fact, dominant sets turn
out to be equivalent to maximal cliques. The definition of the dominant set
simultaneously emphasizes internal homogeneity and together with external in-
homogeneity. Thus it is can be used as a general definition of a "cluster". To
provide an example, assume there are N training samples, each having 5 feature vectors. In order to capture the dominant features from these 5 features
(represented as $F_1, \ldots, F_5$), we construct a graph $G = (V, E)$ with node-set V,
edge-set $E \subseteq V \times V$ and edge weight matrix W whose elements are in the interval $[0, 1]$. Each vertex represents a feature and the edge between two features
represents their pairwise relationship. The weight on the edge reflects the degree
of relevance between two features. Therefore, we represent the graph G with
the corresponding edge-weight or weighted relevance matrix. In our example,
in Fig. 1, features $\{F_1, F_2, F_3\}$ form the dominant set, since the edge weights
"internal" to that set (0.6, 0.7 and 0.9) are larger than the sum of those between
the internal and external features (which is between 0.05 and 0.25).

For the graph $G = (V, E)$ above, we can locate the dominant set by finding
the solutions of a quadratic program that maximizes the functional

$$f(\mathbf{x}) = \frac{1}{2}\mathbf{x}^T \mathbf{W} \mathbf{x} . \tag{1}$$

subject to $\mathbf{x} \in \triangle$, where $\triangle = \{\mathbf{x} \in \mathbb{R}^n : \mathbf{x} \geq 0 \, and \, \sum_{i=1}^{n} x_i = 1\}$ and $\mathbf{W}$ is the
relevance weight matrix between features. The dominant set corresponds in the
strict sense with solutions of the quadratic program. Let u denote a strict local
solution of the above program. It has been proved by [6] that $\sigma(u) = \{i|u_i > 0\}$ is
equivalent to a dominant set of the graph represented by the edge-weight matrix
$\mathbf{W}$. In addition, the local maximum of $f(u)$ indicates the "cohesiveness" of the
corresponding cluster. The replicator equation can be used to solve the program
using the iterative update equation:

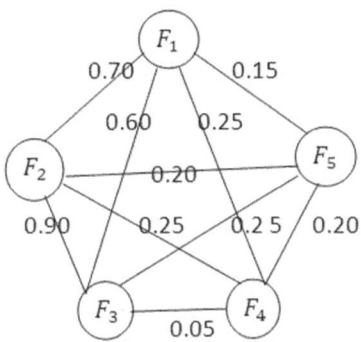

Fig. 1. The subset of features $\{F_1, F_2, F_3\}$ is dominant

$$x_i(t+1) = x_i(t)\frac{(\mathbf{W}\mathbf{x}(t))_i}{\mathbf{x}(t)^T\mathbf{W}\mathbf{x}(t)} \ . \tag{2}$$

where $x_i(t)$ is correspondent to the $i-th$ feature vector at iteration t of the update process.

Dominant-Set Clustering Algorithm: Pavan et al have demonstrated that the concept of a dominant set provides an effective framework for iterative pairwise clustering. Consider a set of features represented by an undirected edge-weighted graph with no self-loops. Let the graph be denoted by $G = (V, E, \omega)$ where $V = 1, \ldots, n$ is the vertex set, $E \subseteq V \times V$ is the edge set, and ω is the weight function. Each vertex represents a feature and the weight residing on the edge between two nodes represents the pairwise affinity of the corresponding features. To cluster the features into coherent groups, a dominant set of the weighted graph is iteratively located, and then removed from the graph. This process is repeated until the node-set of the graph is empty. The main property of a dominant set is that the overall similarity among the internal features is greater than that between the external features and the internal features.

3   Feature Selection Using Dominant-Set Clustering

In this paper we aim to utilize the dominant-set clustering algorithm for feature selection. Using a graph representation of the features, there are three steps to the algorithm, namely a) computing the relevance matrix $\mathbf{W} = (\mathbf{w}_{ij})_{n \times n}$ based on the mutual information between feature vectors, b) dominant-set clustering to cluster the feature vectors and c) selecting the optimal feature set from each dominant set using the multidimensional interaction information (MII) criterion. Fig. 2 shows a schematic view of the proposed method for feature selection. In the remainder of this paper we describe these elements of our feature selection algorithm in more detail.

Computing the Relevance Matrix: In accordance with Shannon's information theory [8], the uncertainty of a random variable Y can be measured by the entropy $H(Y)$. For two variables X and Y, the conditional entropy $H(Y|X)$ measures the remaining uncertainty about Y when X is known. The mutual information (MI) represented by $I(X;Y)$ quantifies the information gain about Y provided by variable X. The relationship between $H(Y)$, $H(Y|X)$ and $I(X;Y)$ is $I(X;Y) = H(Y) - H(Y|X)$.

As defined by Shannon, the initial uncertainty for the random variable Y is expressed as:

$$H(Y) = -\sum_{y \in Y} P(y) \log P(y) \ . \tag{3}$$

where $P(y)$ is the prior probability density function over Y. The remaining uncertainty in the variable Y if the variable X is known is defined by the conditional entropy $H(Y|X)$

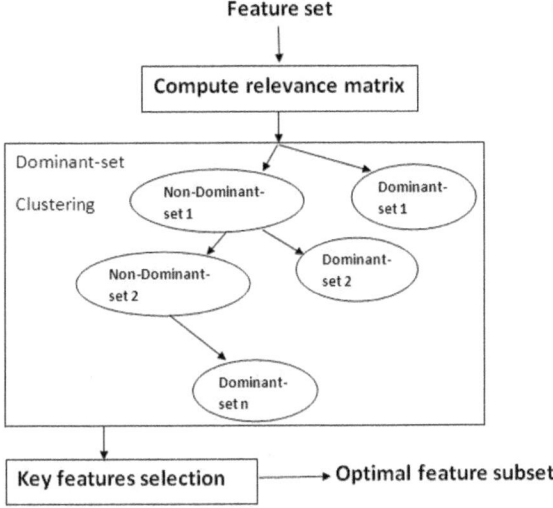

Fig. 2. The flowchart of our approach for feature selection

$$H(Y|X) = -\int_x p(x)\{\sum_{y\in Y} p(y|x)\log p(y|x)\}dx . \qquad (4)$$

where $p(y|x)$ denotes the posterior probability for variable Y given another random variable X. After observing the variable vector x, the amount of additional information gain is given by the mutual information (MI)

$$I(X;Y) = H(Y) - H(Y|X) = \sum_{y\in Y}\int_x p(y,x)\log\frac{p(y,x)}{p(y)p(x)}dx . \qquad (5)$$

From the above definition, we can see that mutual information quantifies the information which is shared by two variables X and Y. When the $I(X;Y)$ is large, this implies that variable X and variable Y are closely related, otherwise, when $I(X;Y)$ is equal to 0, this means that two variables are totally unrelated. Therefore, in our feature selection scheme, the relevance of pairs of feature vectors is computed using mutual information. Suppose there are N training samples, each having K feature vectors. The kth feature vector for the lth training sample is f_k^l, so we can represent the kth feature vector for the N training samples as the long vector $F_k = \{f_k^1, f_k^2, \ldots, f_k^N\}$. The entropy of the feature vector F_k where $(k = 1, 2, \ldots, K)$ can be computed using Equation (3). For two feature vectors F_{k1} and F_{k2}, their mutual information $I(F_{k1}, F_{k2})$ can be computed by Equation (5). The relevance degree between two feature vectors F_{k1} and F_{k2} can be defined as [10]:

$$\mathbf{W}(F_{k1}, F_{k2}) = \frac{2I(F_{k1}, F_{k2})}{H(F_{k1}) + H(F_{k2})} . \qquad (6)$$

where $k1, k2 \in K$ and the higher the value of $\mathbf{W}(F_{k1}, F_{k2})$ the more relevant are the features F_{k1} and F_{k2}. Otherwise, if $\mathbf{W}(F_{k1}, F_{k2}) = 0$, the two features are totally unrelated. In addition, for the above computation, we use Parzen-Rosenblatt window method to estimate the probability density function of random variables F_{k1} and F_{k2} [7]. The Parzen probability density estimation formula is given by: $p(x) = \frac{1}{N}\phi(\frac{x-x_i}{h})$, where $\phi(\frac{x-x_i}{h})$ is the window function and h is the window width. Here, we use a Gaussian as the window function, so $\phi(\frac{x-x_i}{h}) = \frac{1}{(2\pi)^{\frac{d}{2}}h^d|\Sigma|^{\frac{1}{2}}}$ $\exp(\frac{(x-x_i^T)\Sigma^{-1}(x-x_i)}{-2h^2})$, where Σ is the covariance of $(x - x_i)$, d is the length of vector x. When $d = 1$, $p(x)$ estimates the marginal density and when $d = 2$, $p(x)$ estimates the joint density of variables such as F_{k1} and F_{k2}.

Dominant-set Clustering: As illustrated in Fig. 2, the dominant-set clustering algorithm commences from the relevance matrix and iteratively bi-partitions the features into a dominant set and a non-dominant set. It therefore produces the dominant-set progressively and hierarchically. The clustering process stops when all the features are grouped into one of the dominant-sets.

Selecting Key Features: The multidimensional interaction information between feature vector $F = \{f_1, \ldots, f_m\}$ and class variable C is:

$$I(F; C) = I(f_1, \ldots, f_m; C) = \sum_{f_1, \ldots, f_m} \sum_{c \in C} P(f_1, \ldots, f_m; c)$$
$$\times \log \frac{P(f_1, \ldots, f_m; c)}{P(f_1, \ldots, f_m)P(c)} . \tag{7}$$

The main reason for using $I(F; C)$ as a feature selection criterion is that: because $I(F; C)$ is a measure of the reduction of uncertainty in class C due to knowledge of the feature vector $F = \{f_1, \ldots, f_m\}$, from an information theoretic perspective selecting features that maximize $I(F; C)$ translates into selecting those features that contain the maximum information about class C. In practice and as noted in the introduction, locating a feature subset that maximizes $I(F; C)$ presents two problems: 1) it requires an exhaustive "combinatorial" search over the feature space, and 2) it demands large training sample sizes to estimate the higher order joint probability distribution in $I(F; C)$ with a high dimensional kernel [5]. Bearing these obstacles in mind, most of the existing related papers approximate $I(F; C)$ based on the assumption of lower-order dependencies between features. For example, the first-order class dependence assumption includes only first-order interactions. That is it assumes that each feature independently influences the class variable, so as to select the mth feature, f_m, $P(f_m|f_1, \ldots, f_{m-1}, C) = P(f_m|C)$. A second-order feature dependence assumption is proposed by Guo and Nixon [4] to approximate $I(F; C)$, and this is arguably the most simple yet effective evaluation criterion for selecting features. The approximation is given as

$$I(F; C) \approx \hat{I}(F; C) = \sum_i I(f_i; C) - \sum_i \sum_{j>i} I(f_i; f_j)$$
$$+ \sum_i \sum_{j>i} I(f_i; f_j|C) . \tag{8}$$

By using $\widehat{I}(F;C)$ instead of $I(F;C)$, it is possible to locate a subset of informative features by implementing a greedy "pick-one-feature-at-a-time" selection procedure. Given K features, out of which m are to be selected ($m < K$), this involves two steps: 1) select the first feature f'_{max} that maximizes $I(f';C)$, and 2) select $m-1$ subsequent features that maximize the criterion in Equation (8), i.e., select the second feature f''_{max} that maximizes $I(f'';C) - I(f'';f'_{max}) + I(f'';f'_{max}|C)$, select the third feature f'''_{max} that maximizes $I(f''';C) - I(f''';f'_{max}) - I(f''';f''_{max}) + I(f''';f'_{max}|C) + I(f''';f''_{max}|C)$ and so on.

Although an MII based on the second-order feature dependence assumption can select features that maximize class-separability and simultaneously minimize dependencies between feature pairs, there is no reason to assume that the final optimal feature subset is formed by pairwise interactions between features. In-fact, it neglects the fact that third or higher order dependencies can be lead to an optimal feature subset.

The primary reason for using the approximation $\widehat{I}(F;C)$ for feature selection instead of directly using multidimensional interaction information $I(F;C)$ is that $I(F;C)$ requires estimation of the joint probability distribution of features using a large training sample. Consider the joint distribution $P(F) = P(f_1, \ldots, f_m)$, by the chain rule of probability

$$P(f_i, \ldots, f_m) = P(f_1)P(f_2|f_1) \times P(f_3|f_2, f_1) \cdots P(f_m|f_1, f_2 \ldots f_{m-1}) , \quad (9)$$

$$P(F;C) = P(f_1, \ldots f_m; C) = P(C)p(f_1|C)P(f_2|f_1, C)P(f_3|f_1, f_2, C)$$
$$\times P(f_4|f_1, f_2, f_3, C) \cdots P(f_i|f_1, \ldots, f_m, C) . \quad (10)$$

In our feature selection scheme, the original features are clustered into different dominant-sets based on dominant-set clustering and each dominant-set just includes a small set of features. Therefore, for each dominant set, we do not need to use the approximation $\widehat{I}(F;C)$. Instead, we can directly use the multidimensional interaction information $I(F;C)$ criterion for feature selection. Using Parzen windows for probability distribution estimation, we then apply the greedy strategy to select the feature that maximizes the multidimensional mutual information between the features and the output class set. As a result the first feature f'_{max} maximizes $I(f',C)$, the second selected feature f''_{max} maximizes $I(f'', f', C)$, the third feature f'''_{max} maximizes $I(f''', f'', f', C)$, and so on. For each dominant set, we repeat this procedure until $|S| = k$.

4   Experiments and Comparisons

The data sets used to test the performance of our proposed algorithm are the benchmark data sets from the NIPS 2003 feature selection challenge and the UCI Machine Learning Repository. Table. 1 summarizes the properties of these data-sets. Using the feature selection algorithm outlined above, we make a comparison between our proposed feature selection method (referred to as the DSplusMII method) (which utilises the multidimensional interaction information

Table 1. Summary of UCI and NIPS benchmark data sets

Data-set	From	Examples	Features	Classes
Madelon	NIPS	2000	500	2
Breast cancer	UCI	699	10	2
Pima	UCI	768	8	2

Table 2. The experiment results on three data-sets

Method	Madelon	Breast cancer	Pima
MII	$\{f_{476},\ f_{49},\ f_{178},\ f_{131}, f_{491},$ $f_{299},\ f_{283},\ f_{121},\ f_{425}, f_7,$ $f_{385},\ f_{216},\ f_{458},\ f_{237}, f_{310},$ $f_{366},\ f_{98},\ f_{499},\ f_{54}, f_{346},$ $f_{198},\ f_{368}\ \}$	$\{f_3,\ f_7\}$	$\{f_2, f_8, f_6, f_7\}$
DS*plus*MII	$\{f_{476},\ f_{379},\ f_{49},\ f_{330},\ f_{412},$ $f_{137},\ f_{11},\ f_{256},\ f_{135},\ f_{56},$ $f_{138},\ f_{283},\ f_{324},\ f_{425},\ f_{467},$ $f_{62},\ f_{455},\ f_{472},\ f_{208},\ f_{206},$ $f_{169},\ f_{424}\ \}$	$\{f_3,\ f_7\}$	$\{f_2, f_8, f_6, f_1\}$

(MII) criterion and dominant-sets for feature selection) and the use of multidimensional interaction information (MII) using the second-order approximation, see Equation (8).

The experimental results shown in Table. 2 demonstrate that at small dimensionality, i.e. with the Breast cancer data-set(10 features and 699 examples) and the Pima data-set (8 features and 768 examples), the feature subset selected using our proposed method (i.e. DS*plus*MII) is consistent at least to some degree with those obtained using MII with second-order approximation. However, at higher dimensionality (e.g. the Madelon data set with 500 features and 2000 examples), there is a significant difference between the selected feature subsets. There are three reasons for this. The first reason is that dominant-set clustering focuses on the information-contribution of each feature, so the most informative features can be extracted. The second reason is that the multidimensional interaction information (MII) criterion is applied to each dominant set for feature selection, and can consider the effects of third and higher order dependecies between the features and the class. As a result the optimal feature combination can be located so as to guarantee the optimal feature subset. The third and final reason is that multidimensional interaction information (MII) by second-order approximation simply checks for pair-wise dependencies between features and the class, and so only limited feature subsets can be obtained. When the database is large, our proposed method DS*plus*MII shows its advantage.

To illustrate the dominant-set clustering process for feature extraction in more detail, we list the dominant sets for Pima and Breast cancer data-set in Table. 3. By inspection, we can see that the first dominant set includes most of the important features. For example, in the Breast cancer data-set, the final selected

Table 3. The dominant sets for Breast cancer and Pima data-set

Dominant-sets	Breast cancer	Pima
Dominant set 1	$\{f_3,\ f_4,\ f_6,\ f_7$ $f_9,\ f_5,\ f_8\}$	$\{f_5,\ f_2,\ f_4,\ f_8$ $f_3,\ f_6\}$
Dominant set 2	$\{f_1,\ f_2,\ f_{10}\}$	$\{f_7,\ f_1\}$

Table 4. J value comparisons for two methods on three data sets

Method	Madelon	Breast cancer	Pima
MII	1.0867	3.7939	1.3867
DSplusMII	1.1082	3.7939	1.3977

features $\{f_3, f_7\}$ are all from the first dominant-set, the second dominant-set provides no further information relevant to the classication process. For the Pima data-set, most of the final selected informative features are also from the first dominant set. This reveals the advantage of our dominant-set based feature extraction method. It focuses on the information-contribution of each feature which is capable capturing the greatest number of informative features at a low computation cost. Additionally, it also indicates that not all of the dominant-sets located by dominant-set clustering are significant. It is for this reason that we utilize the multidimensional interaction information (MII) criterion for further feature selection.

After obtaining the discriminating features, we compute a scatter separability criterion to evaluate the quality of the selected feature subset. This is a well known measure of class separability introduced by Devijiver and Kittler [3], and given by

$$J(Y) = \frac{|S_w + S_b|}{|S_w|} = \prod_{k=1}^{d}(1 + \lambda_k) \ . \tag{11}$$

where Y denotes the feature set, $tr(S)$ is the sum of the diagonal elements of S, λ_k, $k = 1 \ldots d$, are the eigenvalues of matrix $S_w^{-1}S_b$, and S_w and S_b are the between and within class scatter matrices.

In Table. 4, we compare the the performance of the two methods. At small dimensionality there is little difference between the two methods. However, at higher dimensionality, the features selected by our proposed DSplusMII method are superior to the features selected by MII based on the second-order approximation. This means that our proposed DSplusMII feature selection method can guarantee the optimal feature subset, as it not only focuses on the information-contribution of each feature but also considers its contribution to class.

5   Conclusions

This paper has presented a new graph theoretic approach to feature selection. The proposed feature selection method offers two major advantages. First, dominant-set clustering can capture the most informative features. Second, the MII criteria takes into account high-order feature interactions, overcoming the problem of overestimated redundancy. As a result the features associated with the greatest amount of joint information can be preserved.

References

1. Battiti, R.: Using Mutual Information for Selecting Features in Supervised Neural Net Learning. IEEE Transactions on Neural Networks 5(4), 537–550 (2002)
2. Cheng, H., Qin, Z., Qian, W., Liu, W.: Conditional Mutual Information Based Feature Selection. In: IEEE International Symposium on Knowledge Acquisition and Modeling, pp. 103–107 (2008)
3. Devijver, P., Kittler, J.: Pattern Recognition: A Statistical Approach, vol. 761. Prentice-Hall, London (1982)
4. Guo, B., Nixon, M.: Gait Feature Subset Selection by Mutual Information. IEEE Transactions on Systems, Man and Cybernetics, Part A: Systems and Humans 39(1), 36–46 (2008)
5. Kwak, N., Choi, C.: Input Feature Selection by Mutual Information Based on Parzen Window. IEEE TPAMI 24(12), 1667–1671 (2002)
6. Pavan, M., Pelillo, M.: A New Graph-Theoretic Approach to Clustering and Segmentation. In: IEEE Computer Society Conference on Computer Vision and Pattern Recognition, vol. 1. IEEE, Los Alamitos (2003)
7. Peng, H., Long, F., Ding, C.: Feature Selection Based on Mutual Information: Criteria of Max-Dependency, Max-Relevance, and Min-Redundancy. IEEE Transactions on Pattern Analysis and Machine Intelligence, 1226–1238 (2005)
8. Shannon, C.: A Mathematical Theory of Communication. ACM SIGMOBILE Mobile Computing and Communications Review 5(1), 3–55 (2001)
9. Yang, H., Moody, J.: Feature Selection Based on Joint Mutual Information. In: Proceedings of International ICSC Symposium on Advances in Intelligent Data Analysis, pp. 22–25 (1999)
10. Zhang, F., Zhao, Y., Fen, J.: Unsupervised Feature Selection based on Feature Relevance. In: International Conference on Machine Learning and Cybernetics, vol. 1, pp. 487–492. IEEE, Los Alamitos (2009)

Spatio-Temporal Extraction of Articulated Models in a Graph Pyramid

Nicole M. Artner[1], Adrian Ion[1,2], and Walter G. Kropatsch[1]

[1] PRIP, Vienna University of Technology, Austria
{artner,ion,krw}@prip.tuwien.ac.at
[2] Institute of Science and Technology Austria (IST Austria)

Abstract. This paper presents a method to create a model of an articulated object using the planar motion in an initialization video. The model consists of rigid parts connected by points of articulation. The rigid parts are described by the positions of salient feature-points tracked throughout the video. Following a filtering step that identifies points that belong to different objects, rigid parts are found by a grouping process in a graph pyramid. Valid articulation points are selected by verifying multiple hypotheses for each pair of parts.

Keywords: articulated object, model extraction, graph pyramid.

1 Introduction

Tracking articulated objects is an important and active field of research in Computer Vision [10,15,1]. A model of the target (the object to be tracked) is used by tracking methods to detect and associate instances of the object of interest in consecutive frames. This model is at the minimum a rectangle-shaped close-up of the object (called a template) or a color histogram, but can be as sophisticated as an online-trained classifier [11], or a hierarchical description of the objects parts and their salient features [5].

The proposed method automatically builds such a model of an articulated object, from a set of trajectories of feature-points in an initialization video:

1. build a triangulated graph on the positions of the points in the first frame;
2. label each triangle as "relevant" (on an object) or "separating" (connecting objects/parts) based on the variation of its edge-lengths over the video;
3. group *relevant* triangles in a graph pyramid framework based on their orientation variation, and obtain the "rigid" parts (build on [4]);
4. localize and verify points of articulation by observing the articulated movement of the parts.

The presented approach is related to the work done in *video object segmentation* (VOS), where the task is to separate foreground from background with the help of a video sequence. VOS methods can be divided into two categories [7]: (1) Two-frame motion/object segmentation [3,8] and (2) Multi-frame spatio-temporal

X. Jiang, M. Ferrer, and A. Torsello (Eds.): GbRPR 2011, LNCS 6658, pp. 215–224, 2011.

segmentation/tracking [7,14]. With some exceptions (e.g. [8]), the output of VOS methods is a pixel-level assignment to foreground (FG) and background (BG). Less attention is given to modeling the FG into its constituent rigid parts and joints. Most VOS methods work on the pixel level.

Motion segmentation (MS) [13,16] works on the basis of trajectories of features. Here the main concern is with the segmentation of trajectories to objects and not with detecting their parts. An advantage is that many MS methods can deal with trajectories that do not span the whole video.

The work in factorization methods for *structure from motion* (SfM) [20,19,9,17] is probably the most related to ours. These methods use a factorization technique based on singular value decomposition to detect the linear subspaces in which trajectories of feature-points of rigid/non-rigid parts lie. Like in our case, articulation points/axis are computed in a step following part detection. Trajectories of feature-points that span the whole video are required.

Our approach analyzes the trajectories of features on a higher abstraction level – in a triangulation. The used triangulated graphs encode spatial relationships resulting out of spatial proximity between features, and are the basis for all processes and decisions. The advantage of using a triangulation is the additional information about the motion and behavior of features in this relationship. The motions of parts are not constrained to linear subspaces, however, in this work, parallel projection is assumed and only 2D motions are considered.

1.1   Paper Outline

This paper is organized as follows: Sec. 2 recalls irregular graph pyramids, which are later used for the grouping process. Sec. 3 describes the spatio-temporal filtering and the grouping process, where the "rigid" parts are identified. Sec. 4 explains how the points of articulation are determined. Sec. 5 presents the experiments and in Sec. 6 conclusions are given.

2   Recall: Irregular Graph Pyramids

An irregular graph pyramid is a stack of successively reduced planar graphs $P = \{G_0, \ldots, G_n\}$. A pyramid is typically build in a bottom-up manner using only local operations. Each level $G_k, 0 < k \leq n$ is obtained by first *contracting* edges in G_{k-1}, if their vertices have the same label (regions should be merged), and then *removing* edges in the obtained intermediate graph to simplify the structure. In each G_{k-1} contracted edges form trees called *contraction kernels*. One vertex of each contraction kernel is called a *surviving vertex* and is considered to have been "survived" to G_k. The *receptive field* $F(v)$ of v is the (connected) set of vertices from level 0 that have been "merged" to v over levels $0 \ldots k$. Higher in the pyramid, the receptive fields cover more of the base level and decisions gradually change from local to global. Compared to regular pyramids, irregular graph pyramids have the advantage that their structure is not fixed, it adapts to the data. For more details about graph pyramids see for example [12].

3   Rigid Part Extraction

The input of the method consists of trajectories of feature-points from a training video. Two points can lie on: different objects, the same articulated object but on different rigid parts, or on the same rigid part of an object. To detect the feature points located on the same rigid part we proceed in two steps: (1) spatio-temporal filtering of edges connecting feature points and (2) grouping of triangles formed by the edges. In Sec. 4 we discuss the detection of points of articulation based on the rigid parts found by the method described in the following.

Detect points on different objects. A triangulated graph T is build by a Delaunay triangulation of the positions of the feature-points in the first frame. This step creates a neighborhood for the points and produces entities (edges, triangles), which have not just position, but also size and orientation. For a discussion on the robustness of graphs built on sets of points see [18].

In the graph T a high variation of the length of an edge over the video indicates that its end-points are very likely not located on the same rigid part. Based on this observation, the triangles (faces of T) are labeled as *relevant* if all three edges have the *maximum variation* $\Delta(e) = \max_{0 \leq t_1, t_2 < t_F} \{||e_{t_1}|| - ||e_{t_2}||\}$ below a defined threshold ϵ_r, and *separating* otherwise. Here $||e_{t_1}||$ is used to denote the length of edge e at time t_1 and F is the number of frames in the input sequence.

Connected components made of only *relevant* triangles of T identify detected objects (see Sec. 5, Fig. 5).

Detect points on the same rigid part. This step groups *relevant* triangles to identify the rigid parts. Only triangles that share an edge are grouped thus the obtained parts are guaranteed to be connected.

True rigid motion is rarely observed, e.g.: the skin of a human is elastic, and tracked feature positions are affected by noise. Thus a local decision (e.g. global threshold on $\Delta(e)$) cannot robustly determine which triangles belong to the same "rigid" part (see Fig. 4). Following the approach in [4], the grouping is done using a graph pyramid. Every vertex in the base level G_0 identifies a relevant triangle. Every vertex in the obtained top level G_n identifies a rigid part. Using a graph pyramid allows to make *local-to-global* decisions by using only local operations in a representation that is shift and rotation invariant.

The *orientation variation* $O_e(t)$ of an edge e over time is a 1D signal that encodes at each time t the accumulated orientation change relative to the orientation at frame 0. More formally, $O_e(t) = O_e(t-1) + \theta_e(t)$, where $\theta_e(t)$ is the relative change in orientation (signed angle) of the edge e between frames t and $t-1$. For example, turning around the x axis once will give a value of $360°$ degrees and turning twice in the same direction will give $720°$, not $0°$. The direction of rotation is encoded by the sign: counter clockwise (CCW) is positive, and clockwise (CW) is negative.

The *orientation variation* $O_r(t)$ of a triangle r, at time t is defined as the average of the orientation variations of its edges, at time t. The *maximal relative orientation change* of two triangles is the highest absolute difference of their orientation changes over the input sequence $\Delta(r_1, r_2) = \max_{0 \le t < t_F} \{|O_{r_1}(t) - O_{r_2}(t)|\}$. The value $\Delta(r_1, r_2)$ is used as a cue (grouping criterion) for the triangles r_1, r_2 belonging to the same part.

The graph pyramid groups *relevant* triangles into "rigid" parts such that:

- all the triangles inside a part have a similar O_r,
- the average O_r of triangles in two neighboring parts is different.

Vertices of G_k represent parts consisting of one or more triangles. An edge e in G_k encodes that there exist two triangles r_1, r_2, one belonging to each of the parts represented by the vertices connected by e, such that r_1 and r_2 share an edge in T. Notice the duality between T and G_0: the vertices of G_0 represent the triangles (faces) in T, the edges of G_0 encode adjacency of triangles in T, while the edges in T connect feature-points and make up (the boundaries of) the triangles.

This grouping is similar in spirit to the image segmentation task, where the results should be regions with homogeneous color/texture neighbored with regions that look very different. For more details on the grouping process see [4].

4   Determine Points of Articulation

Articulated motion is a piecewise rigid motion, where the rigid parts conform to the rigid motion constraints, but the overall motion is not rigid [2]. A *point of articulation* connects rigid parts. The parts can move independent of each other, but their distance to the point of articulation remains the same. This paper considers articulation in the image plane (1 degree of freedom). In the following we will call articulated parts, two parts that perform an articulated motion constrained by a point of articulation.

Having the rigid parts in the scene (vertices in the top level G_n) we proceed to discover the parts that move constrained by articulation and to find the corresponding point of articulation. Determining the points of articulation is done in two steps: (1) generate hypotheses for points of articulation (Section 4.1), and (2) verify hypotheses and select valid ones (Sec. 4.2).

4.1   Generation of Hypotheses for Points of Articulation

Given two time steps $0 \le t_1 < t_2 < t_F$ the motion of the points of two articulated parts A and B, can be modeled by considering rotation and translation separately. Using matrices the point correspondence can be written as:

$$\mathbf{p}' = (R * (\mathbf{p} - \mathbf{c}) + \mathbf{c}) + \mathbf{o} \tag{1}$$

where $\mathbf{p}$ is the point at time t_1 and $\mathbf{p}'$ is the same point at time t_2. The point $\mathbf{p}'$ is obtained by first rotating $\mathbf{p}$ around $\mathbf{c}$ with angle θ and then translating it with offset $\mathbf{o}$. R is the 2D rotation matrix with angle θ given by:

$$R = \begin{pmatrix} \cos(\theta) & -\sin(\theta) \\ \sin(\theta) & \cos(\theta) \end{pmatrix}$$

To compute the position of the point of articulation $\mathbf{c}$ at time t_1 it is sufficient to know at times t_1 and t_2 the positions of two points of each of the two rigid parts: $\mathbf{p}_i, \mathbf{p}'_i$, $0 < i \leqslant 4$. Indexes $i \in \{1, 2\}$ are used for the points of part A and $i \in \{3, 4\}$ for the ones of B. These points will be denoted by the term *reference points*. Taking Eq. 1 for the four points produces the following system of equations:

$$\left\{ \mathbf{p}'_i = (R_i * (\mathbf{p}_i - \mathbf{c}) + \mathbf{c}) + \mathbf{o} \qquad i = 1 \ldots 4, \right. \tag{2}$$

where $R_i = R_A$ if $i \leq 2$ and $R_i = R_B$ otherwise. The matrices R_A, R_B are the 2D rotation matrices of the parts A, B with angles θ_A, θ_B, respectively. Solving the system gives the $2D$ coordinates of $\mathbf{c}, \mathbf{o}$ at time t_1 and the values of $\sin(\theta_A), \cos(\theta_A), \sin(\theta_B), \cos(\theta_B)$.

Points of articulation do not have to be visually salient i.e. easily trackable, and are thus not expected to belong to the tracked feature-points. We use Eq. 2 to generate hypotheses for points of articulation for every pair of parts and time stamps t_1, t_2 selected as described in the following. In addition to the detected rigid parts, the grouping step also computes for each part an orientation variation signal as the average of O_r over the contained triangles r. If no rotation exists between frames t_1, t_2 the system in Eq. 2 can have an infinite number of solutions for the point of articulation $\mathbf{c}$. Thus for any two parts A, B, time stamps t_1, t_2 are selected, where the orientation variation signals corresponding to both rigid parts differ between t_1 and t_2 with more than a value ϵ_a. In experiments we divide the sequence in 20 fixed length time windows, and verify the above condition for all of them. For more general purposes, the optimal time steps t_1, t_2 can be found in polynomial time.

The reference points for each rigid part, at time t are the centroid of the part and a point $\mathbf{q}$. The point $\mathbf{q}$ is obtained by translating the centroid over a fixed distance $d > 0$, in the direction given by the corresponding orientation variation of the part, at time t. (This strategy gives a more stable estimate than selecting two of the possibly noisy feature-point positions.) For each rigid part a local coordinate system is derived based on the two reference points. The coordinate system has the origin at the centroid of the part, and the directions of the two axis defined by the motion of the part. Having the coordinates of $\mathbf{c}$ in each of the two local coordinate systems at time t_1, and having the reference points at time t, it is possible to calculate the expected position of $\mathbf{c}$ w.r.t. each adjacent part, at time t. Note that d only acts as a scaling factor for the local coordinate system, and its exact value does not affect the final result. Fig. 1 illustrates the explained concepts.

The output of this step are multiple hypotheses of points of articulation for each pair of rigid parts. Each hypothesis for a point of articulation is described by its position in the local coordinate system of each incident part.

4.2   Verification and Selection of Hypotheses

By definition, rigid parts performing articulated motion keep a constant distance to the point of articulation. This is equivalent to saying that the positions of the

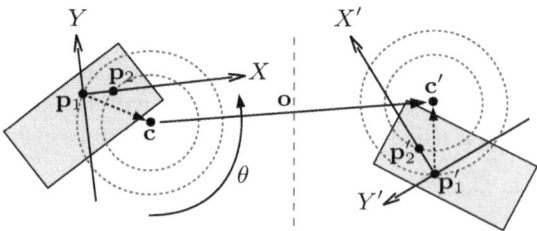

Fig. 1. Determining and encoding the point of articulation in the local coordinate system, during two time steps: left t_1, right t_2

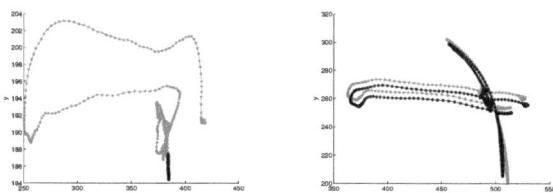

Fig. 2. Verification of hypotheses for points of articulation. Left: invalid hypothesis, right: valid hypothesis. Each curve shows the positions of the hypothesized point of articulation relative to one part. Positions corresponding to consecutive frames have been connected.

points of articulation calculated with the local coordinate systems of the two connected object parts coincide. We use this property to verify the previously generated hypotheses and select the valid ones.

Given a hypothesis for a point of articulation, two positions are computed for each frame of the video – one using the local coordinate system of each connected part. The inaccuracy μ of a hypothesis is the maximum of the distances between the two computed positions over the whole video. If μ is small the hypothesis is considered valid (in practice we take a threshold ϵ_v). If for a pair of rigid parts multiple valid hypotheses exist, the one with the smallest inaccuracy μ is taken. If no hypotheses with a small μ exists, the parts are not considered connected through a point of articulation. Fig. 2 shows positions for hypotheses of points of articulation generated during the verification step.

5   Experiments

In our experiments the Kanade-Lucas-Tomasi tracker [6] is used to track feature points (in this case corners) and supply the necessary trajectories. Only the points which could be tracked successfully over the whole sequence are used.

Sequence 1 is a video with a human, sequence 2 with a finger, and sequence 3 is a synthetic video, all undergoing a globally articulated motion with locally arbitrary deformations (see Fig. 5).

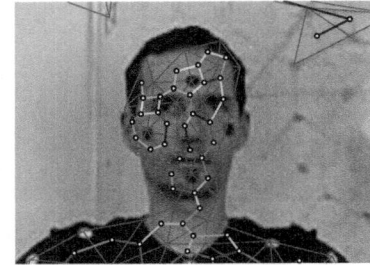

Fig. 3. Left: deformation of the edges over time. Right: dual graph of motion of triangles over time. Color map on the right encodes the degree of deformation and dissimilarity of motion of triangles, where red is high and blue is low.

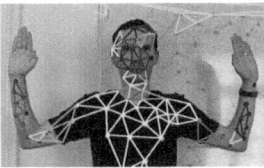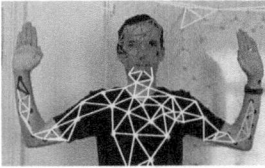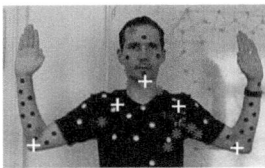

Fig. 4. Left, middle: Grouping result with global thresholds 0.25 and 0.6, respectively (different color means different part). Right: Comparison of identified points of articulation with baseline approach (red stars) and proposed approach (white crosses).

Fig. 3 provides an insight into the motion of the triangles in sequence 1. It visualizes the difference in deformation of edges and motion of triangles over time, which points out the challenge for the grouping process.

In Fig. 4 the results of two baselines are shown: (1) grouping of triangles with a global threshold (criterion: similar orientation of triangles over time) and (2) identification of points of articulation depending on the number of "rigid" edges in a triangle (criterion: two "rigid" edges and one highly deformed over time). Notice that a correct grouping into the whole hand, torso, upper and lower arms is not possible. The number and positions of the articulation points detected with the baseline is incorrect (see Fig. 5 for comparison with the proposed approach).

Fig. 5 collects the results with the proposed approach for sequences 1, 2 and 3 using the parameter values in Table 1. For sequence 1, the torso is connected with the base of the chin, because the features at the base of the chin slide when the head is tilted and remain in the same position in the image, creating a "rigid" triangle. In all three sequences the found points of articulation correctly connect the rigid parts. The threshold ϵ_v is sufficient to separate valid hypotheses from invalid ones, where there is no point of articulation in reality (e.g. head with background in sequence 1).

As in [20], a kinematic model (tree) of the objects can be defined by the found parts and their connections through joints.

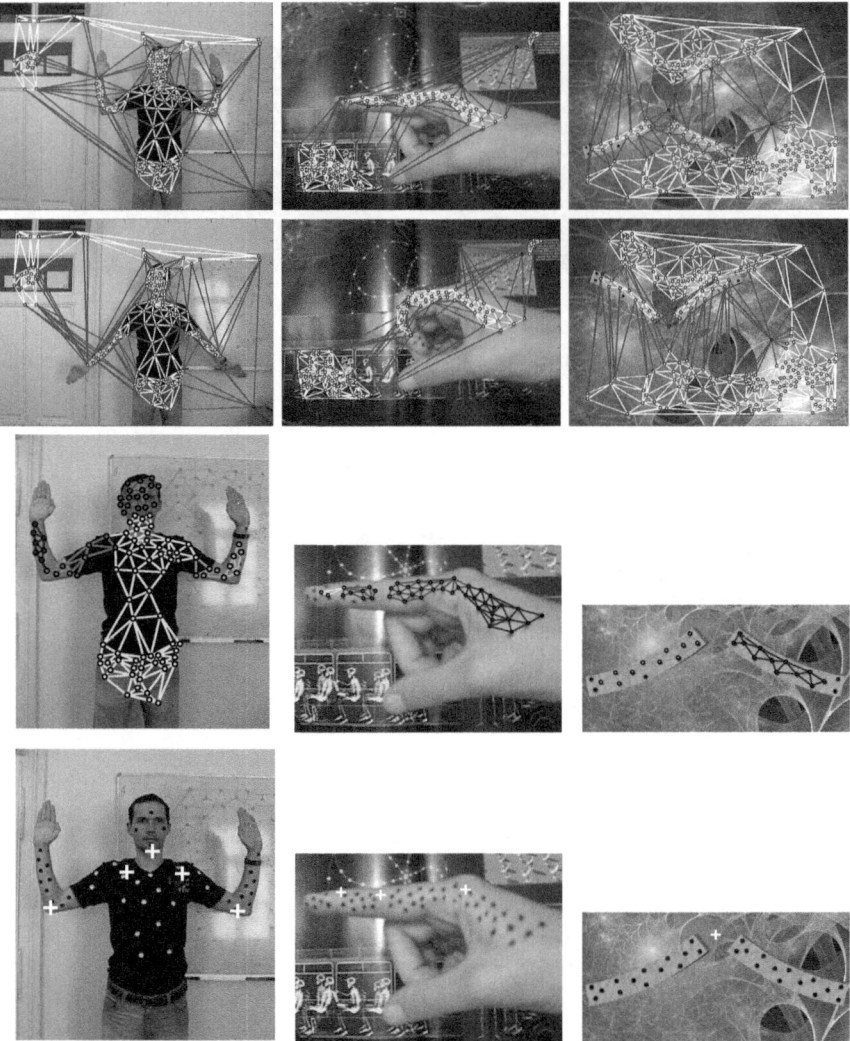

Fig. 5. Results obtained by the proposed method. First and second row: two frames of sequence 1,2, and 3 with spatio-temporal filtering result (white edges: *relevant* triangle and gray edges: *separating* triangle). Third row: Results of grouping process. Forth row: Identified points of articulation.

Discussion: The robustness of the tracker, the presence of salient points on the object(s), and the quality of the video should be sufficient to create the required observations (trajectories of feature-points).

The spatio temporal filtering (Sec. 3) will correctly identify triangles with all vertices on the same object if the motion of the objects relative to each other (distance variation) are larger than the local distance variation between neighboring feature-points of the same object.

Table 1. Values of the used parameters for sequences 1, 2 and 3

Sequence	ϵ_r (*relevant* triangles)	ϵ_a (generate hyp.)	ϵ_v (verify hyp.)
1 (human)	20	0.4	20
2 (finger)	10	0.4	10
3 (synthetic)	15	0.4	20

The grouping into rigid parts gives a correct result if the relative orientation change between two parts is larger than the local differences due to non-rigid deformation (e.g skin) or to imprecisions of the computed feature-point positions.

Points of articulation can be produced between any two pairs of detected rigid parts. To avoid detecting points of articulation between object parts and the background or between unrelated object parts, the unrelated parts (background) should translate with respect to each other.

In the presented approach no prior knowledge is used and it can be applied to videos with any arbitrary articulated or rigid object (i.e.: human, finger, animal, basket ball ...). The approach can only detect points of articulation, when there is articulated motion in the video. If the video contains a rigid foreground object moving in front of a "static" background, the result of the approach is a separation between foreground and background.

6 Conclusion

This paper presented a graph-based approach to identify the rigid parts of articulated objects and find their points of articulation. Trajectories of feature-points are used to describe the motion of objects in the scene. In the first frame a triangulation is built with the positions of the feature-points. A spatio-temporal filtering labels the triangles as *relevant* (object) or *separating* depending on the deformation of the edge lengths over time. A graph pyramid is used to group the *relevant* triangles into rigid parts depending on their orientation change over time. In a following step points of articulation connecting the rigid parts are identified. Experiments on natural and synthetic videos with articulation between quasi-rigid parts (skin, cloth) are used to verify the approach. In future work we plan to deal with input data containing incomplete trajectories and out of the plane motion.

Acknowledgments

This work has been partially supported by the Austrian Science Fund under grants S9103-N13 and P18716-N13.

References

1. Aggarwal, J.K., Cai, Q.: Human motion analysis: A review. Computer Vision and Image Understanding 73(3), 428–440 (1999)

2. Aggarwal, J.K., Cai, Q., Liao, W., Sabata, B.: Articulated and elastic non-rigid motion: A review. In: IEEE Workshop on Motion of Non-Rigid and Articulated Objects, pp. 2–14 (1994)
3. Altunbasak, Y., Eren, P.E., Tekalp, A.M.: Region-based parametric motion segmentation using color information. Graphical Models and Image Processing 60(1), 13–23 (1998)
4. Artner, N.M., Ion, A., Kropatsch, W.G.: Rigid part decomposition in a graph pyramid. In: The 14th Iberoamerican Congress on Pattern Recognition, pp. 758–765. Springer, Heidelberg (2009)
5. Artner, N.M., Ion, A., Kropatsch, W.G.: Multi-scale 2d tracking of articulated objects using hierarchical spring systems. Pattern Recognition 44(4), 800–810 (2010)
6. Birchfeld, S.: Klt: An implementation of the kanade-lucas-tomasi feature tracker (March 2008), http://www.ces.clemson.edu/~stb/klt/
7. Celasun, I., Tekalp, A.M., Gokcetekin, M.H., Harmanci, D.M.: 2-d mesh-based video object segmentation and tracking with occlusion resolution. Signal Processing: Image Communication 16(10), 949–962 (2001)
8. Chen, H.T., Liu, T.L., Fuh, C.S.: Segmenting highly articulated video objects with weak-prior random forests. In: European Conference on Computer Vision, pp. 373–385. Springer, Graz (2006)
9. Drouin, S., Hébert, P., Parizeau, M.: Incremental discovery of object parts in video sequences. Computer Vision and Image Understanding 110, 60–74 (2008)
10. Gavrila, D.M.: The visual analysis of human movement: A survey. Computer Vision and Image Understanding 73(1), 82–980 (1999)
11. Godec, M., Leistner, C., Saffari, A., Bischof, H.: On-line random naive bayes for tracking. In: ICPR, pp. 3545–3548 (2010)
12. Kropatsch, W.G., Haxhimusa, Y., Pizlo, Z., Langs, G.: Vision pyramids that do not grow too high. Pattern Recognition Letters 26(3), 319–337 (2005)
13. Lauer, F., Schnrr, C.: Spectral clustering of linear subspaces for motion segmentation. In: ICCV, pp. 678–685. IEEE, Los Alamitos (2010)
14. Li, H., Lin, W., Tye, B., Ong, E., Ko, C.: Object segmentation with affine motion similarity measure. In: Multimedia and Expo., pp. 841–844 (2001)
15. Moeslund, T.B., Hilton, A., Krüger, V.: A survey of advances in vision-based human motion capture and analysis. Computer Vision and Image Understanding 104(2-3), 90–126 (2006)
16. Nordberg, K., Zografos, V.: Multibody motion segmentation using the geometry of 6 points in 2d images. In: ICPR, pp. 1783–1787. IEEE, Istanbul (2010)
17. Ross, D.A., Tarlow, D., Zemel, R.S.: Learning articulated structure and motion. International Journal of Computer Vision 88(2), 214–237 (2010)
18. Tuceryan, M., Chorzempa, T.: Relative sensitivity of a family of closest-point graphs in computer vision applications. Pattern Recognition 24(5), 361–373 (1991)
19. Walther, T., Würtz, R.P.: Unsupervised learning of human body parts from video footage. In: 2nd Workshop on Non-Rigid Shape Analysis and Deformable Image Alignment, pp. 336–343 (2009)
20. Yan, J., Pollefeys, M.: A factorization-based approach for articulated nonrigid shape, motion and kinematic chain recovery from video. IEEE Trans. Pattern Anal. Mach. Intell. 30(5), 865–877 (2008)

Semi-supervised Segmentation of 3D Surfaces Using a Weighted Graph Representation

Filippo Bergamasco, Andrea Albarelli, and Andrea Torsello

Dip. di Scienze Ambientali, Informatica e Statistica, Universitá Ca' Foscari Venezia

Abstract. A wide range of cheap and simple to use 3D scanning devices has recently been introduced in the market. These tools are no longer addressed to research labs and highly skilled professionals. By converse, they are mostly designed to allow inexperienced users to easily and independently acquire surfaces and whole objects. In this scenario, the demand for automatic or semi-automatic algorithms for 3D data processing is increasing. Specifically, in this paper we concentrate on the segmentation task applied to the acquired surfaces. Such a problem is well known to be ill-defined both for 2D images and 3D objects. In fact, even with a perfect understanding of the scene, many different and incompatible semantic or syntactic segmentations can exist together. For this reasons, we refrain from any attempt to offer an automatic solution. Instead we introduce a semi-supervised procedure that exploits an initial set of seeds selected by the user. In our framework segmentation happens by iteratively visiting a weighted graph representation of the surface starting from the supplied seeds. The assignment of each element is driven by a greedy approach that accounts for the curvature between adjacent triangles. The proposed technique does not require to perform edge detection or to fit parametrized surfaces and its implementation is very straightforward. Still, despite its simplicity, tests made on scanned 3D objects show its effectiveness and easiness of use.

1 Introduction

Segmentation is an important preliminary task in many 2D and 3D data processing pipelines. For instance, splitting an image or a 3D object in smaller parts is very useful to perform high-level recognition [1,2], reverse engineering [3,4] and even tracking [5]. Of course, the expected outcome of a segmentation procedure is different depending on the intended use of the resulting parts. If the goal is to produce a set of image macro pixels, segments will be searched at a purely syntactic level, grouping together pixels of uniform color or texture, regardless of their belonging to one object or another. By contrast, different scenarios require a more semantical splitting, with the aim of separating foreground from background or finding the boundaries of the objects found in the scene. Of course these approaches tend to be more specialized, since the cues to exploit strongly depend on the problem context and on the availability of humans in the loop.

When dealing with the 3D domain, segmentation is mostly targeted at splitting an object or a surface into subdomains that can be later interpreted as

X. Jiang, M. Ferrer, and A. Torsello (Eds.): GbRPR 2011, LNCS 6658, pp. 225–234, 2011.

parametrized primitives. The complexity of such primitives can range from basic items, such as planes, cylinders or spheres, to complete parametrized models, depending on the overall goal of the pipeline. Simple primitives are fitted to the segmented parts mainly for object simplification [6,7], while completed models can be used for direct 3D object recognition with resilience to clutter [8] or invariance to scale [9]. Finally, another important application that needs surface decomposition is the angle and distance-preserving piecewise parametrization needed to apply textures to objects [10].

Surface segmentation can happen through many different methods. Some of them use standard clustering techniques or borrow segmentation procedures from the 2D domain, other exploit graph partitioning algorithms, shape fitting or even the distribution of symmetry planes over watertight objects.

Shlafman at al. propose to use a variation of K-means to group the triangles of the mesh into clusters [11]. This is a quite direct adaptation: first the user specify the desired number of clusters (k), then the process start by randomly selecting a set of k well spaced seed triangles and it iterates by alternating an assignment step (where each non-seed triangle is assigned to the nearest seed) and an adjustment step (where new seeds are selected by picking the triangle nearest to the center of each cluster).

Another classical technique is adapted by Moumoun et al., that suggests the use the Watershed principle [12] on a hierarchical transformation of connected faces structure based on the principal curvature. An interesting perspective on 3D segmentation is supplied by Podolak et al. [13], whose solution exploits the relation between mesh elements and the symmetry planes of the whole object.

Katz et al. [13] split the problem into two separate steps: first a probabilistic clustering is used in order to obtain meaningful but fuzzy components, then exact boundaries are constructed by assigning shared faces to the final cluster. A couple of years later, the same authors present an additional hierarchical method [14] that performs three steps: the transformation of the surface into a pose insensitive representation by mean of Multi-Dimensional Scaling (MDS); the localization of prominent feature points to be used as seeds; the extraction of clusters by refinement of core components obtained using spherical mirroring.

Mortara et al. [15] propose a multi-scale method based on blowing bubbles. The surface segmentation happens by clustering vertices with respect to their morphological behavior at different scales. This is done by centering on each vertex spheres of increasing diameter and using the curves resulting by their intersection with the mesh as a characterizing descriptor for the clustering process. Shapira et al. [16] describe a method that exploits the Shape Diameter Function (SDF), a measure related to the object volume in the neighborhood of each point that is computed for the barycenter of each triangle. The segmentation procedure relies on a two phase process. In the first phase, a Gaussian Mixture Model is used to fit k Gaussians to the histogram of all SDF values in order to produce a probability vector of length k for each triangle indicating its likelihood to be assigned to each of the SDF clusters. In the second step the segmentation is refined using the alpha expansion graph-cut algorithm, which is used to minimize an energy function that combines the vectors obtained in the first phase along with boundary smoothness and concaveness.

Attene et al. [17] introduce the use of geometric primitives to drive a hierarchical segmentation. Specifically, at an initial stage each surface triangle represents a singleton cluster associated to the primitive that best fits it. Such primitives can be planes, spheres and cylinders. At each step, all the adjacent clusters are considered for merging and those that can be better approximated with one of the primitives form a new single cluster. The process stops when the desired number of segments has been obtained. Lai et al. [18] describe a procedure based on random walks that operates in two steps. Initially a set of seeds is chosen and the mesh is over-segmented by assigning each face to the seed that has the highest probability of reaching it by a random walk. The obtained segments are then hierarchically merged until the desired number of cluster is obtained. This is done following an order based on the relative lengths of the intersections and total perimeters of adjacent segments.

Finally, Golovinskiy and Funkhouser [19] use both normalized cuts and randomized cuts. In a similar manner to [17], normalized cut segmentation happens by first assigning each face of the mesh to its own cluster and then by merging them hierarchically in an order determined by the area-normalized cut cost, i.e. the sum of each segment perimeter (weighted by concavity) divided by its area. In this way it is possible to obtain segments that exhibit small boundaries along concave seams while maintaining segments with roughly similar areas. Differently, randomized cuts segmentation is initially applied to a strongly decimated mesh, obtaining very large segments. Those segments are then hierarchically splitted in a top-down manner, starting with all the faces in a single segment. For each split, a set of randomized cuts is computed over the segment, and the cut that is most consistent with others in the randomized set is identified. Among this set of candidates, the one that results in the minimal normalized cut cost is chosen. In both cases, the process stops when the required number of segments has been reached.

In this paper we introduce a novel graph-based segmentation approach. Differently from other previously proposed algorithms we do not adopt any global optimization method and we only rely on surface normals. While an initial seeding is required by the user, our approach is very simple to implement and the experimental validation highlights its speed and its good performance when compared with other graph-based systems.

2   Weighted Graph-Based Seeded Segmentation

Graph-based segmentation has been previously explored by several authors [19,18,20]. Most of the approaches found in literature perform some global computation over the graph in order to evaluate random walk reachability or optimal cuts. The algorithm presented in this paper, after building a weighted dual graph, adopts a straightforward greedy approach that directly extends an initial set of seeds by picking one new vertex at a time.

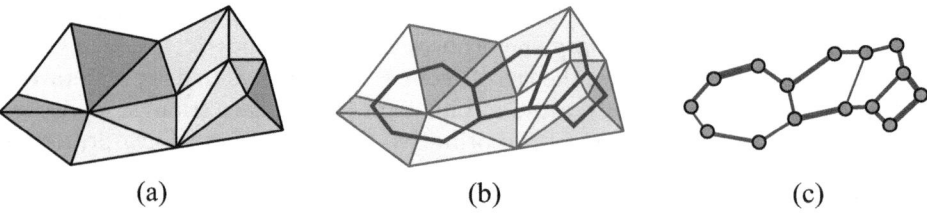

Fig. 1. Steps of the graph creation. From the initial mesh (a) the dual graph is built creating a vertex for each face and connecting adjacent faces (b). Each edge of this graph is then weighted according to the dot product between normals of the connected faces.

2.1   Graph Creation

As for any graph-based mesh segmentation approach our first task is the definition of an apt graphical model for the surface that will be processed. As shown in Fig. 1 each node of this graph corresponds to the a triangle of the mesh. There are no geometrical relations between these node and the absolute position of the triangles in space. For this reason we do not need any attribute on the graph nodes. By converse, we are interested in the relations between adjacent faces, thus we are going to define a scalar attribute for the graph edges. Specifically, we want to assign to each edge a weight that is monotonical with the "effort" required to move between the two barycenters of the faces. This effort should be higher if the triangles exhibit a strong curvature with a short distance between their centers and it should be low if the opposite happens. To this extent, given two nodes of the graph associated to faces i and j, we define the weight between them as:

$$\omega(i,j) = \frac{1- <n_i, n_j>}{|p_i - p_j|} \tag{1}$$

where $\bar{p} = (p_1, p_2...p_k)$ is the vector of the barycenters of the faces and $\bar{n} = (n_1, n_2...n_k)$ are the respective normals. $< \cdot, \cdot >$ denotes the scalar product and $|\cdot|$ the Euclidean norm.

In Fig. 1 (c) edge weight is represented by using a proportional width in the drawing of the line between two nodes. It can be seen how edges that connect faces with stronger curvatures exhibit larger weight.

2.2   Seeding and Greedy Growing

Once the weighted graph has been created, the segmentation can happen. In our framework the surface is segmented starting from one or more hints provided by the user. This human hint expresses a binary condition on the mesh by assigning a small fraction of all the nodes to a set called *user selected green nodes* and another small portion to a set called *user selected red nodes*. We call *green nodes* the faces (nodes) belonging to the segment of interest and *red nodes* the ones that are not belonging to it, regardless of the fact that those nodes have been

Algorithm 2.1: GROW($graph, userSelectGreenNodes, userSelectRedNodes$)

$greenNodes \leftarrow \emptyset$
$redNodes \leftarrow \emptyset$
$unassignedNodes \leftarrow \emptyset$
$seeds \leftarrow \emptyset$

for each $n \in graph.nodes$
  do $\begin{cases} \textbf{if } n \in userSelectGreenNodes \\ \quad \textbf{then } seeds = seeds \cup \{< n, green, 0 >\} \\ \textbf{if } n \in userSelectRedNodes \\ \quad \textbf{then } seeds = seeds \cup \{< n, red, 0 >\} \\ unassignedNodes = unassignedNodes \cup \{n\} \end{cases}$
while $seeds \neq \emptyset$
  do $\begin{cases} s \leftarrow \arg\min_{seed} seed.w, seed \in seeds \\ \textbf{if } s \in unassignedNodes \\ \quad \textbf{then } \begin{cases} \textbf{if } s.type = green \\ \quad \textbf{then } greenNodes = greenNodes \cup \{s.n\} \\ \textbf{if } s.type = red \\ \quad \textbf{then } redNodes = redNodes \cup \{s.n\} \\ unassignedNodes = unassignedNodes \setminus \{s.n\} \\ \textbf{for each } m \in graph.neighbors(s.n) \\ \quad \textbf{do } \begin{cases} \textbf{if } m \in unassignedNodes \\ \quad \textbf{then } seeds = seeds \cup \{< m, s.type, \omega(s.n, m) >\} \end{cases} \end{cases} \end{cases}$

Fig. 2. The simple, yet effective, algorithm proposed to iteratively expand the initial user-specified seeds to cover the whole mesh

manually or automatically labeled. The proposed algorithm distributes all graph nodes in the *green nodes* and *red nodes* sets in a greedy way.

We define a seed as triple $< n, t, w >$ where n is the graph node referred by this seed, t is a boolean flag that indicates if n has to be added to green or red nodes, w is a positive value in $\mathbb{R}^+$. At the initialization step, for each initial green and red node selected by the user, a seed is created and inserted into a priority queue with an initial weight value $w = 0$. All nodes are also added to the *unassigned nodes* set. At each step, the seed $< n, t, w >$ with lowest value of w is extracted from the priority queue and its referred node n is added to *green nodes* or *red nodes* according to the seed's t flag. The node is also removed from *unassigned nodes* to ensure that each node is evaluated exactly once during the execution of the algorithm. For each node $n' \in$ *unassigned nodes* connected to n in the graph, a new seed $< n', t' = t, w' = \omega(n, n') >$ is created and added into the queue. It has to be noted that it is not a direct consequence of such insertion that the final type of n' (either green or red) is determined by the type t' of this seed. At any time multiple seeds referring the same node can exist in the queue, with the only condition that a node type can be set only once. During

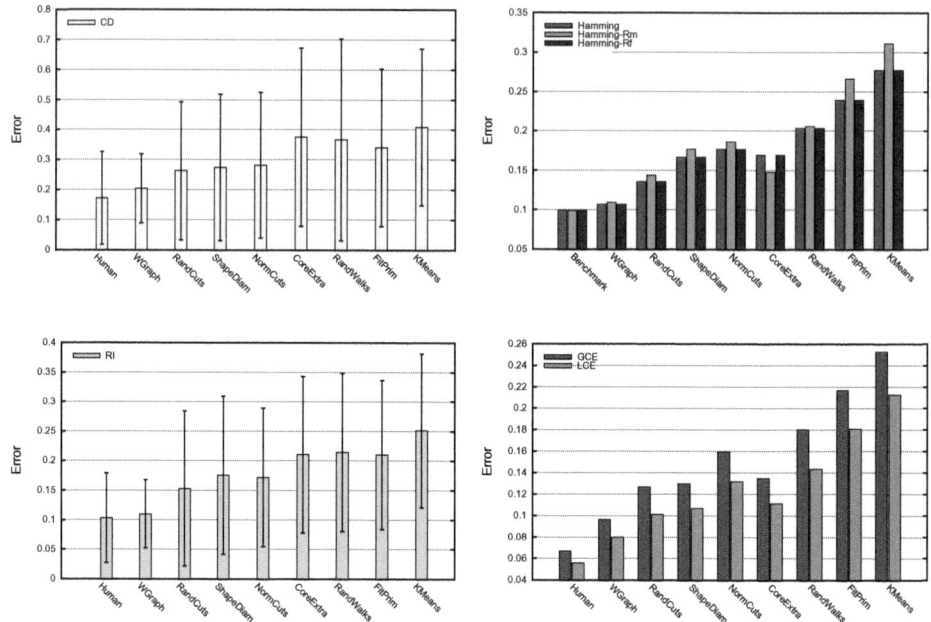

Fig. 3. Benchmark evaluation of our approach (WGraph) with respect to manual segmentation (Human) and other segmentation techniques. The used metrics are respectively the Cut Discrepancy (CD), the Hamming Distance (Hamming), the Rand Index (RI) and the Consistency Error (CE). See the text for details.

the execution of algorithm either the region of green nodes and the region of red ones expands towards the nodes that would require less weight to be reached. Once all nodes in the same connected component are visited, the result of this assignment is shown to the user who can either refine his initial hint or accept the proposed segmentation. Of course the procedure can be iterated to obtain an hierarchical segmentation. In any condition, the algorithm will run in $O(N)$ time since, with the described greedy approach, each node is visited once.

3   Experimental Validation

3.1   Quantitative Evaluation

In order to assess the results obtained by the proposed technique in a quantitative manner some kind of benchmark is needed. We chose to adopt the dataset and metrics proposed in [21]. Namely, the dataset consists in 380 surface meshes, organized in 19 different object categories. A ground truth of 4300 manually generated segmentations is supplied. Since the authors provided both the dataset and the code for running the benchmark, we replicated some of their experiments and added our approach to the set of algorithms to be tested against the ground truth. The evaluation adopts four different metrics. The first one

Fig. 4. Qualitative comparisons between the segmentation obtained with our method (second column) and with a semi-supervised Fitting Primitives respectively with 5 segments (third column) and 10 segments (fourth column). In the first column the ground truth segmentation is shown.

is the *Cut Discrepancy*, that sums the distances from points along the cuts in the obtained segmentation with respect to the closest cuts in the ground truth and vice-versa (CD in Fig. 3). The idea is to measure how well the segment boundaries overlap with the ground truth. The second metric is the Hamming Distance, that measures the overall region-based difference between two segmentations. In particular we evaluate two directional Hamming Distances, the missing rate (Hamming R_m in Fig. 3) and the false alarm rate (Hamming R_f in Fig. 3). In addition also the average between the two is calculated (Hamming R_f in Fig. 3). The third metric is the Rand Index (RI in Fig. 3), that accounts for the likelihood that two triangles belong both to the same o to different clusters in two segmentations. Finally also two Consistency Error metrics are evaluated to measure a triangle-based compatibility between segments that is neutral to differences in hierarchical granularity. Specifically the Global Consistency Error (GCE in Fig. 3), that forces all local refinements to be in the same direction, and the Local Consistency Error (LCE in Fig. 3), that allows for different refinement directions in different parts of the same object (refer to [21] for more details about these metrics). The compared method were Randomized and Normalized Cuts [19], Shape Diameter Functions [16], Core Extraction [14], Random Walks [18], Fitting Primitives [17] and KMeans [11]. While most of these method are not supervised, some required parameters such as the number of segments to extract or initial seeds. For each approach we used the optimal parameter set suggested in [21]. In addition also a set of totally human-supervised segmentation is available in the benchmark. From the results shown in Fig. 3, it is apparent that the proposed method outperforms all the compared approaches. Of course this is somewhat expected since we use an initial set of hints supplied by the user. However the algorithm was always fed with less than 10 seeds, resulting in just a few seconds of operator time. It is interesting to observe that, even with this limited user interaction, the performance obtained is on par with the fully supervised human segmentation.

3.2   Qualitative Evaluation and Running Time

In addition to the quantitative results provided by the benchmark we also present some sample segmentation images in order to give an insight about how the

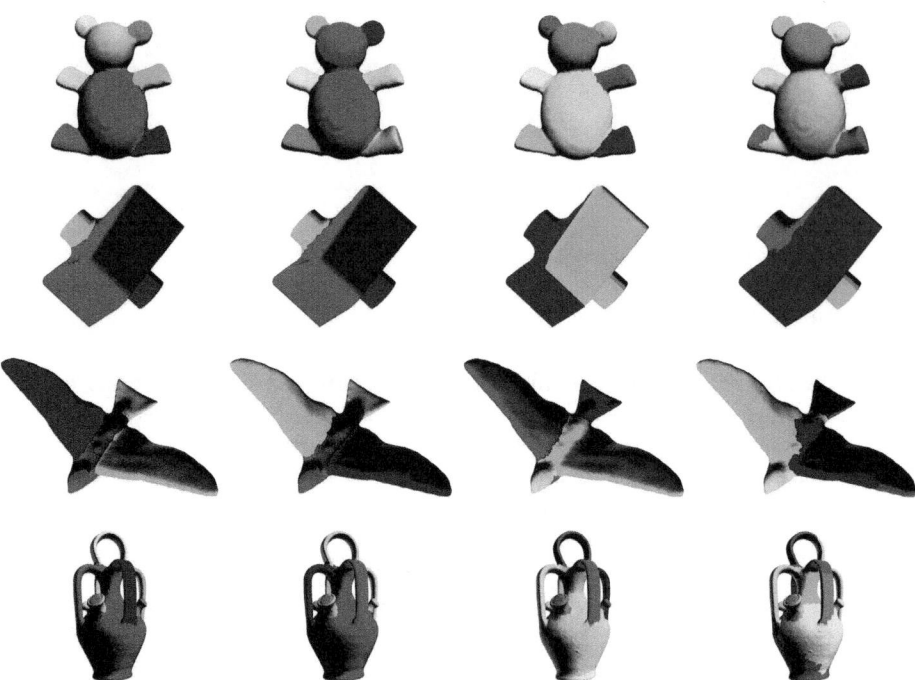

Fig. 5. Qualitative comparisons between the supervised segmentation obtained with our method (second column) and with other state-of-the art unsupervised methods. Respectively Randomized Cuts (third column) and Normalized Cuts (fourth column). In the first column the ground truth segmentation is shown.

different methods actually compare from a qualitative point of view. In Fig. 4 we compare the fully supervised ground truth segmentation (first column) with the result obtained by our method setting just 2 hint seeds for each segment. We tried to replicate this result using the interactive Fitting Primitive tool available from [17], however were not able to obtain a proper segmentation with only 5 clusters (third column), while adding more clusters led the method to over-segmentation (fourth column). In Fig. 5 comparisons are made with the best performing automatic approaches. While objects with sharp edges are usually easily segmented by all methods (teddy bear in the first row), some synthetic shapes can lead to failures for Randomized Cuts and a slight imprecision for Normalized Cuts (CAD model in the second row). Organic shapes (such as the bird in the third row and the vase on the fourth) can make it difficult for automatic algorithm to spot semantically relevant edges.

Regarding the execution time (Fig. 6), the graph building step is the more time consuming, but it can always be performed in less than a second even for large models. The segmentation step itself is very fast and it typically requires less than 50 milliseconds.

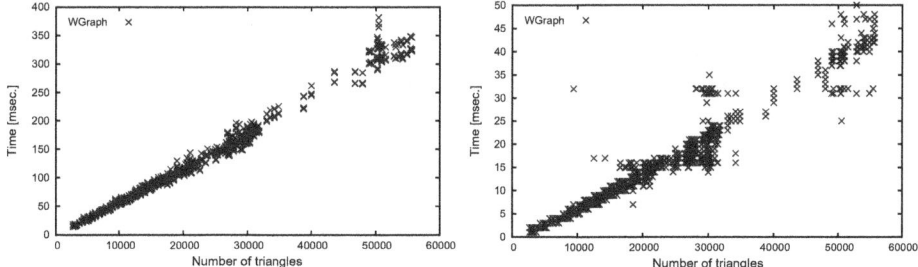

Fig. 6. Evaluation of the time required respectively for creating the weighted graph and to perform a single segmentation on an object. Both scatter-plots relate the execution time to the number of triangles in the mesh.

4 Conclusions

In this paper we introduced a simple yet effective segmentation procedure for 3D objects and surfaces. While our method requires an initial set of user hints, results that are on par with fully supervised segmentations can be obtained even with a very limited amount of seeds placed without special care by the user. Moreover the time required to perform a segmentation is in the order of few millisecond with meshes that count tens of thousands of vertices. This allows for the inclusion of the method in tools that exploit real-time interactive use. Finally, the proposed growing algorithm is very simple and can be easily implemented.

References

1. Kokkinos, I., Maragos, P.: Synergy between Object Recognition and Image Segmentation Using the Expectation-Maximization Algorithm. Pattern Analysis and Machine Intelligence 31, 1486–1501 (2009)
2. Ferrari, V., Tuytelaars, T., Gool, L.: Simultaneous object recognition and segmentation from single or multiple model views. Int. J. Comput. Vision 67, 159–188 (2006)
3. Courtial, A., Vezzetti, E.: New 3d segmentation approach for reverse engineering selective sampling acquisition. The International Journal of Advanced Manufacturing Technology 35, 900–907 (2008), doi:10.1007/s00170-006-0772-3
4. Kim, H.C., Hur, S.M., Lee, S.H.: Segmentation of the measured point data in reverse engineering. The International Journal of Advanced Manufacturing Technology 20, 571–580 (2002), doi:10.1007/s001700200193
5. Colombari, A., Fusiello, A., Murino, V.: Segmentation and tracking of multiple video objects. Pattern Recognition 40, 1307–1317 (2007)
6. Lafarge, F., Keriven, R., Brédif, M., Hiep, V.H.: Hybrid multi-view reconstruction by jump-diffusion. In: The Twenty-Third IEEE Conference on Computer Vision and Pattern Recognition, CVPR 2010, pp. 350–357. IEEE, Los Alamitos (2010)
7. Baillard, C., Zisserman, A.: Automatic reconstruction of piecewise planar models from multiple views. IEEE Computer Society Conference on Computer Vision and Pattern Recognition 2, 2559–2566 (1999)

8. Mian, A.S., Bennamoun, M., Owens, R.: Three-dimensional model-based object recognition and segmentation in cluttered scenes. IEEE Trans. Pattern Anal. Mach. Intell. 28, 1584–1601 (2006)

9. Bariya, P., Nishino, K.: Scale-hierarchical 3d object recognition in cluttered scenes. IEEE Conference on Computer Vision and Pattern Recognition. In: CVPR 2010, pp. 1657–1664 (2010)

10. Wang, Y., Gu, X., Hayashi, K.M., Chan, T.F., Thompson, P.M., Yau, S.T.: Surface parameterization using riemann surface structure. In: Proceedings of the Tenth IEEE International Conference on Computer Vision, ICCV 2005, vol. 2, pp. 1061–1066. IEEE Computer Society, Washington, DC (2005)

11. Shlafman, S., Tal, A., Katz, S.: Metamorphosis of polyhedral surfaces using decomposition. Eurographics 21, 219–228 (2002)

12. Moumoun, L., Chahhou, M., Gadi, T., Benslimane, R.: 3d hierarchical segmentation using the markers for the watershed transformation. International Journal of Engineering Science and Technology 2, 3165–3171 (2010)

13. Katz, S., Tal, A.: Hierarchical mesh decomposition using fuzzy clustering and cuts. ACM Trans. Graph 22, 954–961 (2003)

14. Katz, S., Leifman, G., Tal, A.: Mesh segmentation using feature point and core extraction. The Visual Computer 21, 649–658 (2005)

15. Mortara, M., Patanè, G., Spagnuolo, M., Falcidieno, B., Rossignac, J.: Blowing bubbles for multi-scale analysis and decomposition of triangle meshes. Algorithmica 38, 227–248 (2003)

16. Shapira, L., Shamir, A., Cohen-Or, D.: Consistent mesh partitioning and skeletonisation using the shape diameter function. Vis. Comput. 24, 249–259 (2008)

17. Attene, M., Falcidieno, B., Spagnuolo, M.: Hierarchical mesh segmentation based on fitting primitives. Vis. Comput. 22, 181–193 (2006)

18. Lai, Y.K., Hu, S.M., Martin, R.R., Rosin, P.L.: Fast mesh segmentation using random walks. In: Proceedings of the 2008 ACM Symposium on Solid and Physical Modeling, SPM 2008, pp. 183–191. ACM, New York (2008)

19. Golovinskiy, A., Funkhouser, T.: Randomized cuts for 3D mesh analysis. ACM Transactions on Graphics (Proc. SIGGRAPH ASIA) 27 (2008)

20. Zhang, X., Li, G., Xiong, Y., He, F.: 3d mesh segmentation using mean-shifted curvature. In: Chen, F., Jüttler, B. (eds.) GMP 2008. LNCS, vol. 4975, pp. 465–474. Springer, Heidelberg (2008)

21. Chen, X., Golovinskiy, A., Funkhouser, T.: A benchmark for 3D mesh segmentation. ACM Transactions on Graphics (Proc. SIGGRAPH) 28 (2009)

Convexity Grouping of Salient Contours

Padraig Corcoran[1], Peter Mooney[1], and James Tilton[2]

[1]Department of Computer Science,
National University of Ireland Maynooth
padraigc@cs.nuim.ie
http://www.cs.nuim.ie/~padraigc/
[2]NASA Goddard Space Flight Center

Abstract. Convexity represents an important principle of grouping in visual perceptual organization. This paper presents a new technique for contour grouping based on convexity and has the following two properties. Firstly it finds groupings that form contours of high convexity which are not strictly convex. Secondly it finds groupings that form both open and closed contours of high convexity. The authors are unaware of any existing technique which exhibits either of these properties. Contour grouping is posed as the problem of finding minimum cost paths in a graph. The proposed method is evaluated against two highly cited benchmark methods which find strictly convex contours. Both qualitative and quantitative results on natural images demonstrate the proposed method significantly outperforms both benchmark methods.

Keywords: Salient Contour, Graph Searching, Convexity.

1   Introduction

Contours defining object shape are some of the most important features in visual object recognition. However finding such object contours from real scenes is an extremely difficult task. Most contours are regularly fragmented by occlusion, shadows, and low reflectance contrast. In order to infer shape from contours the human visual system must selectively group contours projecting from a common object while keeping contours from different objects separate [3]. This grouping process in the human visual system is known as perceptual organization and the resulting contours are referred to as salient. Computational modelling of perceptual organization presents the following two challenges [13]. Firstly a function which can determine the saliency of a particular contour grouping must be defined. Secondly an effective algorithm for finding such groupings must be used. Many existing contour grouping techniques are based on local Gestalt properties such as closure, good continuation, and proximity [4]. In this work we propose a technique for finding contour groupings of high convexity. Convexity represents an important factor in grouping for many reasons. It is generally accepted that the parts of an objects contour with high convexity correspond to object parts [8]. Borra and Sarkar [2] compared several grouping methods in the context of object recognition and found that grouping subject to a convex constraint

X. Jiang, M. Ferrer, and A. Torsello (Eds.): GbRPR 2011, LNCS 6658, pp. 235–244, 2011.

Fig. 1. The book in the background is occluded by two foreground objects. The foreground jar is fully visible but is not strictly convex.

gave best performance. They showed in some cases that convexity dominates the effects of Gestalt properties such as good continuation. Convexity is also a nonaccidential property which can distinguish structure from noise in real images [5]. For these reasons convexity, as a grouping property, has been considered by many researchers [16,5].

Despite this progress existing techniques for finding groupings which form convex contours are constrained in two ways. Firstly all techniques can only find closed contours. Although closure is an important grouping property not all salient contours form closed contours. This is case when one object is occluded by another. For example consider the image containing three objects in Figure 1 where a book in the background is occluded by two foreground objects. Finding a complete contour of this occluded book is not possible without prior knowledge of its complete shape. If such knowledge is not available, employing the principle of maximum entropy, we should aim to find the parts of the object's contour which are visible. In this example this would correspond to the open contour of the occluded book which exists between the background and the book in question. Attempting to find a complete contour of the occluded object using exiting techniques would most likely fail [1]. To overcome the problem of occlusion many authors use local features which may still be visible if the object is not completely occluded [9]. The second constraint exhibited by existing techniques is that they can only finding groupings that form strictly convex contours. For example consider the object of the jar in the foreground of Figure 1. This object is not occluded so finding a complete contour is possible. Application of existing techniques will fail to find this contour because it exhibits high convexity but is not strictly convex.

In this paper we propose a new technique for finding open and closed contours of high convexity. This technique involves three steps. We begin by extracting a set of suitable elements for grouping. A graph representation of these elements is then constructed. Finally an efficient graph mining algorithm is applied to find groupings which form salient contours where saliency is defined as a function of convexity. The layout of this paper is as follows. Section 2 describes the method used to extract elements for grouping and the conversion of these elements to a graph representation. Section 3 details the mining algorithm used to group elements into salient contours. Finally sections 4 and 5 present results and draw conclusions respectively.

2   Grouping Element Extraction and Graph Construction

The elements used for grouping are extracted from an image by the following steps. First edge detection is performed. The resulting edge pixels are then grouped into 8-connected contours. Then these contours are segmented into elements of a suitable scale for grouping. We now describe each of these steps.

Colour edge detection is performed using the photometric invariant technique of van de Weijer et al. [18] and the resulting edge image is thinned. The linking of edge pixels into 8-connected contours is performed using a region growing approach. When an edge junction is encountered the contour in question is terminated and separate contours are created for each branch. In many existing grouping algorithms the grouping elements correspond to 8-connected contours [15,16]. Often the scale of such elements is too great. Consider Figure 2a which shows a single 8-connected contour. This contour contains parts of two individual salient contours corresponding to the bird and tree branch. In order to extract these individual salient contours the scale of the original contour must be reduced before a grouping process is applied. To overcome this issue of scale Jacobs [5] proposed the following strategy, which was later used by Stahl et al. [13]. When grouping two contours Jacobs removed the constraint that the entire second contour must be contained in the grouping. This strategy is not suitable in the context of extracting open contours. In cases where the scale of the original contour is too great and no actual grouping is performed it will fail. Consider again the single contour in Figure 2a. A closed contour of the bird cannot be formed due to occlusion by the tree branch. In order to extract the part of this contour corresponding to the bird the scale of the original contour must be reduced. Since no grouping is required the method of Jacobs cannot reduce the scale. To overcome this problem we propose a novel solution. We represent each 8-connected contour as a series of smaller scale line segments. These line segments are generated through a process of line fitting [6]. Grouping is then performed at the line segment scale. Figure 2b shows the result of this line fitting process applied to the contour in Figure 2a. Grouping at this scale allows the extraction of the open contour corresponding to the bird object

To facilitate the search for salient contours we convert the above line segment representation to a graph representation known as an image graph [17,12]. To construct this graph each line segment endpoint is represented as a node in the graph. All nodes have degree 1 or greater. If two endpoints are connected by a line segment a *real edge* is constructed between them [12]. To allow grouping of segments originating from different 8-connected contours *virtual edges* must be constructed between them [12]. A *virtual edge* is constructed between two segment endpoints if and only if both endpoints in question have degree 1 and the spatial distance between them in the image domain is less than a specified threshold. A path in the image graph corresponds to a contour in the image

(a)   (b)

Fig. 2. A single 8-connected edge component is plotted in red in (a). Each connected pair of points in (b) represent a line segment.

domain. This transforms the task of finding salient contour groupings to one of finding suitable paths in the image graph.

3   Salient Contour Grouping

In order to find salient paths in the image graph two elements are required. These are a measure of path saliency and an effective graph searching algorithm. A novel measure of saliency, which is a function of contour convexity, is proposed. The graph searching algorithm used is an implementation of Beam Search [14]. A single application of Beam search starting from a specified initialization or root edge, if successful, will find a single contour. Therefore to find multiple salient contours the search must be repeated multiple times from different root edges. In the following three sections we describe the proposed measure of saliency, the Beam search implementation and the algorithm used to determine a suitable partially ordered set of root edges for searching.

3.1   Contour Saliency Measure

The saliency of a contour i is determined using the function CRL in Equation 1.

$$CRL(i) = C(i) \times R(i) \times L(i) \tag{1}$$

This measure is a product of three terms which we now describe. The function C measures the convexity of the contour in question which lies in the range $[0, 1]$ and approaches 1 for convex contours. This is determined using the convexity measure of Corcoran et al. [11] which has computational complexity of $O(n)$ [7]. The function R in Equation 1 equals the percentage of the contour length which is constructed from real edges and lies in the range $[0, 1]$. This term is important because it models the Gestalt property of good continuation where contours with less breaks appear more salient. It is common for an image to contain many short contours having high convexity which are not perceived by the human visual system as being salient. In order to prevent the extraction of such contours the function L in Equation 1, which is defined by Equation 2, returns a value in the range $[0, 1]$ which approaches 1 for longer contours.

$$L(i) = 1 - \frac{1}{1 + \left(\frac{\Gamma_i}{K}\right)^2} \tag{2}$$

The variable Γ_i equals the total length of the contour in question. The variable K is a specified parameter which determines the point at which this function approaches unity. The shape of this function is such that its first derivative decreases as contour length increases. This models the fact that short contours are generally perceived as being non-salient but when contours are sufficiency long, length becomes a less important factor. All three terms in the product of Equation 1 are in the range $[0, 1]$. The resulting CRL saliency value will also be in this range with higher values signifying greater saliency. Representing a closed contour as an open contour with equal first and last vertices allows CRL to be applied to both types of contour. The time complexity required to compute CRL is $O(n)$. This makes the task of finding image contours of high convexity tractable.

3.2 Beam Search

Beam search uses a breadth-first search to build its search tree. At each level of this tree it generates all successors but only stores a predetermined number of these which are determined to be the best. This number is referred to as the beam size. Due to the fact that a path may be extended in both directions the search tree is bi-directional. A search begins with a path containing two nodes joined by a single edge where this edge is known as the root edge. Consider the image graph in Figure 3a where nodes are represented by circles, real edges by solid lines and virtual edges by dashed lines. A path in this graph, represented by the colour red, contains the four nodes (A, B, C, D, E) of which A and E are leaf nodes in the search tree. The edge connecting nodes B and C is the root edge and is represented by a thicker line. Due to the bidirectional nature of the search the path may be extended along the edge connecting to A to G, and the edge connecting to E to F.

The Beam search terminates when no further extension can be made to any path. The most salient contour encountered during the entire search procedure is then returned. This contour may correspond to only part of a longer path which was discovered. For example the most salient contour encountered during the creation of the path in Figure 3a may correspond to the path containing the nodes (A, B, C, D) which is only a part of an overall longer path discovered. When a path forms a cycle all edges not contained in the cycle are removed. For example consider the path in Figure 3b. If this path was extended by adding the edge connecting H to D this would form a cycle and in turn result in the edge connecting D to E being removed from the path. When a cycle is formed the path in question cannot be extended any further. The proposed Beam search

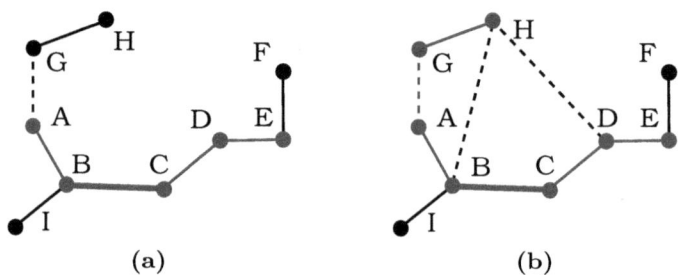

Fig. 3. The paths in (a) and (b) contain the nodes (A, B, C, D, E) and (H, G, A, B, C, D, E) respectively

is subject to two constraints which we now discuss. Any extension to a current path resulting in a contour which self intersects is prohibited. The second constraint imposed is that if an extension results in a cycle, this cycle must contain the corresponding root edge. In the context of Figure 3b extending the path in question by adding the edge connecting H to B would be prohibited. This constraint prevents searches starting at different root edges from consistently finding the same path and was originally proposed by Saund [12].

3.3   Multiple Contour Extraction

The search for salient contours must be performed efficiently. To reduce redundancy Beam searches should commence from root edges, belonging to these contours, as early as possible [12]. Edges corresponding to a single line segment in the image makes it difficult to predict, without considering additional information, if a given edge belongs to a salient contour. To overcome this difficulty we propose the following strategy presented in Algorithm 1. For each line segment we create a corresponding Beam search tree with that line segment as its root edge (line 1). We refer to this set of search trees as ST. Next we increase the depth of each tree to a relatively small size specified by parameter RD (line 2). We then determine the most salient sub-path in each tree and remove all trees from ST for which this value is below a specified threshold RST (line 3). All remaining trees are then ordered by saliency (line 4). The Beam search is then applied to each element of ST in this order until completion (line 6). Upon completion of a given Beam search the following steps are executed: First the most salient contour, MSC, encountered during the search is determined (line 7). If its saliency is greater than a specified threshold CST it is accepted and added to a set entitled SC (line 9). If accepted, the contour MSC is used to reduce the number of elements in ST by deleting any element if its corresponding root edge is contained in MSC (line 10). The most recently processed tree is then removed from ST (line 12). Once all elements in ST have been processed the complete set of contours found is returned (line 14).

Algorithm 1. Multiple Contour Extraction

```
 1: ST = CreateTrees()
 2: increaseDepth(ST, RD)
 3: removeUnsalient(ST, RST)
 4: sort(ST)
 5: while ST.size() > 0 do
 6:    completeSearch(ST[1])
 7:    MSC = SetTrees[1].MSC
 8:    if MSC.Saliency > CST then
 9:       SC.add(MSC)
10:       ST.delete(MSC)
11:    end if
12:    ST.erase(1)
13: end while
14: return SC
```

4   Results

To evaluate the proposed technique for finding salient contour groupings the Berkeley segmentation dataset [10] was used. To allow a comparative study the contour grouping techniques of Wang et al. [16] and Jacobs [5] were used. Both these techniques are constrained by the fact that they can only extract strictly convex closed contours. We qualify this statement with the fact that the method of Jacobs [5] is robust to noise and therefore can extract closed contours which are not strictly but are almost convex. Also the method of Wang et al. [16] is robust to occlusion due to the image boundary. These two techniques represent state of the art convexity based grouping methods. When applied to images containing unoccluded object contours which are strictly convex both techniques perform well. Examples of this can be found in the original publications. It was discovered that when these techniques were applied to most natural images they performed poorly. For example consider Figure 4 which displays the results of applying both techniques to a single image. Both techniques fail to extract the most salient contours in this image which correspond to the contours of the bird and the tree branches. This failure can be attributed to the fact that these contours are occluded by the a tree branch and the image boundary respectively, and are not strictly convex. For example the contour of the bird contains two concave parts just below its head.

Figure 5 displays the contours extracted using the proposed technique for a set images. These results were achieved using a Beam width of 1, RD value of 15, RST value of 0.6, CST value of 0.7 and K value of 20. When the Beam width was greater than 1 the algorithm discovered a greater number of contours which were salient with respect to the proposed metric but these in general did not correspond to genuine salient contours or object parts. From this analysis we drew the conclusion that a best first search performs the task of filtering noise. In all cases the proposed method accurately finds the most salient contours. These include the contours of the bird and deer in Figure 5a and Figure 5b respectively.

Fig. 4. Convex contours extracted using the methods of Wang et al. and Jacobs are shown in (a) and (b) respectively

Fig. 5. Contours extracted using the proposed technique are shown

It is evident from this qualitative evaluation the proposed method outperforms both benchmark methods. To quantify this performance, we compared precision-recall curves on the Berkeley benchmark set [10] which contains 100 test images. The method of Wang et al. [16] is supervised in the sense that the number of contours extracted must be specified by the user. Therefore it cannot be subject to this analysis. The curves in question are shown in Figure 6 and were generated by varying the threshold applied to the gradient image processed by both algorithms. The dip on the left hand side of the curves is due to the fact that with a high threshold the salient contours contain many breaks in the edge image. The curve of the proposed technique is superior to the curve corresponding to

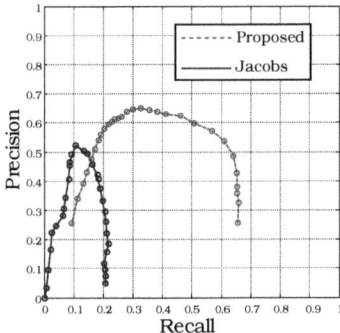

Fig. 6. The PR curves for the proposed and Jacobs method are shown

the method of Jacobs signifying its superior performance. In all test cases the running time of the proposed algorithm was under 1 second.

5   Conclusions

This paper presents a new technique for finding salient image contours of high convexity. The novelty of this technique lies in its ability to find both open and closed contours which exhibit high convexity but may not be strictly convex. Qualitative and quantitative results achieved on natural images show the proposed technique to outperform two current methods for finding convex image contours.

Acknowledgments

Research presented in this paper was part-funded by a Strategic Research Cluster grant (07/SRC/I1168) and a Research Professor Award (07/RPR/I1177) from Science Foundation Ireland under the National Development Plan.

References

1. Ansari, N., Delp, E.: Partial shape recognition: a landmark-based approach. IEEE International Conference on Systems, Man and Cybernetics 2, 831–836 (1989)
2. Borra, S., Sarkar, S.: A framework for performance characterization of intermediate-level grouping modules. IEEE Transactions on Pattern Analysis and Machine Intelligence 19(11), 1306–1312 (1997)
3. Elder, J., Elder, J., Zucker, S., Zucker, S.: A measure of closure. Vision Research 34, 3361–3369 (1994)
4. Geisler, W.S., Perry, J.S., Super, B.J., Gallogly, D.P.: Edge co-occurrence in natural images predicts contour grouping performance. Vision Research 41(6), 711–724 (2001)
5. Jacobs, D.: Robust and efficient detection of salient convex groups. IEEE Transactions on Pattern Analysis and Machine Intelligence 18(1), 23–37 (1996)

6. Jain, R., Kasturi, R., Schunck, B.G.: Machine vision. McGraw-Hill, Inc., New York (1995)
7. Latecki, L.: Shape similarity measure based on correspondence of visual parts. IEEE Transactions on Pattern Analysis and Machine Intelligence 22(10), 1185–1190 (2000)
8. Latecki, L.J., Lakmper, R.: Convexity rule for shape decomposition based on discrete contour evolution. Computer Vision and Image Understanding 73(3), 441–454 (1999)
9. Lowe, D.G.: Object recognition from local scale-invariant features. IEEE International Conference on Computer Vision 2, 1150–1157 (1999)
10. Martin, D., Fowlkes, C., Tal, D., Malik, J.: A database of human segmented natural images and its application to evaluating segmentation algorithms and measuring ecological statistics. In: Proc. 8th Int'l Conf. Computer Vision, vol. 2, pp. 416–423 (July 2001)
11. Corcoran, P., Mooney, P., Tilton, J.: Convexity Measure for Partial Contours by Shape Similarity. In: British Machine Vision Conference. Dundee, Under Review (2011)
12. Saund, E.: Finding perceptually closed paths in sketches and drawings. IEEE Transactions on Pattern Analysis and Machine Intelligence 25(4), 475–491 (2003)
13. Stahl, J., Wang, S.: Convex grouping combining boundary and region information. In: IEEE International Conference on Computer Vision, vol. 2, pp. 946–953 (2005)
14. Russell, S., Norvig, P.: Artificial Intelligence: a Modern Approach, 3rd edn. Prentice-Hall, Upper Saddle River (2009)
15. Wang, S., Kubota, T., Siskind, J., Wang, J.: Salient closed boundary extraction with ratio contour. IEEE Transactions on Pattern Analysis and Machine Intelligence 27(4), 546–561 (2005)
16. Wang, S., Stahl, J., Bailey, A., Dropps, M.: Global detection of salient convex boundaries. International Journal of Computer Vision 71(3), 337–359 (2007)
17. Wang, S., Wang, J., Kubota, T.: From fragments to salient closed boundaries: an in-depth study. In: IEEE Conference on Computer Vision and Pattern Recognition, vol. 1, pp. I–291– I–298 (2004)
18. van de Weijer, J., Gevers, T., Smeulders, A.: Robust photometric invariant features from the color tensor. IEEE Transactions on Image Processing 15(1), 118–127 (2006)

Hierarchical Interactive Image Segmentation Using Irregular Pyramids*

Michael Gerstmayer, Yll Haxhimusa, and Walter G. Kropatsch

Pattern Recognition and Image Processing Group
Institute of Computer Graphics and Algorithms, Vienna University of Technology
{mig,yll,krw}@prip.tuwien.ac.at

Abstract. In this paper we describe modifications of irregular image segmentation pyramids based on user-interaction. We first build a hierarchy of segmentations by the minimum spanning tree based method, then regions from different (granularity) levels are combined to a final (better) segmentation with user-specified operations guiding the segmentation process. Based on these operations the users can produce a final image segmentation that best suits their applications. This work can be used for applications where we need accuracy in image segmentation, in annotating images or creating ground truth among others.

1 Introduction

Image segmentation cannot produce a perfect final segmentation, only by using low-level visual cues. The reason is the *intrinsic ambiguity* in the exact location of region boundaries in digital images. In general, homogeneity of low-level cues will not map to the semantics [12], and the degree of homogeneity of a region is in general quantified by threshold(s) for a given measure [6]. To avoid problems with the automatic segmentation methods one can use human help to guide segmentation methods, producing results acceptable by users/practitioners. Most interactive or semi-automatic segmentation algorithms make use of this external knowledge, some of them e.g. Snakes [11], Live Wire (or Intelligent Scissors) [17] and recent approaches based on the Graph Cuts formalism [1,18,16] are well known. They are often used in e.g. medical image segmentation, image or video object extraction and to refine or improve results from automatic methods.

But the notion of 'interactive' is ambiguous and not very well defined. Some of the methods are initialized (e.g. statistic shape models, rule sets, training sets) others use seed points or strokes for guiding and limiting a segmentation process [9]. Besides initialization, existing methods can also be categorized either as optimizing (manually guided, influenced) or post-processing methods (manually corrected, modified) [10]. The work presented in this paper is located between the last two categories. It uses user-interaction to guide the minimum spanning tree (MST) based pyramid segmentation [8]. The MST-based segmentation method produces a stack of (dual) graphs (a graph pyramid) [8] on each level of the

---

* This paper has been supported by the ASF under grant FWF-P20134-N13.

X. Jiang, M. Ferrer, and A. Torsello (Eds.): GbRPR 2011, LNCS 6658, pp. 245–254, 2011.
© Springer-Verlag Berlin Heidelberg 2011

pyramid (Fig. 3). Each level of this hierarchy corresponds to one image segmentation. Regions from different levels of the segmentation pyramid are combined by user interaction resulting in a modified pyramid containing the 'final good' segmentation at top. In the regions where the user did not set any modification operation, the algorithm will be guided automatically by a pairwise comparison of region similarity [5].

Meine et al. [15] use the topological GeoMap representation and user interaction to guide a segmentation method in the medical image analysis. The use of a topologically correct representation has a major impact on the processing and its results. This motivated us to use combinatorial maps, as a topological representation. The authors in [13] interactively modify a hierarchical watershed segmentation, which is similar to our user modification(s) of the MST based segmentation hierarchy. In our case we can, if needed, access each pixel, which is not the case in the work of [13].

The paper is structured as follows. We first give a short overview of the combinatorial maps and combinatorial pyramids (Section 2). After that the operations that a user can set are presented in Section 3.1, in Section 4 we show some results and conclude the paper.

2   Combinatorial Image Pyramids

In this section a short overview of the most important concepts of combinatorial image pyramids are given. Combinatorial maps and generalized combinatorial maps define a general framework which allows to encode any subdivision on nD topological spaces orientable or non-orientable with or without boundaries [2]. Images and its structure are represented in this work as weighted combinatorial maps. Using $2D$ images, combinatorial maps may be understood as a particular encoding of a planar graph, where each edge is split into two half-edges called darts. Since each edge connects two vertices, each dart belongs to only one vertex. A $2D$ combinatorial map is formally defined by the triplet $G = (\mathcal{D}, \sigma, \alpha)$ [3] where $\mathcal{D}$ represents the set of darts and $\sigma(d)$ is a permutation on $\mathcal{D}$ encountered when turning clockwise around each vertex. Finally $\alpha(d)$ is an involution on $\mathcal{D}$ which maps each of the two darts of one edge to the other one. Given a combinatorial map $G = (\mathcal{D}, \sigma, \alpha)$, its dual is defined by $\overline{G} = (\mathcal{D}, \varphi, \alpha)$, with $\varphi = \sigma \circ \alpha$. The cycles of permutation φ encode the faces/regions of the combinatorial map. In the following cycles of α, $\sigma(d)$ and φ containing a dart d will be respectively denoted by $\alpha^*(d)$, $\sigma^*(d)$ and $\varphi^*(d)$ (an example of a combinatorial map is shown in Fig. 1). Thus all graph definitions used in irregular pyramids are analogously defined. A *combinatorial pyramid* is a stack of combinatorial maps successively reduced by the set of contraction and removal operations, i.e. $(G_0, ..., G_k)$ where k represents the levels of the pyramid. Each map $k+1$ is built from the one below, k, by selecting a set of contraction kernels $K_{k,k+1}$ and applying it to a given combinatorial map G_k to get the reduced map $G_{k+1} = \mathcal{C}[G_k, K_{k,k+1}] = G_k \setminus K_{k,k+1}$. More on removal of the redundant edges can be found in [2].

Region adjacency graphs (RAG), dual graphs [8], combinatorial maps [7], and GeoMap [15] have been used before [2] to represent the partitioning of $2D$

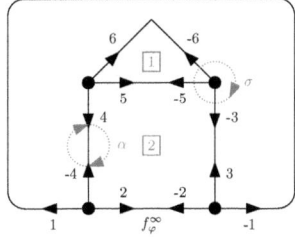

$$\mathcal{D}\,(1,-1,2,-2,3,-3,4,-4,5,-5,6,-6):$$

$$\alpha = (1,-1)(2,-2)(3,-3)(4,-4)(5,-5)(6,-6)$$
$$\sigma = (1,-4,2)(-2,3,-1)(-3,-5,-6)(4,6,5)$$
$$\varphi = (5,-6)(2,3,-5,4)(-1,-4,6,-3)(-2,1)$$

Fig. 1. The house and its region relations are described with permutations on the dartset $\mathcal{D}$ of the combinatorial map. The infinite face f_φ^∞ is encoded with the orbit $\varphi^*(-2) = (-2,1)$.

space. From these structures, we use the combinatorial maps because *RAG*s cannot correctly encode multiple boundaries and inclusions. Dual graphs lack the explicit encoding of edge orientation around vertices, which is present in a combinatorial map [2](e.g. Fig. 1). Moreover with combinatorial maps, its dual must not be explicitly represented. One combinatorial map is enough to fully characterize the partition and to deduce the dual graph.

3 Interactive Operations on Pyramids

Usually, automatic segmentation methods will not be able to deliver a final segmentation that is acceptable by the users (see Fig. 2). Thus there is a need to perform a user interaction such that one can produce a better image segmentation. We have chosen a (hierarchical) pyramid based segmentation method where we define user operations which will guide the merging and division by using regions in different level of the pyramid to a final acceptable image segmentation. The irregular (combinatorial) pyramid [7] produces automatically a stack of segmented images (only some levels are shown in Fig. 2). The segmentation results are produced automatically by merging processes that take low-level cues (in this example RGB color values) into consideration [5].

In this work, the user can set focus on region(s) lying in different levels of the pyramid (having different granularities). One can apply the process of manually changing the image segmentation by merging and/or division operations in any level within the stack of segmentations. Note that in our pyramid all these manual operations defined on the regions will change the merging tree (Fig. 3a). Moreover they also guide the processes in changing it, resulting in a stack of segmented images where the final (wished) segmentation is at top of the pyramid. Because we keep the hierarchy it is always possible to decompose the object into its subparts or restart the process for further refinement. Instead of doing this with unpredictable result (e.g. effect of a stroke in Graph Cuts) we can explicitly address each region in the merging tree (until the pixel level if needed) while non restricting the flexibility of the algorithm in merging (other) non selected (or not in the focus) regions automatically. E.g. Fig. 3b) shows (two) final segmentations (level k and $k+1$). The user has decided first to put focus on parts of the face

bottom level   mid level   top level of the pyramid

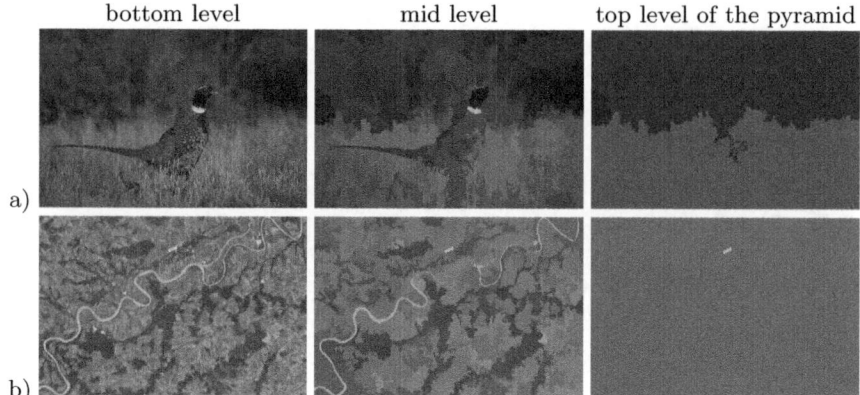

Fig. 2. Image segmentation with automatic segmentation method [8];
a) Due to lack of contrast bird is merged with meadow (Image 43074 from [14]);
b) Due to the thin structure the Danube river disappears.

like eyes, mouth, nose, ears etc. (lv. k), which then got manually merged with
the face (lv. $k + 1$). The inclusion relations of the eyes, mouth, nose and ears,
within the face is correctly encoded in our merging tree. As it is shown, note
that in our pyramid we can keep many 'final' segmentations.

3.1   Modifying Operations

The manual image segmentation process consists of two parts: (1) building the
irregular pyramid based on the MST [8,7], and (2) a user interface where the
user places its modification operations[1]. The framework uses as input the original
image, its stack of automatically generated segmented images at different levels
(e.g. images in Fig. 2a) and the hierarchical information of the merging tree.

Modifying relations between regions, requires the creation of a correspondence
between the user-operations based on regions in the user interface (visual representation) and their combinatorial map correspondent in the same level of the
pyramid. This information is given by the structural description of the image
relations, inherently encoded in its primal and dual combinatorial maps. To get
the inter-pixel boundary [4] in the primal or the edge representing the adjacency
in its dual, a calculation is done through permutations on the darts in $\mathcal{D}$ (e.g.
boundary between the roof (region 1) and the wall (region 2) in Fig. 1). Each
region is represented in the primal by a dart $\in \mathcal{D}$ and its orbit φ^* describing the
boundary. These darts (aligned around a vertex corresponding to this region)
encode also the relations to the neighboring regions in the dual. Therefore the
joining edge, for e.g. merging two regions, is calculated through iterating the
φ^* permutations for each region until both darts belong to the same cycle of α
and hence to the same edge. This transformation from regions to edges is necessary because the operations placed in the user interface implicitly describe what

---

[1] Can be found at http://www.prip.tuwien.ac.at/

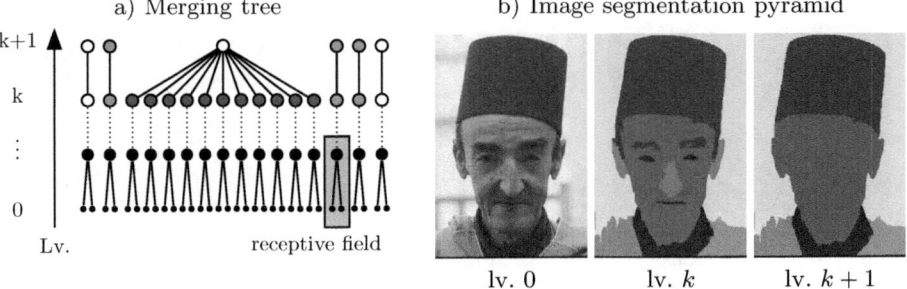

Fig. 3. a) Discrete levels with the merging tree. Interactive operations combine regions from different segmentations in different levels of the pyramid. Green vertices represent regions inhibited from merging in further processing, and red vertices are chosen to get merged. b) User segmentation of Image 189080 from [14].

to do and the representation in terms of combinatorial maps explicitly defines what to be modified. In the presented work we need only the concept of regions' boundary in order to define the user operations. Thus, the idea is general and can be implemented also on other topological representations like region adjacency graphs, dual graphs [8] etc. Therefore this work is not limited to the image representation with combinatorial maps even though it strongly benefits from them.

The operations for the purpose of modifying the relations between the regions are placed in the user interface either by separate selection or brushing over them (e.g. Fig. 4). The two fundamental operations that can be applied on adjacent regions r_1 and r_2 for guiding the segmentation process are:

- $mrg(r_1, r_2)$: merging regions r_1 and r_2,
- $imrg(r_1)$: inhibition of merging r_1 with other regions in the levels above

One can define other operations as a combination of these two basic modifying operations. These operations exert influence on the merging tree. But merging solely does not implicate that the resulting region will be inhibited automatically since the algorithm can decide to merge it with other regions in higher levels of the pyramid. Through nesting other combinations are possible, for e.g.:

- $imrg(mrg(r_1, r_2))$: merge and inhibit the resulting region,
- $imrg(r_1, r_2)$: inhibit both from merging with other regions, etc.

Since we use a hierarchical representation, it is also possible to select regions at different granularity from multiple levels. The basic way of doing this is traversing through all the segmentation levels of the pyramid. The other way is the 'explore' mode, intended to traverse down (toward higher resolution) only within the receptive field of a single region (Fig. 4c). This can be used as a way of applying operations in higher levels on previously merged regions (at lower level of granularity). Its main purpose is dividing them again while e.g. inhibiting the

a) Brush mode   b) Merge mode   c) Explore mode

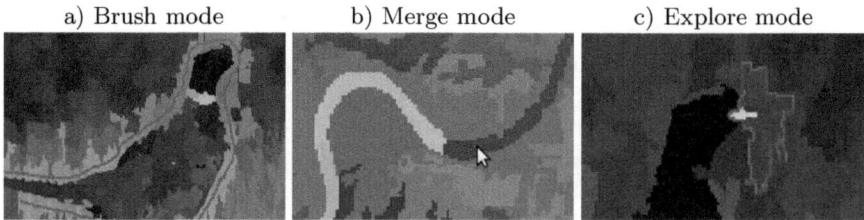

Fig. 4. User interactions. a) Using brush mode (red line) for effectively applying merge operations (encircle bird), b) Merging regions by selection (yellow - first selected region, violet - region to be selected), c) Dividing a region (surrounded red) into its components within its receptive field and restoring a single region from a lower level (indicated with yellow) e.g. the birds beak.

current state. It is also possible that some regions (or their receptive fields) overlap which can lead to conflicts in operations e.g. merging two regions and one of them is inhibited at the same time. We resolve these conflicts by analyzing the combinatorial map and the relations of region(s) under operation i.e. the affected edges (see Section 3.2).

The listing above of operations should be understood as an outline of the versatility of the framework, since many other operations are possible. Finally a set of operations is passed on to the framework, to perform the final segmentation. Each operation entry is of the form $\{\text{lv.}, \text{op.}, \text{r}_1[, \text{r}_2]\}$ where lv. is the level, op. the operation, r_1 the first region and r_2 the second one (optional).

3.2   Building Segmentation

In the bottom-up automatic building of the pyramid [8] candidate edges for contraction are chosen to be the smallest weighted edges. The decision whether two regions (vertices connected by edges) are merged is guided by comparison of region similarity [5]. The set of user operations are not immediately applied upwards in the merging tree. Since we allow operations to be selected on different levels of the pyramid we need to down-project these operations on a common starting level. A common starting level is the lowest level in which at least one operation is defined. In this common starting level the candidate-edges affected from operations (contraction or inhibition) are determined, allowing rebuilding the pyramid from this level upwards. More precisely, deleting all older levels above the common starting level and recalculating new levels, i.e. the merging tree is changed only from the common starting level above. That implies the recalculation of the operations to an equivalent instruction set, which is achieved with the hierarchical information, through permutations on $\mathcal{D}$ with φ and set operations as shown in the following example.

Fig. 1 and Fig. 5 are two consecutive levels, $k + 1$ and k in the pyramid. Let Fig. 5 be a common starting level. The boundary of the roof of the house (region 1) remains the same in both of the levels. However the wall of the house (region 2) in level k is divided into two subregions $2'$ and $2''$. When applying

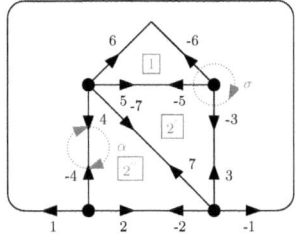

$$R_{k+1,1} \xrightarrow{div} R_{k,1}$$

$$R_{k+1,2} \xrightarrow{div} R_{k,2} = R_{k,2'} \wedge R_{k,2''}$$

Darts for $R_{k,1} = \varphi(5, -6)$

Darts for $R_{k,2} = \varphi(3, -5, -7) \wedge \varphi(2, 7, 4)$

Fig. 5. The combinatorial map from a segmentation in a lower level (than Fig. 1) of the hierarchy

e.g. the operation $\mathtt{imrg}(\mathtt{mrg}(1, 2))$ in level $k + 1$, meaning merge region 1 (roof) and 2 (wall) in Fig. 1 and inhibit the resulting region from merging' we can see that this operation leads to a different non corresponding result $\mathtt{imrg}(\mathtt{mrg}(1, 2'))$ when applied on the combinatorial map in the level below (shown in Fig. 5).

Beginning from the relation between the regions in the common starting level k and the level above $k + 1$, its φ permutation, the corresponding darts (or edges) can be reconstructed as shown in Fig. 5 (right). There are two different categories and collections of edges (inhibition and removal) necessary to process the operation correctly:

1. **Former Contractions:** in each set of $R_{k,1}$ and $R_{k,2}$ some darts belong to the same cycle of α. This edge indicates the adjacency/subdivision from region $R_{k+1,2}$ in level k and is marked for **removal** e.g. $(7, -7)$.
2. **User Operations:**
   - all edges of $R_{k,1}$ and $R_{k,2}$ without the marked edges for removal from former contractions represent the outer borders of regions $R_{k,1}$ and $R_{k,2}$. All these edges are marked for **inhibition**, e.g. darts $(3, -3)(5, -5)$, $(2, -2)(4, -4)$ and $(6, -6), (5, -5)$
   - the darts from the outer borders of $R_{k,1}$ and $R_{k,2}$ of each region are compared to find the edge in common. This edge is marked for **removal** e.g. $(5, -5)$.

Selection of edge $(5, -5)$ would lead to a conflict because it appears in both inhibition and removal edge sets. Because user intention is to merge, a rule decides that this edge has to be removed. Similar to above example we have correctly defined the assignment to the collections of user selected edges for processing (removal or inhibition) for all other user operations. Rebuilding the pyramid from the common starting level, is done as follows: first all edges in this collection of edges are processed first, then all other edges not in this collection are automatically processed by [8]. The processing is based on dual contraction [3] and guarantees a consistent state of the combinatorial map. Note that all edges marked for inhibition will not be deleted by the automated process and survive until the top of the pyramid. The resulting segmentation pyramid considers the selected regions and operations containing a new final segmentation at top, where a final state is reached (e.g. Fig. 3, lv. k).

This framework can be used easily for creating a nested image annotation tree, e.g. by locking all levels below the top of the pyramid and iteratively merge the remaining regions (e.g. Fig. 3, lv. $k + 1$) to build a new hierarchy. In contrast to other tools (e.g. labelme[19]) where each level of abstraction has to be created separately, this is an enormous simplification. Furthermore it could be easily used in creating ground truth in image and video databases.

4   Segmentation Results

We show by means of some images the applicability of the framework. The user has to define what is the region of focus and which objects are of interest. To produce the result in Fig. 6a three pyramid rebuilds are necessary. A detailed set of user operations applied are shown in Table 1. The brush was used to encircle the bird (Fig. 4a) in a level of the pyramid with a fine segmentation. This causes that a limiting boundary is created. In the second run errors in segmentation were removed. Finally in the third run the outside area is inhibited and all inner regions are automatically merged to produce a final segmentation (Fig. 6a).

In an aerial image of the Danube river (Fig. 2b) the user wants to segment the river properly. Some parts of the river were correctly segmented automatically in higher levels, thus these regions are inhibited in the first manual interaction. In the thinner branches correction at pixel level is necessary, thus 36 clicks are needed (Table 1, Fig. 6b). The last image segmented (Fig. 6c) by a user has multiple regions of interest at different granularities (the rock, the person on the rock and the lake/river). Within multiple levels and two rebuilds regions are merged together and inhibited from merging. The processing takes in general per rebuild couple of seconds on images with 500×500 on a PC (2 GHz processor with 1 GB RAM). Since each user segmentation is intended for different applications, a straightforward comparison with other frameworks is difficult.

a) 43074: Focus on the bird   b) Focus on the river   c) 14036: Focus on details

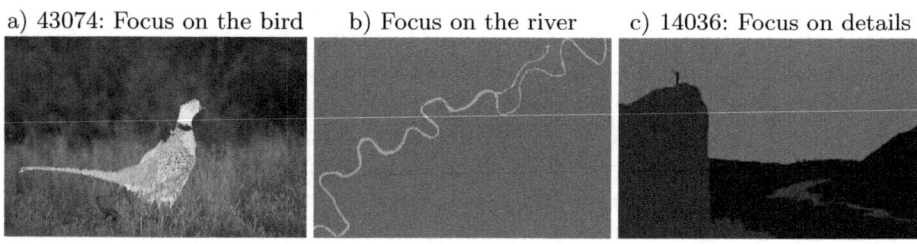

Fig. 6. Final segmentation after user interaction on the segmentation hierarchy

Table 1. Segmentation results in Fig. 6. # Lv.: number of different levels modified, # Op.: sum of operations, Time: *min*:sec, interaction time contains time for placing operations/int. time overall, computation time: time for result.

Image	Run	Input	#	Op.	# Lv.	Clicks	Interaction Time	Comp. Time
43074	1.	brush	104	imrg(mrg)	1	1	01:09/01:41	00:32
	2.	select	7	imrg	5	7	00:13/03:51	00:55
		select	3	imrg(mrg)		6		
	3.	select	1	imrg(mrg)	1	2	00:02/00:18	00:10
River	1.	select	14	imrg(mrg)	3	28	00:49/06:24	00:43
		select	5	imrg	5	5		
14036	2.	select	3	imrg	1	3	00:02/00:33	00:10
	1.	select	2	imrg	3	2	00:09/01:20	01:35
		select	2	imrg(mrg)		4		
	2.	select	3	imrg	1	3	00:14/01:00	00:15
		select	2	imrg(mrg)		4		

5 Discussion

Depending on the object(s)/region of interest where the user wants to set the focus on, the following ways can lead to the same final segmentation result: (1) select different regions from different levels, (2) encircle object of interest with the brush (or selection), (3) limit/fill-out object of interest using the brush (or selection), (4) start from a coarse segmentation of the object and then split up or (5) start from a fine segmentation of the object and merge. One can choose to combine one of the above to have a hybrid approach. Explicitly defining what to be done might cause that an object not denoted, but correctly segmented before will get lost. As shown in Figure 6, it is possible to force a segmentation, in the worst case by defining operations using segments at pixel level. Undoing operations relates to modify the instruction set, hence it is also possible to start from or correct an existing segmentation produced by other segmentation methods. A solution might be to reconstruct the hierarchical merging tree by using the segments of the output of the segmentation method. Out of the receptive fields of each segment, operations can be created to recalculate the intermediate levels.

The user interface interacts with the segmentation framework but is clearly separated and no knowledge about the underlying data-structure is necessary. We will further evaluate this framework in the terms of usability.

6 Conclusion

The approach of interactively modifying an irregular pyramid by guiding it with user-specified operations is introduced. In contrast to other methods, the presented framework delivers a general purpose but versatile method for creating user-guided segmentation hierarchies. The various strategies of interactions and the solutions developed for effective processing have been analyzed and discussed.

References

1. Arbelaez, P., Maire, M., Fowlkes, C., Malik, J.: Contour detection and hierarchical image segmentation. IEEE Transactions on Pattern Analysis and Machine Intelligence (99) (2010)
2. Brun, L., Kropatsch, W.: Contraction kernels and combinatorial maps. Pattern Recognition Letters 24(8), 1051–1057 (2003)
3. Brun, L., Kropatsch, W.G.: Dual contraction of combinatorial maps. Tech. Rep. PRIP-TR-54, Institute of Computer Graphics and Algorithms 186/3, Pattern Recognition and Image Processing Group, TU Wien, Austria (1999)
4. Brun, L., Vautrot, P., Meyer, F.: Hierarchical watersheds with inter-pixel boundaries. In: Campilho, A.C., Kamel, M.S. (eds.) ICIAR 2004. LNCS, vol. 3211, pp. 840–847. Springer, Heidelberg (2004)
5. Felzenszwalb, P.F., Huttenlocher, D.P.: Efficient graph-based image segmentation. International Journal of Computer Vision 59(2), 167–181 (2004)
6. Fuh, C.S., Cho, S.W., Essig, K.: Hierarchical color image region segmentation for content-based image retrieval system. IEEE Transactions on Image Processing, 9(1), 156–162 (2000)
7. Haxhimusa, Y., Ion, A., Kropatsch, W.G., Brun, L.: Hierarchical image partitioning using combinatorial maps. In: Chetverikov, D., Czuni, L., Vincze, M. (eds.) Proceeding of the Joint Hungarian-Austrian Conference on Image Processing and Pattern Recognition, Hungary, May 2005, pp. 179–186 (2005)
8. Haxhimusa, Y., Kropatsch, W.G.: Segmentation graph hierarchies. In: Fred, A., Caelli, T.M., Duin, R.P.W., Campilho, A.C., de Ridder, D. (eds.) SSPR&SPR 2004. LNCS, vol. 3138, pp. 343–351. Springer, Heidelberg (2004)
9. Heimann, T., et al.: Comparison and evaluation of methods for liver segmentation from ct datasets. IEEE Transactions on Medical Imaging 28(8), 1251–1265 (2009)
10. Hug, J.M.: Semi-Automatic Segmentation of Medical Imagery. Ph.D. thesis, Swiss Federal Institute of Technology Zürich (2000)
11. Kass, M., Witkin, A.P., Terzopoulos, D.: Snakes: Active contour models. International Journal of Computer Vision 1(4), 321–331 (1988)
12. Keselman, Y., Dickinson, S.: Generic model abstraction from examples. IEEE Transactions on PAMI 27(5), 1141–1156 (2005)
13. Klava, B., Sumiko, N., Hirata, T.: Interactive image segmentation with integrated use of the markers and the hierarchical watershed approaches. In: VISSAPP (1), pp. 186–193 (2009)
14. Martin, D., Fowlkes, C., Tal, D., Malik, J.: A database of human segmented natural images and its application to evaluating segmentation algorithms and measuring ecological statistics. In: Proc. 8th Int'l Conf. Computer Vision, vol. 2, pp. 416–423 (2001)
15. Meine, H., Köthe, U., Stiehl, H.: Fast and accurate interactive image segmentation in the geomap framework. In: Tolxdorff, T. (ed.) Bildverarbeitung für die Medizin 2004, pp. 60–65. Springer, Heidelberg (2004)
16. Micusík, B., Hanbury, A.: Automatic image segmentation by positioning a seed. In: ECCV (2), pp. 468–480 (2006)
17. Mortensen, E.N., Barrett, W.A.: Interactive segmentation with intelligent scissors. Graphical Models and Image Processing 60(5), 349–384 (1998)
18. Rother, C., Kolmogorov, V., Blake, A.: "grabcut": interactive foreground extraction using iterated graph cuts. ACM Trans. Graph. 23(3), 309–314 (2004)
19. Torralba, A., Russell, B., Yuen, J.: Labelme: Online image annotation and applications. Proceedings of the IEEE 98(8), 1467–1484 (2010)

Tiled Top-Down Pyramids and Segmentation of Large Histological Images

Romain Goffe[1], Luc Brun[1], and Guillaume Damiand[2]

[1] GREYC, ENSICAEN, CNRS, UMR6072, F-14050, Caen, France
{romain.goffe,luc.brun}@greyc.ensicaen.fr
[2] LIRIS, Université de Lyon, CNRS, UMR5205, F-69622, Villeurbanne, France
guillaume.damiand@liris.cnrs.fr

Abstract. Recent microscopic imaging systems such as whole slide scanners provide very large (up to 18GB) high resolution images. Such amounts of memory raise major issues that prevent usual image representation models from being used. Moreover, using such high resolution images, global image features, such as tissues, do not clearly appear at full resolution. Such images contain thus different hierarchical information at different resolutions. This paper presents the model of tiled top-down pyramids which provides a framework to handle such images. This model encodes a hierarchy of partitions of large images defined at different resolutions. We also propose a generic construction scheme of such pyramids whose validity is evaluated on an histological image application.

Keywords: Irregular pyramid, Topological model, Combinatorial map.

1 Introduction

The increasing amount of high resolution images raises new and major issues in the field of image analysis and processing. For instance, microscopic scanners have recently been improved to the point where whole slide imaging techniques may offer a $\times 40\,000$ magnification. However, the segmentation of such images requires to handle a hierarchy of large partitions defined on up to 18GB data volumes. A suitable model to encode segmentation of such images should be compatible with memory constraints induced by this large amount of data and should allow to design segmentation algorithms based on the same top-down analysis scheme than pathologists.

In this regard, we identified two key steps during pathologists' manual analysis of pyramids of histological images. First, an identification of histological components for each resolution is performed and regions of interest are determined according to topological or geometric features. Second, these regions of interest are used within a hierarchical scheme from the lowest to the highest resolution (top-down), each region being analyzed in the context defined by its ancestors.

As a consequence, a model with geometrical, topological and hierarchical features is required to provide a segmentation of large histological images. Usual

X. Jiang, M. Ferrer, and A. Torsello (Eds.): GbRPR 2011, LNCS 6658, pp. 255–264, 2011.

non hierarchical models are devoted to few partitions' properties and do not provide a full encoding of geometrical and topological properties of a partition. For example, RAG data structures do not provide an efficient access to the geometry of regions' borders. This drawback is addressed by more sophisticated models such as topological maps [2,3] which encode both geometric and topological properties of a partition.

Hierarchical models such as quadtrees or regular pyramids are commonly used for multi-resolution images representation and segmentation. Yet, both frameworks entail several drawbacks [1]. For example, they may not preserve adjacency of connected regions through different levels of a pyramid. Those drawbacks lead to the design of irregular pyramids [2,3] in order to take advantage of the efficiency of graphs and topological maps for geometric and topological operations while keeping the advantages of their regular ancestors.

We have previously introduced the tiled maps model [4], defined as a topological map decomposed into topological tiles to encode partitions of large images. In [5], we have proposed an efficient construction scheme of tiled top-down pyramids. Using such a construction scheme, each partition defined at a given resolution is initialised by the projection of the partition defined at the previous (lower) resolution and then further refined by split and merge operations. In this paper, we focus on three main points. We first provide an improved formalism to define our model of tiled top-down pyramid (Section 2). We propose in Section 3 a new projection step of each partition which takes into account the additional information provided by the current resolution. We finally show in Section 4 the efficiency of our model with the segmentation of an histological image.

2   Top-Down Framework

2.1   Topological Maps

A topological map model [2,3] encodes both topological and geometrical properties of a partition. It combines three distinct models: a 2-map that encodes topological relationships, a matrix of interpixel elements [8] that encodes the geometry of the partition and a tree of regions for inside relationships (Figure 1).

A combinatorial map is defined from a set of abstract basic elements called *darts*. Two operators denoted by β_i, $i \in \{1,2\}$ that apply on darts allow to represent adjacency relationships between cells (Definition 1).

Definition 1 (Combinatorial map). *In two dimensions, a combinatorial map M is a triplet $M = (\mathcal{D}, \beta_1, \beta_2)$ where:*

1. *$\mathcal{D}$ is a finite set of darts;*
2. *β_1 is a* permutation[1] *on $\mathcal{D}$;*
3. *β_2 is an* involution[2] *on $\mathcal{D}$.*

---

[1] A *permutation* is a one to one mapping from S onto S.
[2] An *involution* f is a one to one mapping from S onto S such that $f = f^{-1}$.

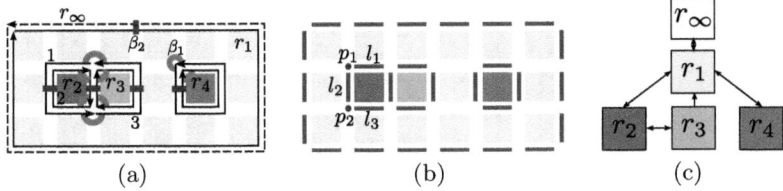

Fig. 1. Topological maps: a combination of three models for the representation of a partition. (a) Combinatorial map for adjacency relationships. Dotted darts belong to the infinite region. (b) Interpixel representation for geometrical borders: bounding pointels and linels are drawn as circles and bold segments. (c) Tree of regions for inside/contains relationships.

Intuitively, a combinatorial map may be considered as a planar graph whose edges are decomposed into half edges called darts. Each dart is incident to a single vertex and belongs to a single edge and a single face. Sets of darts defining high level entities such as vertices, edges and faces are retrieved using cycles[3] of β_i permutations. In practice, the β_1 permutation connects a dart of a face to the next dart encountered when turning clockwise around it. The involution β_2 separates two adjacent faces and maps a dart of an edge to the only dart with an opposite orientation which belongs to the same edge. For instance, in Figure 1(a), $\beta_1(1) = 3$ and $\beta_2(1) = 2$. Vertices, edge and faces of a combinatorial map are respectively encoded by the cycles of the permutations $\beta_1 \circ \beta_2, \beta_2$ and β_1. All faces of a combinatorial map but one encode finite faces (faces corresponding to regions with a finite area). The face which does not satisfy this property encodes the background of the partition encoded by the combinatorial map and is called an infinite face. A combinatorial map is called *minimal* in number of cells if it does not contain any vertex with a degree lower or equal to 2 and removing any dart would change the topology.

Matrix of interpixel elements. Using interpixel elements [8], the geometry of an an $n \times m$ image partition is encoded by an $(n + 1) \times (m + 1)$ array of marks. Each entry of this array encodes the existence of a given linel (or crack) element within the set of linels encoding image's boundaries and called bounding linels (Figure 1(b)). Note that both pointels incident to a same bounding linel are considered as bounding elements. We call *embedding* the association of bounding cells with a topological element (vertex, edge or face). In Figure 1(b), embedding of dart 1 is defined as the sequence of bounding elements $(l_1, p_1, l_2, p_2, l_3)$. A region corresponds to the embedding of a face and is a set of adjacent pixels.

Tree of regions. The tree of regions describes inside/contains relationships: a region is the father of the regions it contains. In Figure 1(c), r_1 contains $\{r_2, r_3, r_4\}$. The root of the tree encodes the background of the image and is called the infinite region (denoted by r_∞).

---

[3] A cycle of a permutation is defined as its restriction to the set of images of an element.

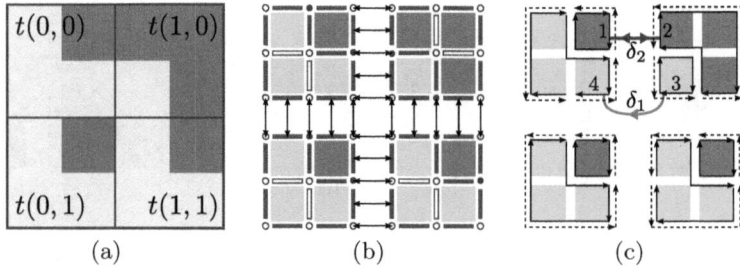

Fig. 2. Tiled maps: an extension to the topological map model designed for large partitions. (a) Original image decomposed into four tiles. (b) Interpixel representation: arrows map equivalent bounding cells. (c) Tiled map: each tile is encoded by a topological map. Darts are connected between tiles through δ_1 (arcs) and δ_2 (segments) operators.

2.2   Tiled Topological Maps

A tiled extension of the topological map model which decomposes a map into smaller elements called *tiles* has been proposed to overtake memory issues induced by the representation of large partitions [4]. Indeed, tiles may be swapped on disk when they are not being processed. A *topological tile* is a geometric subdivision of a topological map whose interpixel matrix only encodes a subdivision of the image. Adjacency relationships between topological tiles are encoded by equivalent interpixel cells on tiles' shared borders: two pointels are equivalent if they have the same coordinates in the image referential and two linels are equivalent if their incident pointels are equivalent (Figure 2(b)). Darts' embeddings are encoded within the interpixel framework as sequences of pointels and linels. Consequently, two darts are said to have an equivalent embedding if their embeddings correspond to equivalent sequences of pointels and linels.

Let $\mathcal{D}^t$ be the set of darts that belong to a tile t and $\mathcal{D}_{\partial}^{t,t'}$ be the subset of $\mathcal{D}^t$ composed of darts that are incident to the border of t and whose embedding has equivalent cells in t'. Connections between darts of two adjacent tiles t and t' are represented through a bijection $\varphi^{t,t'} : \mathcal{D}_{\partial}^{t,t'} \rightarrow \mathcal{D}_{\partial}^{t',t}$ that maps each dart d of $\mathcal{D}_{\partial}^{t,t'}$ to a dart d' of $\mathcal{D}_{\partial}^{t',t}$ such that d and d' have equivalent embeddings (Figure 2(c)). Note that such a bijection can always be obtained if we consider a decomposition of the tiles' borders into basic darts whose embedding is a single couple (pointel, linel).

Besides, the decomposition in tiles entails that some darts of a tile's border may encode the tile's decomposition of the image without encoding a real border between regions. In this case, the border is considered as a *fictive border*. Otherwise, the border is called a *real border*. Thus, two additional operators δ_1 and δ_2 are introduced in order to skip darts encoding fictive borders which leads to the definition of the tiled map model (Definition 2).

Definition 2 (Tiled topological map). *Let T be a set of topological tiles. Let $\mathcal{D}$ be the subset of $\mathcal{D}^T$ whose darts belong to a real border. A tiled topological map (or tiled map) is a triplet $M = (\mathcal{D}, \delta_1, \delta_2)$ where:*

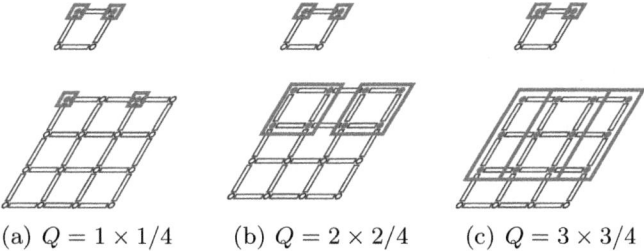

(a) $Q = 1 \times 1/4$ (b) $Q = 2 \times 2/4$ (c) $Q = 3 \times 3/4$

Fig. 3. Reduction windows of pointels within the interpixel representation of a pyramid of images. (a) Non-overlapping holed pyramid. (b) Non-overlapping pyramid without holes. (c) Overlapping pyramid.

- δ_2 is an involution on $\mathcal{D}$ such as: $\delta_2(d) = \begin{cases} \varphi^{t,t'}(d) & \text{if } d \in \mathcal{D}_{\partial}^{t,t'} \\ \beta_2(d) & \text{otherwise} \end{cases}$
- δ_1 is a permutation on $\mathcal{D}$ such as: $\delta_1(d) = \beta_1 \circ (\delta_2 \circ \beta_1)^k(d)$
  with $k = min\{p \in \mathbb{N} \mid \beta_1 \circ (\delta_2 \circ \beta_1)^p(d) \in \mathcal{D}\}$

Since δ_1 and δ_2 operators respectively define a permutation and an involution on $\mathcal{D}$ [5], the triplet $M = (\mathcal{D}, \delta_1, \delta_2)$ is a combinatorial map.

2.3   Tiled Top-Down Pyramids

Tiled maps allow to represent partitions from large images. However, such images usually contain different information at different resolutions hence requiring the design of a multi-resolution hierarchical data structure, such as the tiled top-down pyramidal model [5].

In this paper, we propose a novel approach for the definition of tiled top-down pyramids based on the combination of three hierarchical data structures: a pyramid of images encoding a multi-resolution image, a pyramid of tiles with a constant size and a pyramid of tiled maps.

A regular pyramid of images [1] is a sequence $(I^0, \ldots, I^n)$ of images with exponentially increasing resolutions such that the *reduction factor* r encoding the ratio between the size of two successive images remains constant along the pyramid. Each pixel in an image I^k is related to a connected set of pixels in I^{k+1} called its *reduction window* and encoded as a $M \times N$ window. The value of a pixel in I^k is deduced from the values of pixels within its reduction window using a *reduction function*. Different types of pyramids may be distinguished according to the ratio $Q = M \times N/r$: holed pyramids ($Q < 1$), non overlapping pyramids without holes ($Q = 1$) and overlapping pyramids ($Q > 1$).

Such a pyramid may be implicitly associated to a stack of matrices of inter pixel elements, also defining a regular pyramid with a same reduction factor. The notion of reduction window on such a pyramid may be adapted as follows: the *reduction window* of a pointel p (or $\mathcal{RW}(p)$) at level k ($0 \leq k < l$) is an $M' \times N'$ window defined at level $k + 1$ and corresponding to the set of pointels encoded by p at level k. A hierarchical relationship is thus induced between pointels, each pointel being the father of the pointels within its reduction window. Different types of pyramids of pointels may be distinguished according to the ratio $M' \times N'/r$ (Fig. 3). In the following, we use non overlapping pyramids without holes both for images and pointels pyramids.

Let us now consider a given tile's size and perform a tiling of each image of a non overlapping image pyramid without holes $\mathcal{I} = (I^0, \ldots, I^n)$. Each tile defined in an image I^k is expanded into $r \times r$ tiles in I^{k+1}, where r denote the reduction factor of $\mathcal{I}$. Hierarchical relationships established within an image pyramid induce thus hierarchical relationships between tiles decomposing each image of the pyramid. This last hierarchy is called a pyramid of tiles (Definition 3).

Definition 3 (Pyramid of tiles). *Let I be an image. A pyramid of tiles associated with I is a couple $(\mathcal{I}, \mathcal{T} = (T^0, \ldots, T^n))$ where $\forall k$, $0 \leq k < n$:*

- *$\mathcal{I} = (I^0, \ldots, I^n)$ is a non-overlapping pyramid of images without holes associated with I;*
- *$T^k = \{t(i,j,k)\}_{(i,j) \in [0, l^k/l[\times [0, h^k/h[}$ is a rectangular tiling of the $l^k \times h^k$ image I^k by a set of $l \times h$ tiles $t(i,j,k)$;*
- *$\mathcal{RW}(t(i,j,k)) = \{t(i',j',k+1) \mid \lfloor i'/r \rfloor = i, \lfloor j'/r \rfloor = j\}$ is the reduction window of tile $t(i,j,k)$, where $\lfloor . \rfloor$ denotes the floor operator.*

While both pyramids of images and tiles are regular hierarchical data structures, our aim is to define an irregular pyramid of tiled maps. A tiled top-down pyramid (Definition 4) is thus a stack of finer and finer partitions, defined at different resolutions and encoded by tiled topological maps. The hierarchy is represented by *up/down* relationships between darts and regions [4]. Moreover, since each map G^{k+1} of the pyramid is finer than G^k, a border defined at a given level of the pyramid cannot be removed at levels below. A tiled top-down pyramid is thus a *causal structure* [6]. Consequently, we propose a new definition for tiled top-down pyramids (Definition 4) where the tiled maps (levels) are based on the tiling of a given pyramid of tiles.

Definition 4 (Tiled top-down pyramid). *Let I be an image. A tiled top-down pyramid is a triplet $(\mathcal{I}, \mathcal{T}, \mathcal{G} = (G^0, \ldots, G^n))$ where $\forall k$, $0 \leq k < n$:*

- *$(\mathcal{I}, \mathcal{T})$ is a pyramid of tiles composed of $n + 1$ levels and associated with I;*
- *G^k is a tiled map corresponding to the tiling of T^k;*
- *G^{k+1} is defined on (I^{k+1}, T^{k+1}) and is deduced from G^k by decomposition operations (e.g. splitting of faces of G^k) in order to be finer than G^k.*

The construction scheme of a tiled top-down pyramid may rely on the notion of focus of attention which allows to refine only regions of interest identified in the upper levels of the pyramid. The advantage of such a construction scheme is twofold since it may reduce memory usage and processing time while imitating experts' analysis scheme. Finally, note that within a given application field, only some specified levels of the pyramid may need to be explicitly encoded.

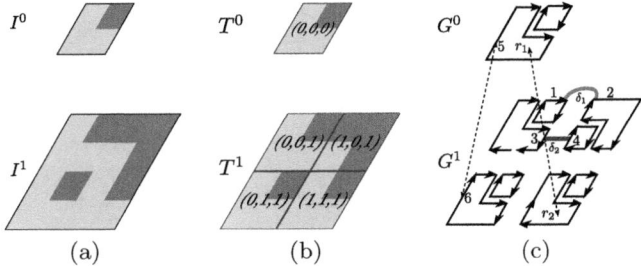

Fig. 4. Tiled top-down pyramids: a combination of three hierarchical models for the representation of large multi-resolution images. (a) Pyramid of images ($r = 2$). (b) Pyramid of tiles. (c) Pyramid of tiled maps. Dotted arrows illustrate *up/down* relationships between darts and regions. Darts that belong to r_∞ are not represented.

3 Segmentation Scheme

We propose a generic segmentation scheme for large multi-resolution images based on the tiled top-down pyramid framework. Our segmentation process relies on the definition of an *Oracle* that may combine several criteria of different natures to take advantage of geometrical, topological and hierarchical features. Since this segmentation step is part of the global construction scheme of the tiled top-down pyramid, it must neither impact the causality of the pyramid nor its refinement by focus of attention. The process assumes an initial partition G^0 which may be a single region or which may result from an extraction algorithm [2,3] at resolution I^0. In both cases, a partition at level $k + 1$ is deduced from level k by applying the following two steps procedure:

- a projection step of the regions' borders of G^k onto G^{k+1} which preserves the topology of G^k and expands the geometry of G^{k+1} borders according to the reduction factor.
- an *Oracle*-based refinement step, restricted to regions whose father is a region of interest in the previous level (focus of attention).

Figure 5 illustrates the main steps of our projection procedure. The main issue is to project the former borders onto the current resolution while preserving the same topology. Intuitively, the projection of a border $b = (p_1, \ldots, p_m)$ in G^k is a border b' in G^{k+1} contained within the strip formed by the reduction windows of the pointels $(p_i)_{i=1,m}$ of b. This strip is a connected set since the pyramid of images contains no holes. Moreover, the pyramid is not overlapping and therefore, two projected borders cannot intersect, except on their extremities which correspond to the intersection of several borders. Let us consider a pointel p_1 in G^k which corresponds to an intersection of borders and its projection $P(p_1)$ in G^{k+1}. In order to keep the same topology in G^k and G^{k+1}, we define the projection of each linel in G^k incident to p_1 as an horizontal or vertical straight line of length r in G^{k+1}, incident to $P(p_1)$, and with the same direction than the initial linel. These straight lines correspond to a standard linel's projection

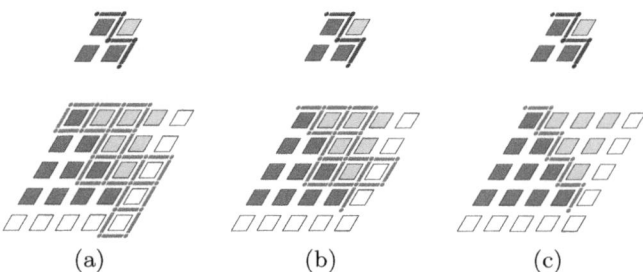

Fig. 5. Projection of regions borders. (a) Strip composed by the reduction windows of the pointels that belong to the border of the previous level. (b) Restrictions on borders extremities. (c) Shortest path according to a given energy.

which guarantees that $P(p_1)$ is the only intersection point of projected borders within p_1's reduction window. This standard projection is also performed on p_m and the connection between the two straight lines incident to $P(p_1)$ and $P(p_2)$ is achieved using a Dijkstra algorithm within the strip formed by the reduction windows of pointels $(p_i)_{i=2,m-1}$. Dijkstra algorithm is applied with an edge weight proportional to the inverse of the gradient computed in I^{k+1}, hence taking into account the additional information provided by I^{k+1}.

The refinement step relies on the definition of the *Oracle* proposed in Algorithm 1: according to a segmentation criterion (**line 3**), our scheme refines a region if its father in the previous level is a region of interest (**line 2**) while preserving the causality of the model (**line 1**).

Algorithm 1. `Oracle`

> **Data:** Two adjacent pixels p and p' of a tile $t \in G^{k+1}$.
> **Result:** true if p and p' belong to the same region.
> $r \leftarrow region(up(p))$; $r' \leftarrow region(up(p'))$;
> $l \leftarrow$ incident linel to p and p';
> 1 **if** l *is a bounding linel* **then**
>   └ **return** *false*;
> 2 **if** $up(r)$ *is not a region of interest in* G^k **then**
>   └ **return** *true*;
> 3 **return** `segmentation_criterion`(region(p), region(p'));

4   Application for Large Histological Images Segmentation

We propose a practical use case[4] of the top-down framework for the segmentation of large histological images. Our objective is to demonstrate the generic aspect of our model with the integration of an existing segmentation scheme. In this

---

[4] Our model is implemented in C++ and computations are carried out on an Intel E5300@2GHz with 2GB RAM.

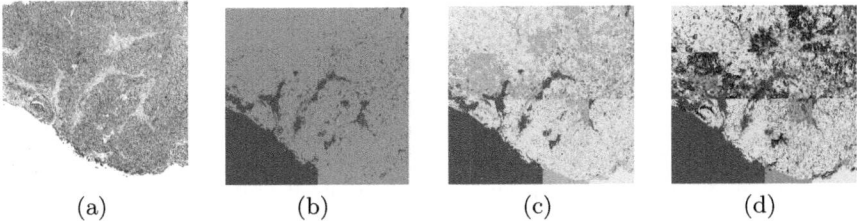

(a)   (b)   (c)   (d)

Fig. 6. Classification scheme of histological images. (a) Original image. (b) Resolution 1: distinction between tissue and background. (c) Resolution 2: distinction between lesions and tissue. (d) Resolution 3: distinction between cancer *in situ* and stroma.

Table 1. Segmentation of an histological image within our top-down framework

level	size (pixels)	number of tiles	number of regions	runtime (s) model	segmentation	ram (MB)
1	$2\,500 \times 2\,500$	16	57 435	9.8	35.2	104
2	$5\,000 \times 5\,000$	64	132 402	43.6	186.6	101
3	$10\,000 \times 10\,000$	256	225 672	267.8	645.0	106

application, histological images are produced as a stack of three images with increasing resolutions. Their manual analysis by pathologists are performed using a focus of attention over regions of interest up to the full resolution image where cells' mitosis are enumerated. This estimation is then used for breast cancer diagnosis grading.

Our top-down segmentation algorithm uses the same scheme than [9] with the additional use of a tiled top-down pyramid (Figure 6). Successive k-means based classifications [7] allow to label and refine regions of interest. At resolution 1, the background is separated from tissue. Then, a classification in two classes of tissue at resolution 2 allows to extract lesions and a last classification distinguishes the cancer *in situ* from stroma at resolution 3.

Table 1 presents experimental results for the segmentation of a representative histological image. Column 2, 3 and 4 respectively indicate the size of the images, the number of tiles and the number of regions of their associated partitions. In Column 5 and 6, we provide runtimes for both pyramid construction scheme (model) and our segmentation algorithm (segmentation). These results demonstrate that our model does not introduce an important additional cost while providing an efficient access to hierarchical, topological and geometrical features. Note that runtimes for levels' extraction are linear with the size of the image but slightly increase with the number of tiles due to disc access delays. Finally, tiled maps allow to preserve a constant memory usage around 100MB (Column 7) with a tile size of 625 × 625 pixels.

5   Conclusion

In this paper, we have proposed a new approach to the definition of tiled top-down pyramids which emphasizes the combination of regular and irregular hierarchical data structures. We have described a segmentation scheme with a new projection step which takes into account the additional information provided by each image of the pyramid. Finally, we have demonstrated the efficiency of our framework with an application on histological images.

This work opens interesting perspectives such as the definition of topological criteria that could be combined to the present segmentation process. Those criteria may enhance results by taking further advantage of our top-down framework. Finally, a medical evaluation should be performed to confirm the accuracy of the partitions provided by our method.

References

1. Bister, M., Cornelis, J., Rosenfeld, A.: A critical view of pyramid segmentation algorithms. Pattern Recognition Letters 11(9), 605–617 (1990)
2. Brun, L., Mokhtari, M., Domenger, J.P.: Incremental modifications on segmented image defined by discrete maps. Journal of Visual Communication and Image Representation 14, 251–290 (2003)
3. Damiand, G., Bertrand, Y., Fiorio, C.: Topological model for two-dimensional image representation: definition and optimal extraction algorithm. Computer Vision and Image Understanding 93(2), 111–154 (2004)
4. Goffe, R., Brun, L., Damiand, G.: Tiled top-down combinatorial pyramids for large images representation. International Journal of Imaging Systems and Technology 21(1), 28–36 (2011)
5. Goffe, R., Damiand, G., Brun, L.: A causal extraction scheme in top-down pyramids for large images segmentation. In: Hancock, E.R., Wilson, R.C., Windeatt, T., Ulusoy, I., Escolano, F. (eds.) SSPR&SPR 2010. LNCS, vol. 6218, pp. 264–274. Springer, Heidelberg (2010)
6. Guigues, L., Cocquerez, J.P., Le Men, H.: Scale-sets image analysis. International Journal of Computer Vision 68(3), 289–317 (2006)
7. Kanungo, T., Mount, D.M., Netanyahu, N.S., Piatko, C.D., Silverman, R., Wu, A.Y.: An efficient k-means clustering algorithm: Analysis and implementation. IEEE Trans. on PAMI 24(7), 881–892 (2002)
8. Kovalevsky, V.A.: Finite topology as applied to image analysis. Computer Vision, Graphics, and Image Processing 46(2), 141–161 (1989)
9. Roullier, V., Lézoray, O., Ta, V.-T., Elmoataz, A.: Mitosis extraction in breast-cancer histopathological whole slide images. In: Bebis, G., Boyle, R., Parvin, B., Koracin, D., Chung, R., Hammoud, R., Hussain, M., Kar-Han, T., Crawfis, R., Thalmann, D., Kao, D., Avila, L. (eds.) ISVC 2010. LNCS, vol. 6453, pp. 539–548. Springer, Heidelberg (2010)

Segmentation of Similar Images Using Graph Matching and Community Detection

Charles Iury Oliveira Martins[1,*], Roberto Marcondes Cesar Jr.[1,*],
Leonardo Ré Jorge[2], and André Victor Lucci Freitas[2]

[1] Institute of Mathematics and Statistics, Department of Computer Science,
University of São Paulo, São Paulo, Brazil, CEP: 05508-090
{charles,cesar}@ime.usp.br
[2] Institute of Biology, Department of Animal Biology,
State University of Campinas, Campinas-SP, Brazil, CP 6109, CEP: 13083-970
leonardorejorge@gmail.com, baku@unicamp.br

Abstract. In this paper, we propose a new method to segment sets of
similar images using graph matching and community detection algorithms.
The images in a database are represented by Attributed Relational Graphs,
allowing the analysis of structural and relational information of the regions
(objects) inside them. The method gathers such information by matching
all images to each other and stores them in a single graph, called Match
Graph. From it, we can check the obtained pairwise matchings for all images of the database and which objects relate to each other. Then, with the
interactive segmentation from one single image from the dataset (e.g. the
first one) we can observe these relationships between them through a color
label, thus leading to the automatic segmentation of all images. We show
an important biological application on butterfly wings images and a case
using images taken by a digital camera to demonstrate its effectiveness.

1 Introduction

Image segmentation methods using interactive approaches (also known as semi-
automatic) have been widely described in literature (e.g. [1,2,3]). These approaches require some kind of a priori knowledge, which is provided by the
user as an input information to the program. This is used to initiate and guide
the segmentation task.

Most interactive methods only take into account the gray tone (or color) of the
pixels, or regions, in the images and do not establish any kind of neighborhood relationship between them. Thus, although the semi-automatic approaches, as the
automatic ones, have raised important tools for image segmentation literature,
most of the proposed methods are not designed to take into account the overall
structure of the image in order to guide this process. Furthermore, the segmentation of (possibly large) sets of similar images is a problem that has been overlooked

---

* The authors are grateful to CAPES, CNPq, FAPESP and FINEP for the financial
support. Roberto M. Cesar-Jr is also with the e-Science Lab - CTBE.

X. Jiang, M. Ferrer, and A. Torsello (Eds.): GbRPR 2011, LNCS 6658, pp. 265–274, 2011.
© Springer-Verlag Berlin Heidelberg 2011

in the literature. In fact, most interactive segmentation methods would require user interaction for each image in the set. An important application would be to consider the automatic segmentation of sets of similar images that should contain the same structure, rather than segmenting one by one, manually. These two issues are addressed by the method introduced in the present paper.

Representations through graphs are widely used to treat structural information in different applications such as networking, psycho-sociology, image interpretation and pattern recognition [4]. In several approaches, graphs are used to represent knowledge and information extracted from images, where the vertices represent pixels, or regions, and edges define the relationship between them.

Therefore, for any image representation through graphs, we can gather not just vertices information, representing regions or objects, and edges representing the connections between them, but also the characteristics of each component. We can thus obtain a powerful image representation that allows their exploration in different ways [5] as graphs are a very flexible representation of data capable of representing a large set of problems related to pattern recognition [7].

Image representation using graphs and the application of techniques of graph matching in image processing and pattern recognition have been explored in works such as Bunke [8], Conte et al. [9], Felzenszwalb and Huttenlocher [10] and Wilson and Hancock [11].

In recent works [2,3], where image segmentation is seen as a model-based recognition problem, a model of an image to be segmented is provided by the user and a graph representation is extracted from the model. In the input image we initially apply an over-segmentation process (using Watershed [12]) and a graph representation is achieved. The final result of the segmentation is accomplished by matching the input and model graphs. Therefore, we explicitly manipulate the structural information present in the image to produce the final result.

The works of Ribalta and Serratosa [6,7] present new algorithms to find an optimal Common Labelling (CL) of a set of atributted graphs to compute a representative of a set of vetices. First, they compute all possible isomorphism between the attributed graphs and then compute de CL. Although these are interesting works, this is still a NP-problem with an exponential computational cost depending on the number of vertices and edges of all graphs.

In this work, we aim to gather information for a particular set (database) of images in an efficient data structure capable of representing it entirely. Each set must contain similar images and its representation will be given as a graph or, more specifically, as an Attributed Relational Graph (ARG), called Match Graph (MG). The MG contains all information of the location of the regions (objects) belonging to the set of images and the relations between them. From the interactive segmentation of a single image of this set (reference image), we can use these relations to propagate an obtained color label for each region of interest, thus achieving the automatic segmentation of all images in this set. We used the algorithm proposed by Noma et al. [3] to constitute the MG, associated to the community detection algorithm proposed by Clauset et al. [13] to achieve our objective.

This paper is organized as follows. The proposed methodology is presented in Sec. 2 where each step is discussed. After, we present the results obtained from a set of real butterfly wings images in Sec. 3. A different application for the segmentation of series of similar images obtained in burst mode of a digital camera is also presented. We conclude with a discussion of the results and prospects of future work in Sec. 4.

2   Methodology

Figure 1 displays the representation and the processes at each stage of the proposed method. The following sections explain each step represented in this figure.

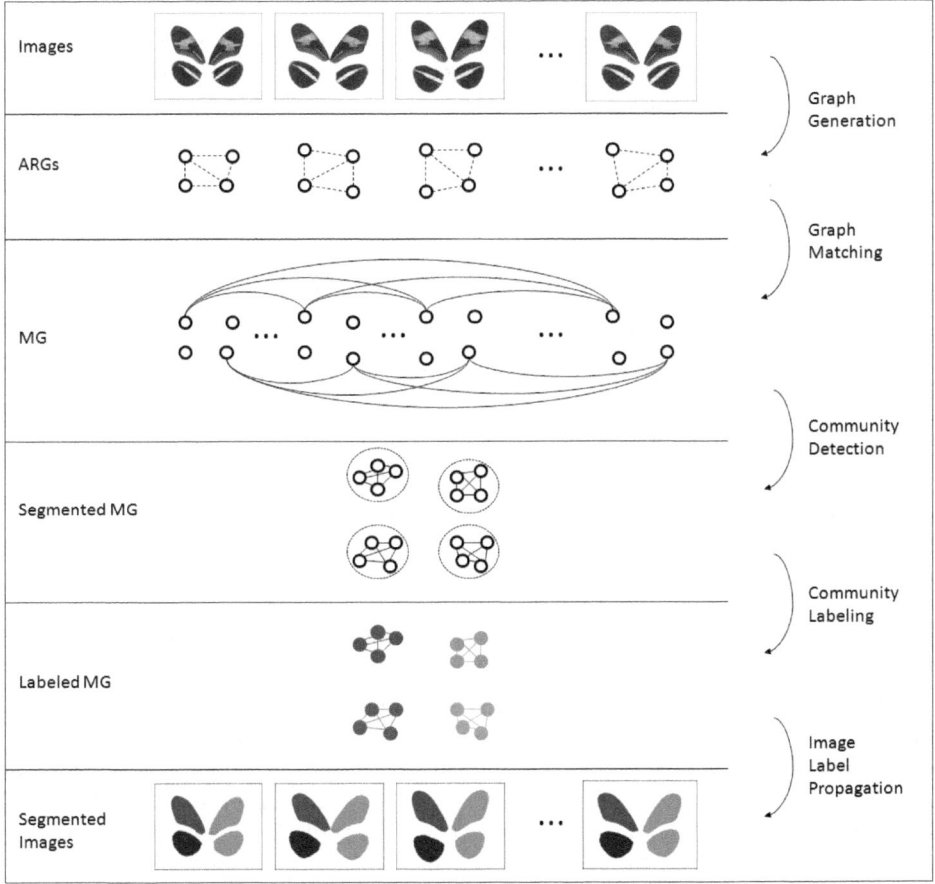

Fig. 1. (a) Illustrated representation of the proposed method

2.1 Graph Generation

Fristly, in the Graph Generation process, all images of a given database are represented by ARGs. In this work, an ARG is a directed graph expressed as a tuple $G = (V, E, \mu, \nu)$, where V is its set of vertices and $E \subseteq V \mathrm{x} V$ is its set of edges. In Fig. 1 a vertex (circles) represents an image region (subset of pixels) and an edge (black dotted line) is placed between vertices that represent two regions in the image. The vertices are obtained by the Watershed transformation [12], where each vertex represents a catchment basin (region), and the edges are obtained by Delaunay triangulation [14]. $\mu : V \rightarrow L_V$ associates an object attribute vector to each vertex of V, while $\nu : E \rightarrow L_E$ associates a relational attribute vector to each edge of E. These attributes are defined in details in [3].

2.2 Graph Matching

The Graph Matching process starts by taking all pairs of images within the set. Let A and B be two images in the dataset, represented by two ARGs G_A and G_B, respectively. The outcome corresponds to a mapping function $f : V_A \rightarrow V_B$, such that the vertices of G_A are associated with a vertex in G_B. Thus, for a vertex $v_a \in V_A$ we have $|V_B|$ possibilities to perform the matching and the right decision depends on an optimization procedure.

To help making such decisions, we used the graph matching algorithm proposed by Noma et al. [3], which takes into account the differences of structural information between images to perform it. Although the images are similar their structures are not the same, thus leading to an inexact graph matching. This algorithm runs in time proportional to $O(|V_A||V_B|)$ per iteration of the algorithm.

2.3 Community Detection

Ideally, the MG should contain a well defined community structure[1] [15]. However, in real images the objects are susceptible to noise and object variatons (e.g. color and shape). Thus, undesirables edges between communities may appear in the MG. In order to cope with this situation, we used the algorithm proposed by Clauset et al. [13] to eliminate these edges from the MG and help to identify the groups of vertices that correspond to the associated objects into the image database.

The algorithm is based on the greedy optimization of the quantity known as modularity [13]. It is a property of a network and leads to a specific proposed division of that network into communities. It measures when the division is good, in the sense that there are many edges within communities and only a few between them. This algorithm is fast and suitable for very large networks, thus its use is required for the proposed method since the number of vertices and edges in the MG increases according to the number of images in the set. This algorithm runs in essential linear time $O(n \log^2 n)$, where n is the number of vertices of the graph.

---

[1] Densely connected groups of vertices with only sparser connections between groups.

2.4 Image Label Propagation

Having defined the communities in the MG, the next process consists in labeling them. Thus, we apply a different color label to each community, obtained by the interactive segmentation of some single image from the set (reference image). This image is obtained using the software avaiable by the authors in [3] or can be drawn manually. The vertices in the MG that correspond to the regions in the reference image are labeled as those respective regions in the image. Then, the label is propagated to all other vertices of each community, so that each one will be represented by a specific color defined by the user.

The final segmentation result of all the images in database is hence accomplished by this Image Label Propagation process. Since all vertices in each community have a color label, it is easy to identify all the corresponding regions by propagating them with the respective color into the images.

3 Experimental Results

In order to demonstrate the effectiveness of the proposed method over real images, we aim to segment the different color structures within the wings of the butterfly *Heliconius erato* (Nymphalidae). To perform this test we will use an image database[2] obtained for a previous work [16], gathered to study the role of the host-plant used as food during the larval stage on wing shape in this species. A small part is illustrated in Fig. 2 where eight images are displayed.

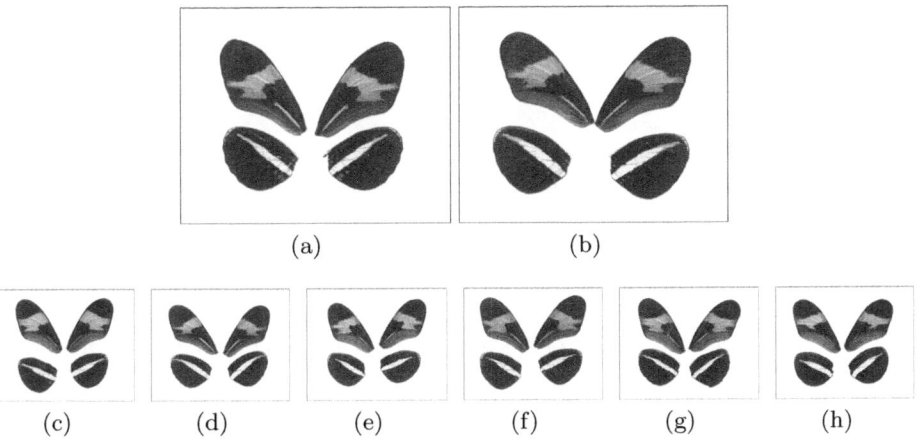

Fig. 2. (a)-(h): Wings of butterfly species *Heliconius Erato* feed with the *Passiflora edulis* plant

---

[2] This database contains 289 images obtained by removing the wings from the butterflies, and capturing them of the ventral side of the four wings of each individual using a flatbed scanner (HP Scanjet 3800; Hewlett-Packard).

The segmentation result of the butterflies wings in the eight images in this set is shown in Fig. 3, where the results demonstrate the ability of the method in discriminating the objects. Initially, we identified all the different wings (top and bottom, left and right) using different color labels, according to the reference image (first image in the set, segmented interactively). All groups and wings were identified through the MG (Fig. 4) and the labeling was performed.

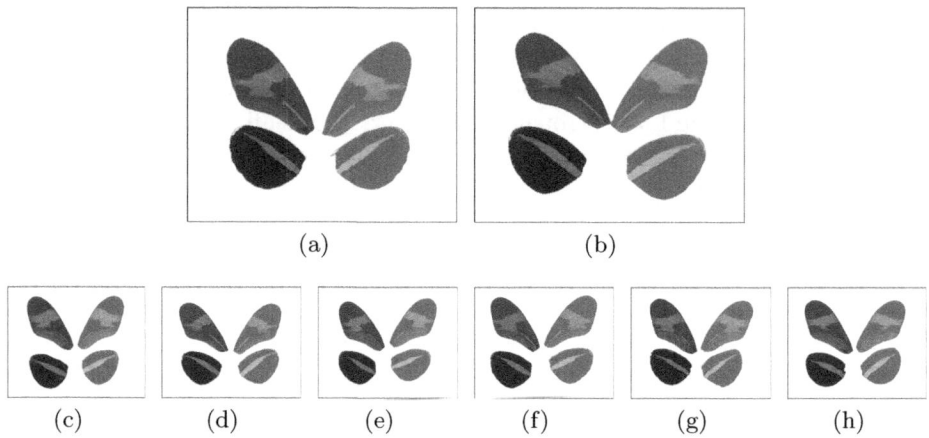

Fig. 3. (a)-(h): Segmentation results by the proposed method applied to images of butterfly wings in Figs. 2(a)-2(h), respectively

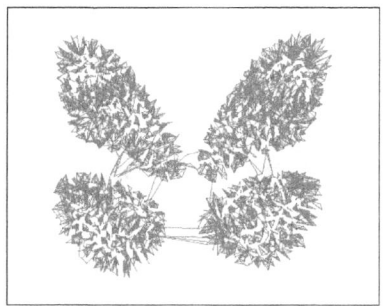

Fig. 4. MG obtained for the set of images of Fig. 3 containing 4092 vertices and 57288 edges

In order to provide a more detailed segmentation of the color patterns that compose the butterfly wings, we applied to the method a reference image containing more details. This is important to identify all the color patterns on the wings of this species in order to study, in future works, the influence that they suffer in relation to their feeding [16].

Therefore, from the interactive segmentation of the first image of this set, we achieve the automatic segmentation of all other images (only four were listed in

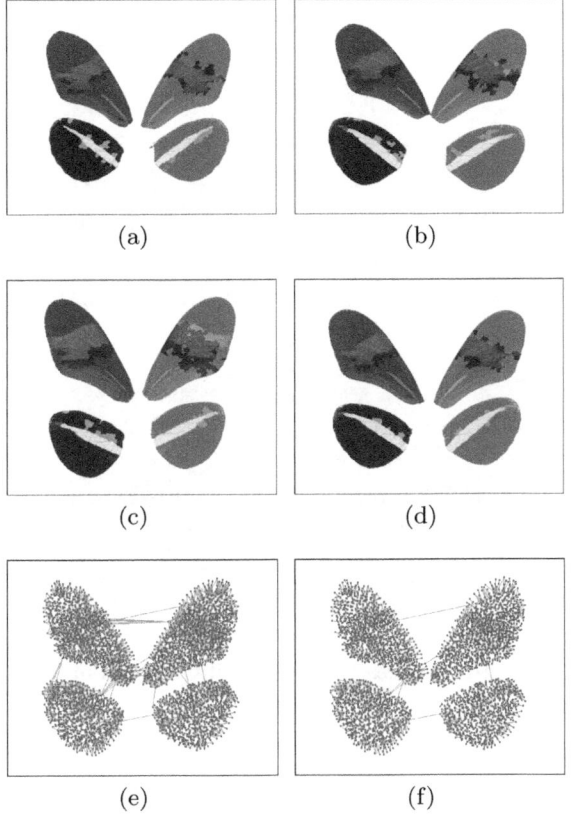

Fig. 5. (a)-(d): Segmentation results by the proposed method applied to images of butterfly wings in Figs. 2(a)-2(d), respectively; (e) MG before-CD; (f): MG after-CD

this example), showed in Figs. 5(a) to 5(d). Figure 5(e) depicts the MG obtained before the community detection algorithm. It is worth noting that its number of edges is greater, due to the many matchings, but then corrected after this process, as seen in Fig. 5(f).

As the wings on the images of Fig. 2 are close together and contain differences in shape and size, some error of matching may occur. Two vertices near each other and that belong to different wings can be joined, thus creating a community whose vertices represent regions of different wings. This fact is illustrated in the images of Figs. 3(a)-3(h). Accordingly, a small labeled region in a different color appears in the wings, in the central part of the image, indicating that there was a spread of the label of one wing to another. In the MG illustrated in Fig. 4, we see this problem through the vertices that were joined by edges that arise from distinct regions of wings. However, it is worth noting that user intervention would be required just to correct such small details, as most structures of all images in the collection would be correctly identified. This is much easier than requiring the user to segment each image from scratch.

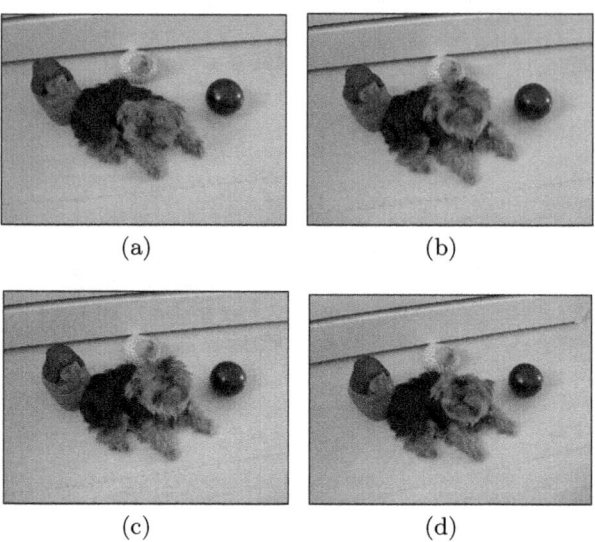

Fig. 6. (a)-(d): Images taken by digital camera in burst mode

Fig. 7. (a)-(d): Segmentation results by the proposed method applied to images in Figs. 6(a)-6(d), respectively; (e) MG after-CD containing 2918 vertices and 13706 edges

In Fig. 5, we can observe in detail the effect of applying the community detection algorithm. Clearly we note that the number of edges is greater, in Fig. 5(e), than the generated MG before the process of CD. This is due to the matching between vertices that represent neighboring regions and are similar in color or gray level but which are not a desirable solution to the problem. In this case, after applying the CD algorithm, there is a large reduction in the number of edges and a better definition of the number of communities is obtained, as illustrated in the MG of Fig. 5(f).

In order to test the proposed method using other types of images, we considered its application to a set of images taken by a digital camera in burst mode. Figure 6 depicts four images of a residential environment containing an animated element, which is a dog, and other static objects. In this example, it is interesting to segment each of the highlighted parties as the dog and the three objects around it.

The automatic segmentation of the four images in Figs. 6(a)-6(d) is illustrated in Figs. 7(a)-7(d). The obtained MG, depicted in Fig. 7(e), is naturally more dense due to the large number of regions in the images, thus increasing the number of added vertices.

4   Conclusion

We have presented a new method to segment similar images in a group using graph matching and community detection algorithms. Our goal consists in gathering information about the objects (regions) of all the images in a single graph, named Match Graph. It is obtained by the pairwise matching of all image pairs in this group and post-processed with cuts of undesirable matching edges, provided by community detection. A color label from an interactive segmentation of some single image of the group is associated to its respective vertices and propagated to the other vertices of their communities. Then, the label can also be associated to all regions and an automatic segmentation of the entire group of images is provided.

We intend to continue working to have a more robust method for identifying each type of situation that leads to an incorrect matching. For example, by eliminating undesirable edges between regions according to valid criteria established by the program (such as proximity and color or gray levels). In order to improve our method, we are also working to improve the computational performance in order to treat large image sets.

References

1. Klava, B., Hirata, N.S.T.: Interactive image segmentation with integrated use of the markers and the hierarchical watershed approaches. In: International Conference on Computer Vision Theory and Applications - VISSAPP, pp. 186–193 (2009)
2. Consularo, L.A., Cesar Jr., R.M., Bloch, I.: Structural image segmentation with interactive model generation. IEEE International Conference on Image Processing, ICIP 2007 (2007)

3. Noma, A., Graciano, A.B.V., Consularo, L.A., Cesar Jr., R.M., Bloch, I.: A new algorithm for interactive structural image segmentation. Computing Research Repository - CoRR 0805.1854 (2008)
4. Cesar-Jr., R.M., Bengoetxea, E., Bloch, I., Larrañaga, P.: Inexact Graph Matching for Model-Based Recognition: Evaluation and Comparison of Optimization Algorithms. Pattern Recognition 38(11), 2099–2113 (2005)
5. Haxhimusa, Y.v., Kropatsch, W.G.: Segmentation graph hierarchies. In: Fred, A., Caelli, T.M., Duin, R.P.W., Campilho, A.C., de Ridder, D. (eds.) SSPR&SPR 2004. LNCS, vol. 3138, pp. 343–351. Springer, Heidelberg (2004)
6. Solé-Ribalta, A., Serratosa, F.: On the computation of the common labelling of a set of attributed graphs. In: Bayro-Corrochano, E., Eklundh, J.-O. (eds.) CIARP 2009. LNCS, vol. 5856, pp. 137–144. Springer, Heidelberg (2009)
7. Sol-Ribalta, A., Serratosa, F.: Graduated Assignment Algorithm for Finding the Common Labelling of a Set of Graphs. In: Hancock, E.R., Wilson, R.C., Windeatt, T., Ulusoy, I., Escolano, F. (eds.) SSPR&SPR 2010. LNCS, vol. 6218, pp. 180–190. Springer, Heidelberg (2010)
8. Bunke, H.: Recent Developments in Graph Matching. International Conference on Pattern Recognition 2, 117–124 (2000)
9. Conte, D., Foggia, P., Sansone, C., Vento, M.: Thirty years of graph matching in pattern recognition. International Journal of Pattern Recognition and Artificial Intelligence 18(3), 265–298 (2004)
10. Felzenszwalb, P.F., Huttenlocher, D.P.: Efficient Graph-Based Image Segmentation. International Journal of Computer Vision 59(2), 167–181 (2004)
11. Wilson, R.C., Hancock, E.R.: Structural matching by discrete relaxation. IEEE Transactions on Pattern Analysis and Machine Intelligence 19, 634–648 (1997)
12. Soille, P.: Morphological Image Analysis. Springer, Heidelberg (2003)
13. Clauset, A., Newman, M.E.J., Moore, C.: Finding community structure in very large networks. Physical Review E 70(6), 66111 (2004)
14. da Costa, L.F., Cesar Jr, R.M.: Shape Analysis and Classification: Theory and Practice. CRC Press, Inc., Boca Raton (2000); ISBN 0849334934
15. Newman, M.E.J.: Modularity and community structure in networks. Proceedings of the National Academy of Sciences 103(23), 8577–8582 (2006)
16. Jorge, L.R., Cordeiro-Estrela, P., Klaczko, L.B., Moreira, G.R.P., Freitas, A.V.L.: Host-plant dependent wing phenotypic variation in Heliconius erato. Biological Journal of the Linnean Society (2011)

Automatic Street Graph Construction in Sketch Maps

Klaus Broelemann[1], Xiaoyi Jiang[1], and Angela Schwering[2]

[1] Department of Mathematics and Computer Science, University of Münster,
Germany
[2] Institute for Geoinformatics, University of Münster, Germany

Abstract. In this paper we present an algorithm for automatic street
graph construction of hand-drawn sketch maps. This detection is important for a subsequent graph matching in order to align the sketch
map with another map. Our algorithm detects a number of street candidates and selects street lines by rating the candidates and their neighbors
in the street candidate graph. To evaluate this approach, we manually
generated a ground truth for some maps and conducted a preliminary
quantitative performance study.

1 Introduction

During the last years, Geographic Information Systems (GIS) became a widely-
used technology in people's daily life. While the abilities and complexity of GI
systems are continuously increasing, there is still an absence of easy-to-use interaction methods. According to Schlaisich and Egenhofer [8] hand-drawn sketch
maps are an intuitive way to interact with a GIS.

There are two aspects that separate hand-drawn maps from other map types.
One is the fact that they are drawn on a paper, a whiteboard or a tablet PC by
hand. The other aspect is the difference between metric maps and sketch maps.
Metric maps display a region accurate based on a mathematical projection. In
contrast to metric maps, sketch maps are not drawn accurate. They do not show
an accurate image of a region but a conceptual one, which can be drawn by
everyone.

One application for sketch maps is the route-finding to an unknown destination. Nowadays, navigation systems can guide people to each destination,
provided that the address is entered. But these systems fail without knowing
the address. Let us assume some participants of a conference want to meet at a
restaurant for dinner. One of them has seen a nice restaurant before, but does
not remember its name. He can sketch the way to the restaurant on a sheet of
paper in order to share the way with others. In order to find the restaurant,
people can take a photo of this sketch map and an automatic system under-
stands the map and finds the sketched destination on a metric map as input for
a navigation system.

Beside the navigation task, there are other applications for automatic processing of sketch maps, like volunteer geographic information (VGI) systems, which

X. Jiang, M. Ferrer, and A. Torsello (Eds.): GbRPR 2011, LNCS 6658, pp. 275–284, 2011.

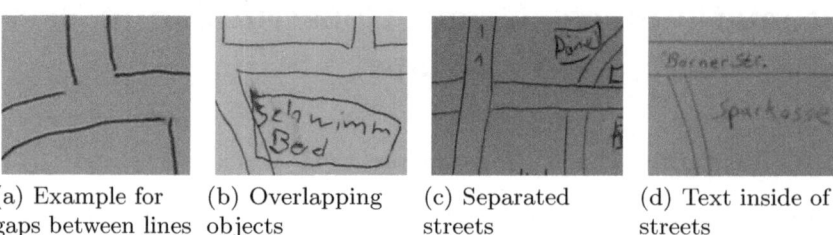

(a) Example for gaps between lines   (b) Overlapping objects   (c) Separated streets   (d) Text inside of streets

Fig. 1. Examples inaccuracies of hand-drawn sketches

allow users to annotate, add and modify content of maps. Using sketch maps as input would enable users to provide their knowledge in a natural way.

All such applications need to align different maps with each other. For the navigation task, the system has to align the sketched map with a stored map in order to identify the location of the drawn destination.

There are two ways to process sketch maps in an automatic way: online with pen-based input devices and offline using images of sketch maps. While most sketch processing systems use pen-based input devices, we deal with photos of sketch maps. Thus, our approach is not restricted to special devices, but can run on any smartphone with a build-in camera. Although the available resources on smart phones are increasing, mobile multimedia processing [5] still has to deal with low memory and computational power. Thus, our algorithm tries to meet the needs of mobile devices.

Previous research has shown that sketch maps are distorted in several ways: the drawn lengths do not correspond to the lengths in metric maps, angles tend to be 90° and curved streets often are drawn straight. Even though the drawn street network is subject to granularity (which means in this context that small or unimportant streets might be missing in sketch maps), connections that are drawn between streets exist in reality, too. Thus, we believe that street graph matching is a good way to match sketch maps against both metric maps and other sketch maps.

The detection of street networks of sketch maps is an important step for automatic matching of sketch maps. In this paper, we present an algorithm for such an automatic detection of street graphs from camera images of sketch maps. The algorithm has to deal with inaccuracies of hand-drawn sketches: lines that should be connected have gaps in between, objects can overlap each other, due to drawing order of streets, connected streets can be separated by street border lines and text might be drawn inside the streets. Figure 1 shows examples for these inaccuracies. In addition, there can be inhomogeneous illumination effects caused by the mobile camera.

We structure this paper as follows: In section 2 we will give a brief description of related work. Section 3 is used to describe the basic idea of our algorithm as well as the details of its parts. Some results of our algorithm are presented in section 4. Finally, in section 5 we will give a conclusion and an outlook on future work.

2   Related Work

Sketch recognition has been subject of research for several years. Most of the previous work concentrates on pen-based sketch recognition [1,3]. These methods process the sketches online and, thereby, can use a segmentation into single strokes and the drawing order of strokes. For our offline approach, this information is not available. Thus, different methods have to be applied to recognize elements of the sketch.

Previous methods that were proposed to detect pairs of parallel or equidistant lines in hand-drawn images [3,10] base on the angle of lines. Thus, these methods can only be applied to straight lines.

The detection of street networks is not only interesting for the domain of sketch maps. Some work has been done to detect street networks in aerial and satellite images [7]. Hu *et al.* [4] presented a system for finding streets in aerial images that follows the streets from starting points. The idea is to measure the distance from a point to the boarder as a function of the angle. The local maximums will belong to the main directions of the street. For sketch maps, the problem of such a method is to find a scale-independent way to detect starting points. As people differ in their sketching style, including the dimension of objects, some kind of scale invariance is necessary for street detection algorithms.

3   Street Graph Detection

The first proposed step for sketch map registration is to detect the street graph of the sketch map. In this section we present our algorithm for offline detection of street graphs from images of sketch maps. There are two ways of drawing streets in sketch maps: as single lines or as pairs of parallel lines. The aim of our algorithm is to detect double-lined streets and build the graph of these streets. The core idea of the algorithm is to produce a large number of middle lines and select the ones that belong to a street segment. These streets can be combined into a street graph.

Our algorithm finds a number of street segment candidates using well-known techniques like thinning. The novel approach is to select candidates that are likely to be street segments based on a rating and the rating of their neighbors in the candidate graph. By using ratings, the algorithm can easily be extended to create a probabilistic graph suitable for probabilistic graph matching.

The algorithm can roughly be divided into three parts: preprocessing, line segment detection, and graph creation. The goal of the first part is to binarize the image and to enhance the quality of the image for further processing. By doing so the algorithm removes artefacts, fills gaps between drawn lines and separates text and graphic elements.

The second part finds middle lines by thinning the background. The resulting line segments are rated according to their likelihood to be the middle line of a street. The basic idea for the rating are the two criteria that the width of streets does not change abruptly and that streets are much longer than wide.

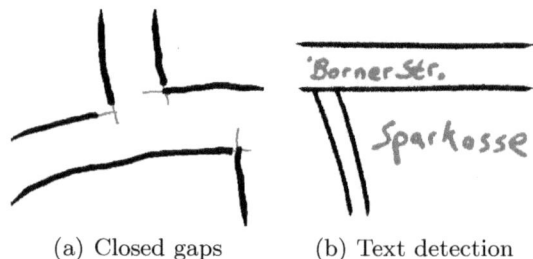

(a) Closed gaps   (b) Text detection

Fig. 2. Examples for results of preprocessing operations

In a third part, the algorithm selects lines to be composed to a street graph. This selection is based on the ratings of line segments and their neighbors.

3.1   Preprocessing

The input for the algorithm is a photo of a hand-drawn sketch map. That means that the algorithm has to deal with both the inaccuracies of human drawings and the typical effects of camera images like inhomogeneous illumination.

To meet the latter effect, the image binarization is done using adaptive thresholding methods. For this purpose each pixel is compared to the average gray level of a given neighborhood. Artefacts that might occur can be detected by considering the size of connected components. To reduce the computation costs, the removal of artefacts will be done in combination with a later preprocessing step.

One consequence of inaccurate drawings are small gaps between lines. An example of such gaps can be seen in Fig. 1(a). As streets will be detected by thinning the background, such gaps lead to undesirable lines in the resulting skeleton. To close these gaps, lines are extended by a fixed number N of pixels. This can be done by thinning the drawn lines and repeating the last N pixel of a thinned line along the contour direction. Figure 2(a) shows an example for such extended lines.

As we use a thinning algorithm to detect streets, all elements within streets can disturb the street detection. Beside previously mentioned artefacts, streets can contain written text. Figure 1(d) shows an example for such text within streets. To remove the text, we use a algorithm similar to the text / graphic separation algorithm of Tombre *et al.* [9] to detect text and remove it from the image. Figure 2(b) shows the detected text for the previous given example. The text separation algorithm is based on connected component analysis. It detects text based on size and ratio of the bounding box as well as the density of pixels. Since the text separation algorithm calculates the connected components of the image, it is possible also to remove binarization artefacts by deleting components below a given size.

3.2   Line Segment Detection

The main idea for our algorithm is to find a huge number of potential middle line segments of streets, rate these line segments and build a street graph based

on these ratings. In this subsection we describe how to find middle line segments
and how to rate these segments.

To detect the line segments, we apply a thinning algorithm on the background
pixels. For further details of the used thinning algorithm see the first algorithm
proposed by Guo and Hall [2]. Figure 3(a) shows the thinning result for a typical
sketch map. By the thinning, streets are approximately reduced to their middle
lines. Beside the middle lines of the streets, the thinning algorithm also creates
lines in city blocks and between objects like houses.

As can be seen in Fig. 3(a), the skeleton contains some "twigs". To remove the
twigs, we calculate the distance transform for the preprocessed image to get the
distance of each line pixel to the background. Later this distance can be used to
calculate the line segment width. We define line segments as "twigs" if they have
an open endpoint and if the length is less than double of the maximum distance.
This definition is meant to find line segments going from a middle point to the
border. Since these line segments do not belong to street middle lines, we remove
all twigs from the skeleton. As can be seen in Fig. 3(b) this step removes some
long background segments, too.

For the construction of the street graph, it is necessary to distinguish between
street line segments and background line segments. We rate all line segments in
order to measure the likelihood of being a street line segment. For this rating
two characteristics of streets are used:

– Streets are normally much longer than width.
– The width of streets does not change rapidly.

For the measurement of both characteristics it is necessary to know the width of
the street. Thus, we calculate a distance transform for the preprocessed image
to get the width for each line pixel. Given a line segment of n pixels with the
distances $(d_i)_{i=1,\ldots,n}$, we calculate the quotient q of length to average width of
the segment:

$$q = \frac{n}{\frac{1}{n}\sum_{i=1}^{n} d_i} = \frac{n^2}{\sum_{i=1}^{n} d_i} \qquad (1)$$

(a) Skeleton

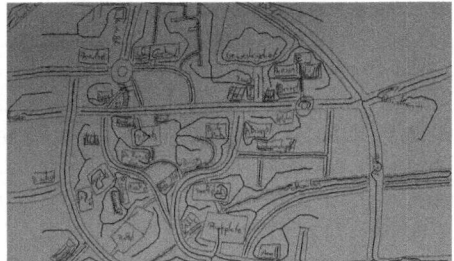

(b) Skeleton without twigs

Fig. 3. Line segment detection

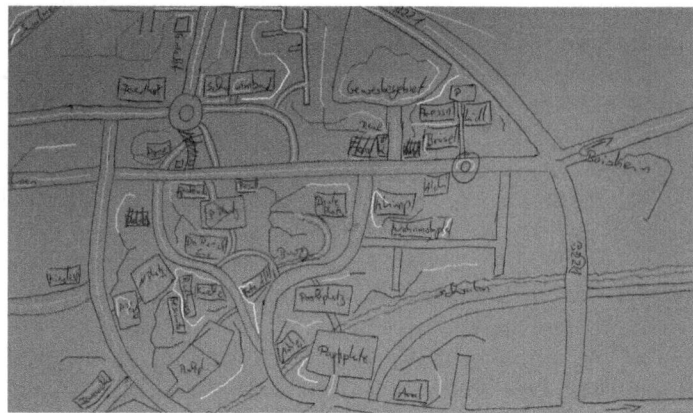

Fig. 4. Line segments colored by their ratings. Rating less than 0.5 (red), less than 0.75 (orange), less than 1 (yellow), less than 1.5 (light green) and 1.5 and greater (green).

To measure the change of width of line segments, we perform the derivation of distances d_i and smooth the result with a Gaussian filter in order to reduce the effect of local changes. With the smoothed derivations m_i the average change of width can be calculated by $\bar{m} = \frac{1}{n} \sum_{i=1}^{n} |m_i|$. Instead of defining hard threshold we define a rating function

$$r(\bar{m}, q) = \exp\left(-\frac{\bar{m}^2}{2\sigma^2}\right) \cdot \sqrt{q} \qquad (2)$$

with the parameter σ to control the tolerance of street width changes.

3.3   Graph Creation

As result of the previous part we get a set of line segments, each segment with a rating. From this set, we select the segments that are likely to be street line segments. Some parts of the sketch map can have a street-like appearance without being a street and, thus, get a high rating. In this subsection we will describe how considering the neighbors of a line segment can help to distinguish between real street line segments and non-street line segments.

We perform several tests to accept line segments as street line segments. At first we want to accept highly rated line segments that are not isolated. This means that the lines should be connected to at least one other highly rated line. To do so, we initially create a connectivity matrix $C = (c_{ij})$ for the segments with $c_{ij} = 1$ if segments i and j are connected.

As mentioned in the introduction, there might be connected streets that are separated by a line (see Fig. 1(c)). Though the line segments are not connected at such junctions, we want to consider them for the connectivity matrix. Therefore, we elongate line segments at endpoints that are not connected to any other segments. We use the method that is also applied in the preprocessing part to fill the gaps, with the difference that the length is proportional to the width of the street segment. For these new connections, we set a connectivity value in the connectivity matrix. This value is a function of the elongation that is necessary to connect both lines, which can be described by $c_{ij} = 1 - \frac{d}{d_{max}}$ with elongation d and maximal elongation d_{max}.

We use the connectivity matrix C to calculate weighted ratings for the neighbors of a segment. In order to accept non-isolated, highly rated line segments, we define two thresholds T_1 and T_2 and accept only segments s with a rating $r(s) \geq T_1$ and a maximal weighted rating of neighbor segments $r_m \geq T_2$. Our experiments have shown that $T_1 = 1$ and $T_2 = 0.75$ are good choices.

At some points like crossroads, there can be small line segments connecting two nearby nodes. Due to the small length, these parts have a low rating, but are still part of the street. To detect these segments, we accept line segments that connect two previously accepted lines and have a length below a maximal length L_{max}. L_{max} is chosen that two third of all segments are below it. We had to introduce L_{max} because some segments at the border of the image can go around the sketch and, thus, connect two segments unintentionally.

The rating and selection is done scale invariant. Beside the advantage of independence to sketch size, image resolution and distance of the camera to the sketch, there is the disadvantage that some line segments that belong to city blocks get a high rating and are accepted as streets. Figure 4 also shows some examples for such segments. To remove these segments, we build the average width of all accepted segments and remove all segments that are more than double the width.

All accepted segments build a street graph. We perform a last selection step by considering the connected components of this graph. Though the graph contains two or more disconnected components, we want to remove small ones. Therefore, we measure the size of these components and remove all components that are less than half the size of the biggest component.

4   Results

For testing our algorithm, we used sketch maps that were created in previous experiments to analyze people's sketching habits. These experiments were made without paying attention on drawing style or accuracy. Thus, these sketch maps contain *natural* drawing styles with typical errors. To simulate mobile usage, we took photos of the sketch maps with the camera of a *HTC Desire* smart phone. Figure 5 shows an example sketch map with the detected street graph.

As can be seen in the figure, most of the streets could be detected, but there are still some problems. Elongating line segments is a way to deal with connected streets that are separated by a line, but it also connects background lines with street lines. Especially small space between buildings and streets can be detected as street in this way. Further work has to find a way to distinguish between both. As can be seen in Fig. 4, in some cases text separation does not work well. This leads to some wrong detection. One example is the street at the upper right

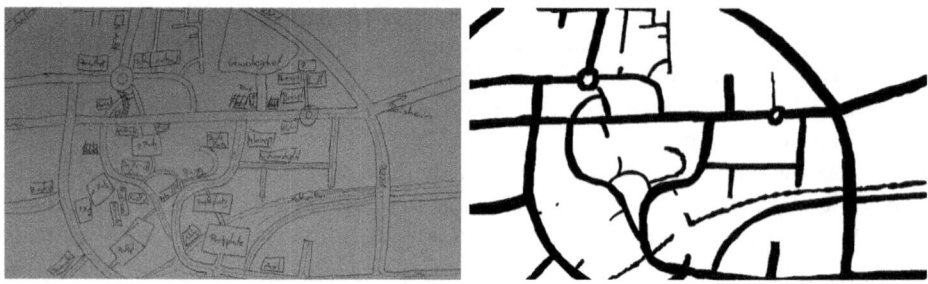

Fig. 5. Algorithm result for map 1 (left) and the manually generated ground truth (right)

border. A last element disturbing the detection are traffic circles that can not be well detected as streets, yet.

To evaluate the results of our algorithm, we produced a ground truth for some sketch maps by manually marking the streets. For comparing we used a method to compare curvilinear structures [6]. This method compares two binary images, one showing the ground truth, the other one showing the detected streets. The evaluation method measures two error values: the false negative rate (FNR) and the false positive rate (FPR). The first one is calculated from all streets that were not recognized, the latter one from all line segments that were considered as streets without being streets. Given a binary ground truth GT and its skeleton GT_S as well as the detected streets DS and their skeleton DS_S. Using the complements GT^c and DS^c, the two rates can be calculated by:

$$FNR = \frac{|GT_S \cap DS^c|}{|GT_S|}, \qquad FPR = \frac{|DS_S \cap GT^c|}{|DS_S|} \qquad (3)$$

Since we only detect the thinned streets, we have to reconstruct the streets from the detected line segments by using the previously calculated distance map. Table 1 shows both values for the maps from Fig. 6. For some of the tested maps we obtain an error rate of more than 20 percent. This is due to similar reasons as discussed above for the first map. Furthermore, some of these maps contain small streets between areas like parking places. Due to the restriction on bigger connected components in the street graph, these streets can not be detected by our algorithm. We do not consider this fact as an disadvantage of our algorithm: In the context of sketch maps, where positions are heavily distorted, isolated streets cannot be used for the subsequent graph matching.

In addition, we transformed the maps using a set of 35 similarity and 10 perspective transformations to measure the stability of our algorithm. For the similarity transformations we combined 5 scales (0.5, 0.75, 1, 1.5, and 2) with 7 rotations (0°, 15°, ..., 90°). For the perspective transformations we altered the angle of view between 5° and 50°. In each transformed map a graph is extracted using our method, which is back transformed according to the known transformation. Then, the back transformed graph is compared to the original

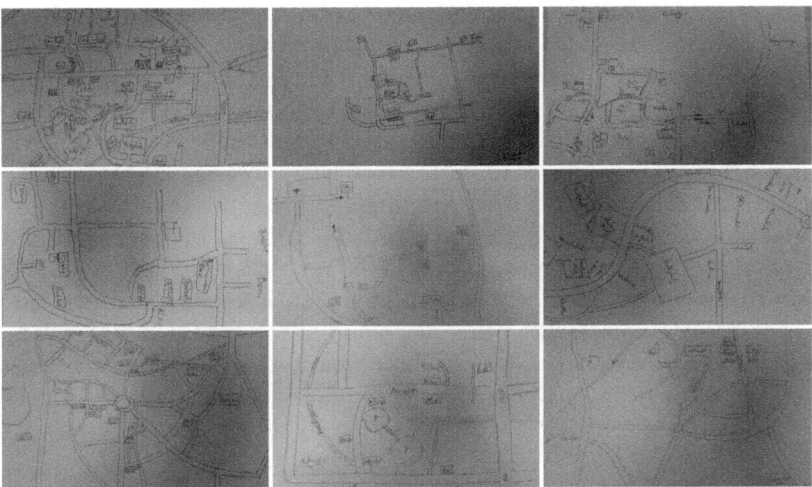

Fig. 6. Maps used for the performance test

Table 1. Error rates and stability of the algorithm

Map	FNR	FPR	Similarity FNR	FPR	Perspective FNR	FPR
Map 1	19.96%	10.35%	17.73%	8.76%	7.72%	21.82%
Map 2	12.57%	11.10%	16.84%	8.71%	19.44%	11.84%
Map 3	21.39%	31.52%	35.15%	21.46%	30.88%	28.80%
Map 4	2.25%	13.10%	13.24%	6.43%	5.08%	19.75%
Map 5	4.84%	6.21%	6.73%	5.63%	8.18%	11.46%
Map 6	17.13%	0.31%	2.86%	5.67%	5.01%	19.44%
Map 7	5.62%	13.10%	12.76%	3.63%	8.18%	11.97%
Map 8	20.10%	12.07%	28.26%	15.96%	20.84%	28.58%
Map 9	16.71%	8.29%	12.32%	9.98%	9.54%	24.26%

graph. Table 1 also shows the mean errors for both similarity and perspective transformations and thus indicates the stability of our algorithm.

Sketch maps are not an exact image of a region and do not display every detail of a region. Often small or unimportant streets are missing or just indicated at a junction. Thus, graph matching algorithms for sketch maps have to be able to deal with missing streets, even in higher numbers than the FNR of our algorithm. Depending on the matching algorithm, it might even be good to choose more restrictive parameters to increase the FNR and decrease the FPR.

5 Conclusion and Future Work

In this paper we have presented an algorithm that is able to extract street graphs from the images of sketch maps. The algorithm is able to deal with different distortions and inaccuracies that can occur in sketch maps. The goal is to create graphs that can be used for graph matching in order to align sketch maps with both metric maps and other sketch maps. Although the algorithm still produces some errors, we believe that the results could be good enough for graph matching methods that are also able to deal with errors from the sketching process.

The presented algorithm deals with streets that are drawn by pairs of parallel lines. An algorithm to deal with single lined streets and a method to automatically distinguish between both styles will be subject of future work. Beside broadening the class of supported maps, the next task is a graph matching algorithm that is able to deal with sketch maps. Such an algorithm will allow the alignment of sketch maps with other maps and in this way enable sketch maps as input method for automatic systems. Some basic research on sketch maps that can be used to develop such a matching method has previously been done [11].

References

1. Davis, R.: Magic paper: Sketch-understanding research. Computer 40(9), 34–41 (2007)
2. Guo, Z., Hall, R.W.: Parallel thinning with two-subiteration algorithms. Commun. ACM 32, 359–373 (1989)
3. Hammond, T., Davis, R.: Ladder, a sketching language for user interface developers. Computers & Graphics 29(4), 518 (2005)
4. Hu, J., Razdan, A., Femiani, J.C., Cui, M., Wonka, P.: Road network extraction and intersection detection from aerial images by tracking road footprints. IEEE Transactions on Geoscience and Remote Sensing 45(12), 4144–4157 (2007)
5. Jiang, X., Ma, M.Y., Chen, C.W. (eds.): Mobile Multimedia Processing: Fundamentals, Methods, and Applications. Springer, Heidelberg (2010)
6. Jiang, X., Mojon, D.: Supervised evaluation methodology for curvilinear structure detection algorithms. In: Proceedingd of International Conference on Pattern Recognition, Los Alamitos, CA, USA, vol. 1, pp. 103–106 (2002)
7. Mena, J.B.: State of the art on automatic road extraction for GIS update: a novel classification. Pattern Recognition Letters 24(16), 3037–3058 (2003)
8. Schlaisich, I., Egenhofer, M.: Multimodal spatial querying: What people sketch and talk about. In: 1st International Conference on Universal Access in Human-Computer Interaction, pp. 732–736 (2001)
9. Tombre, K., Tabbone, S., Pélissier, L., Dosch, P.: Text/Graphics separation revisited. In: Lopresti, D.P., Hu, J., Kashi, R.S. (eds.) DAS 2002. LNCS, vol. 2423, pp. 615–620. Springer, Heidelberg (2002)
10. Varley, P.A.C., Martin, R.R.: A system for constructing boundary representation solid models from a two-dimensional sketch. In: Proceedings of the Geometric Modeling and Processing, GMP, Washington, DC, USA, pp. 13–32 (2000)
11. Wang, J., Mülligann, C., Schwering, A.: An empirical study on relevant aspects for sketch map alignment. In: Proceedings of the 14th AGILE International Conference on Geographic Information Science (to appear, 2011)

People Re-identification by Graph Kernels Methods

Luc Brun[1], Donatello Conte[2], Pasquale Foggia[2], and Mario Vento[2]

[1] GREYC UMR CNRS 6072
ENSICAEN-Université de Caen Basse-Normandie,
14050 Caen, France
luc.brun@greyc.ensicaen.fr
[2] Dipartimento di Ingegneria dell'Informazione e di Ingegneria Elettrica,
Università di Salerno, Via Ponte Don Melillo, 1 I-84084 Fisciano (SA), Italy
{dconte,pfoggia,mvento}@unisa.it

Abstract. People re-identification using single or multiple camera acquisitions constitutes a major challenge in visual surveillance analysis. The main application of this research field consists to reacquire a person of interest in different non-overlapping locations over different camera views. This paper present an original solution to this problem based on a graph description of each person. In particular, a recently proposed graph kernel is used to apply Principal Component Analysis (PCA) to the graph domain. The method has been experimentally tested on two video sequences from the PETS2009 database.

1 Introduction

Over a couple of decades, visual surveillance gained more and more interest due to its important role in security. Fundamental research issues in this context are object detection, tracking, shadow removal, an so on. However, recently, researchers draws much attention to high-level event detection, such as behaviour analysis, abandoned object detection, etc. An important task within this research field is to establish a suitable correspondence between observations of people who might appear and reappear at different times and across different cameras. This kind of problematic is commonly known as "people re-identification".

Several applications using single camera setup may benefit from information induced by people re-identification. One of the main appllications is loitering detection. Loitering refers to prolonged presence of people in an area. This behaviour is interesting in order to detect, for example, beggars in street corners, or drug dealers at bus stations, and so on. Beside this, information on these re-occurrences is very important in multi-camera setups, such as the ones used for wide area surveillance. Such surveillance systems create a novel problem of discontinuous tracking of individuals across large sites, which aims to reacquire a person of interest in different non-overlapping locations over different camera views.

X. Jiang, M. Ferrer, and A. Torsello (Eds.): GbRPR 2011, LNCS 6658, pp. 285–294, 2011.

Re-identification problem has been studied for last five years approximately. A first group [9,17,2,3] deals with this problem by defining a unique signature which condenses a set of frames of a same individual; re-identification is then performed using a similarity measure between signatures and a threshold to assign old or new labels to successive scene entrances. In [9] a panoramic map is used to encode the appearance of a person extracted from all cameras viewing it. Such a method is hence restricted to multicamera systems. The signature of a person in [17] is made by a combination of SIFT descriptors and color features. The main drawback of this approach is that people to be added into the database are manually provided by a human operator. In [2] two human signatures, which use haar-like features and dominant color descriptor (DCD) respectively, are proposed while in [3] the signature is based on three features, one capturing global chromatic information and two analyzing the presence of recurrent local patterns.

A second group ([25,4]) deals with re-identification of people by means of a representation of a person in a single frame. Each representation corresponds to a point in a feature space. Then a classification is performed by clustering these points using a SVM ([25]) or a correlation module ([4]). Both [25,4] use the so-called "color-position" histogram: the silhouette of a person is first vertically divided into n equal parts and then some color features (RGB mean, or HSV mean, etc.) are computed to characterize each part.

This paper can be ascribed to the second group but with some significant novelty: first, we have a structural (graph-based) representation of a person; second, our classification scheme is based on *graph kernels*. A graph kernel is a function in graph space that shares the properties of the dot-product operator in vector space, and so can be used to apply many vector-based algorithms to graphs.

Many graph kernels proposed in the literature have been built on the notion of *bag of patterns*. Graphlets kernels [21] are based on the number of common sub-graphs of two graphs. Vert [14] and Borgwardt [22] proposed to compare the set of sub-trees of two graphs. Furthermore, many graph kernels are based on simpler patterns such as walks [13], trails [8] or paths.

A different approach is to define a kernel on the basis of a graph edit distance, that is the set of operations with a minimal cost transforming one graph into another. Kernels based on this approach do not rely on the (often simplistic) assumption that a bag of patterns preserves most of the information of its associated graph. The main difficulty in the design of such graph kernels is that the edit distance does not usually corresponds to a metric. Trivial kernels based on edit distances are thus usually non definite positive. Neuhaus and Bunke [15] proposed several kernels based on edit distances. These kernels are either based on a combination of graph edit distances (trivial kernel, zeros graph kernel), use the convolution framework introduced by Haussler [11] (convolution kernel, local matching kernel), or incorporate within the kernel construction schemes several features deduced from the computation of the edit distance (maximum similarity edit path kernel, random walk edit kernel). Note that a noticeable exception to this classification is the diffusion kernel introduced by the same authors [15] which defines the gram matrix associated to the kernel as the exponential of a similarity matrix deduced from the edit distance.

We propose in this paper to apply a recent graph kernel [5,10] based on edit distance, together with statistical machine learning methods, to people re-identification. The remaining of this paper is structured as follows: we first describe in Section 2 our graph encoding of objects within a video. Moving objects are acquired from different view points and are consequently encoded by a set of graphs. Given such a representation we describe in Section 3 an algorithm which allows to determine if a given input graph corresponds to a new object. If this is not the case, the graph is associated to one of the objects already seen. The different hypotheses used to design our algorithm are finally validated through several experiments in Section 4.

2   Graph-Based Object Representation

The first step of our method aims to separate pixels depicting people on the scene (foreground) from the background. We thus perform a detection of moving areas, by background subtraction, combined with a shadow elimination algorithm [6]. This first step provides a set of masks which is further processed using mathematical morphology operations (closing and opening) (Fig. 1a). Detected foreground regions are then segmented using Statistical Region Merging (SRM) algorithm [16] (Fig. 1c). Finally, the segmentation of the mask within each rectangle is encoded by a Region adjacency Graph (RAG). Two nodes of this graph are connected by an edge if the corresponding regions are adjacent. Labels of a node are: the RGB average color, the area, and the size η normalized with respect to the overall image (Fig. 1d).

3   Comparisons between Objects by Means of Graph Kernels

Objects acquired by multiple cameras, or across a large time interval, may be subject to large variations. Common kernels [13] based on walks, trails or paths are quite sensitive to such variations. On the other hand, graph edit distances correspond to the minimal overall cost of a sequence of operations transforming two graphs. Within our framework, such distances are parametrized by two sets of functions $c(u \to v), c(u \to \epsilon)$ and $c(e \to e'), c(e \to \epsilon)$ encoding respectively the substitution, and deletion costs for nodes and edges. Using such distances, small graph distortions may be encoded by small edit costs, hence allowing to capture graph similarities over sets having important within-class distance. Unfortunately, the computational complexity of the exact edit distance is exponential in the number of involved nodes, which drastically limits its applicability to databases composed of small graphs.

This paper is based on a sub optimal estimation of the edit distance proposed by Riesen and Bunke [18,19]. In this estimation, first a cost for matching two

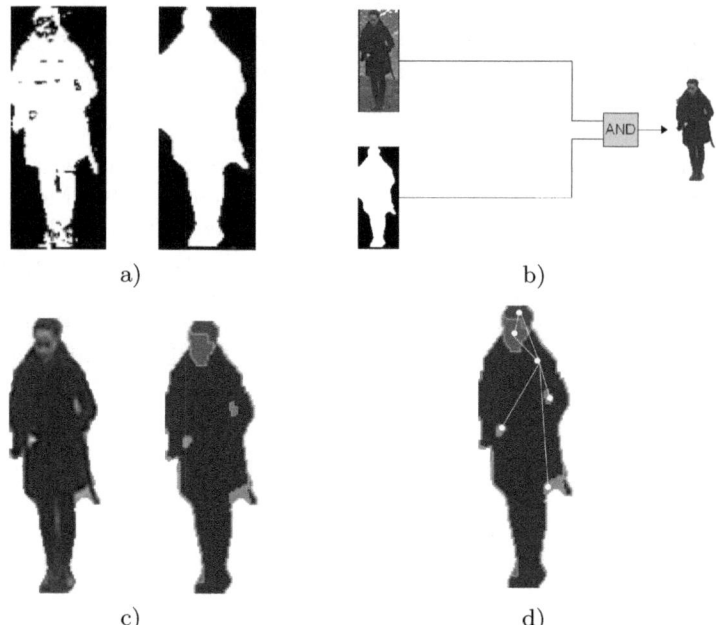

a)   b)

c)   d)

Fig. 1. a) Application of a suited morphological operator; b) Extraction of person appearance; c) Image segmentation; d) RAG construction

nodes is defined, taking into account also the edges incident to the nodes. This cost can be computed in a polynomial time (with respect to the number of incident edges) using the Hungarian Algorithm [19]. The cost for matching a pair of nodes depends on $c(u \to v)$, $c(e \to e')$ and $c(e \to \epsilon)$.

Then, the edit distance is estimated by finding a mapping between the nodes of the two graphs that minimizes the total cost of the mapped pairs, attributing the deletion cost $c(u \to \epsilon)$ to unmapped nodes. This problem can be solved in polynomial time with the Hungarian algorithm. It can be demonstrated that the estimate computed in this way lies below the true value of the edit distance, since it computes the edge-related costs in an optimistic way, assuming that the edge mapping can be performed using only local information, ignoring global constraints.

Now we will discuss the four cost functions used for defining the edit distance. Within our framework, each node u encodes a region and is associated to the mean color (R_u, G_u, B_u) and to the normalized size η_u of the region (Section 2). We experimentally observed that small regions have larger chances to be deleted between two segmentations. Hence, the normalized size of a region can be used as a measure of its relevance within the whole graph.

The cost of a node substitution is defined as the distance between the mean colors of the corresponding regions. We additionally weigh this cost by the maximum normalized size of both nodes. Such a weight avoids to penalize the matching of small regions, which should have a small contribution to the global similarity of both graphs. Also, a term is added to account for the size difference between the regions:

$$c(u \rightarrow v) = \max(\eta_u, \eta_v) \cdot d_c(u, v) + \gamma_{NodeSize} \cdot |\eta_u - \eta_v|$$

where $d_c(u, v)$ is the distance in the color space, and $\gamma_{NodeSize}$ is a weight parameter selected by cross validation. The distance $d_c(u, v)$ is not computed as the Euclidean distance between RGB vectors, but uses the following definition that is based on the human perception of colors:

$$d_c(u, v) = \sqrt{(2 + \frac{\overline{r}}{2^k})\delta_R^2 + 4\delta_G^2 + (2 - \frac{(2^k - 1) - \overline{r}}{2^k})\delta_B^2}$$

where k is the channel depth of the image, $\overline{r} = \frac{R_u + R_v}{2}$ and δ_R, δ_G and δ_B encode respectively the differences of coordinates along the red, green and blue axis.

The cost of a node deletion should be proportional to its relevance encoded by the normalized size, and is thus defined as:

$$c(u \rightarrow \epsilon) = \gamma_{NodeSize} \cdot \eta_u$$

Using the same basic idea, the cost of an edge removal should be proportional to the minimal normalized size of its two incident nodes.

$$c((u, u') \rightarrow \epsilon) = \gamma_{Edge} \cdot \gamma_{EdgeSize} \cdot \min(\eta_u, \eta_{u'})$$

where $\gamma_{EdgeSize}$ encodes the specific weight of the edge removal operation while γ_{Edge} corresponds to a global edge's weight.

Within a region adjacency graph, edges only encode the existence of some common boundary between two regions. Moreover, these boundaries may be drastically modified between two segmentations. Therefore, we choose to base the cost of an edge substitution solely on the substitution's cost of its two incident nodes.

$$c((u, u') \rightarrow (v, v')) = \gamma_{Edge} \cdot (c(u \rightarrow v) + c(u' \rightarrow v'))$$

Note that all edge costs are proportional to the weight γ_{Edge}. This last parameter allows thus to balance the importance of node and edge costs.

3.1   From Graph Edit Distance to Graph Kernels

Let us consider a set of input graphs $\{G_1, \ldots, G_n\}$ defining our graph test database. Our person re-identification is based on a distance of an input graph G from the space spanned by $\{G_1, \ldots, G_n\}$. Such a measure of novelty detection requires to embed the graphs into a metric space. Given our edit distance (Section 3), one may build a $n \times n$ similarity matrix $W_{i,j} = exp(-EditCost(G_i, G_j)/\sigma)$ where σ is a tuning variable. Unfortunately, the edit distance does not fulfill all the requirements of a metric; consequently, the matrix W may be not semi-definite and hence does not define a kernel.

As mentioned in Section 1, several kernels based on the edit distance have been recently proposed. However, these kernels are rather designed to obtain a definite positive matrix of similarity than to explicitly solve the problem of kernel-based classification or regression methods. We thus use a recent kernel construction scheme [5,10] based on an original remark by Steinke [23]. This scheme [5,10] exploits the fact that the inverse of any regularised Laplacian matrix deduced from W defines a definite positive matrix and hence a kernel on $\{G_1, \ldots, G_n\}$. Thus, our kernel construction scheme first builds a regularised Laplacian operator $\tilde{L} = I + \lambda L$, where λ is a regularisation coefficient and L denotes the normalized Laplacian defined by: $L = I - D^{-\frac{1}{2}} W D^{-\frac{1}{2}}$ and D is a diagonal matrix defined by $D_{i,i} = \sum_{j=1}^{n} W_{i,j}$. Our kernel is then defined as: $K = \tilde{L}^{-1}$. Using a classification or regression scheme, such a kernel leads to map graphs having a small edit distance [5,10] (and thus a strong similarity) to close values.

3.2   Novelty Detection and Person Re-identification

Within our framework, each reappeared person is represented by a set of graphs encoding the different acquisitions of this person. Before assigning a new input graph to an already created class, we must determine if this graph corresponds to a person already encountered. This is a problem of novelty detection, with the specific constraint that each class of graphs encoding an already encountered person has a large within-class variation. Several methods, such as one class SVM [20] or support vector domain description [24] have been used for novelty detection. However, these methods are mainly designed to compare an incoming data with an homogeneous data set. The method of Desobry [7] has the same drawback and is additionally mainly designed to compare two sets rather than one set with an incoming datum.

The method introduced by Hoffman [12] is based on kernel Principal Component Analysis (PCA). An input datum is considered as non belonging to a class if its squared distance from the space spanned by the first principal components of the class is above a given threshold. Note that this method is particularly efficient using high dimensional spaces such as the one usually associated to kernels. This method has the additional advantage of not assuming a strong homogeneity of the class.

Given an input graph G and a set of k classes, our algorithm first computes the set $\{d_1(G), \ldots, d_k(G)\}$ where $d_i(G)$ is the squared distance of the input graph G from the space spanned by the first q principal component of class i. Our novelty decision criterion is then based on a comparison of $d(G) = \min_{k=1,n} d_k(G)$ against a threshold.

If $d(G)$ is greater than the specified threshold, G is considered as a new person entering the scene. Otherwise, G describes an already encountered person, which is assigned to the class i that minimizes the value of $d_i(G)$.

4   Experimental Results

We implemented the proposed method in C++ and tested its performance on two video sequences taken from the PETS2009 [1] database (Fig. 2). Each video

a) View 001   b) View 005

Fig. 2. Sample frames from the PETS2009 dataset

sequence is divided in two parts so as to build the training and test sets. In this experiment we have used one frame every 2 seconds from each video, in order to have different segmentations of each person. The training set of the first sequence (View001) is composed of 180 graphs divided into 8 classes, while the test set contains 172 graphs (30 new and 142 existing). The second sequence (View005) is composed of 270 graphs divided into 9 classes for the training set, and 281 graphs (54 new and 227 existing) for the test set.

In order to evaluate the performances of the algorithm, we have used the following measures:

– The **true positives rate (TP)**, i.e the rate of test patterns correctly classified as novel (positive): TP = true positive/total positive
– The **false positives rate (FP)**, i.e the rate of test patterns incorrectly classified as novel (positive): TP = false positive/total negative
– The **detection accuracy (DA)**:

$$DA = (\text{true positive} + \text{true negative})/(\text{total positive} + \text{total negative})$$

– The **classification accuracy (CA)**, i.e the rate of samples classified as negatives which are then correctly classified with multi-class SVM
– The **Total Accuracy**: $TA = DA \times CA$.

As shown on Fig. 3 we obtained around 85% of novelty detection accuracy, and 70% of total accuracy for both View001 and View005 sequences. These results were obtained with the Graph Laplacian Kernel using $\sigma = 4.7$ and $\lambda = 10.0$.

These results appear very promising. For a certain interval of threshold values the classification accuracy rate keeps to the value of 100% Furthermore the True Positive Rate curve has an high slope in correspondence of an high value of the threshold: this means that the method is quite robust. Finally, ROC curves (Fig. 4) show that the algorithm have a good true positives rate, with a quite low false positives rate.

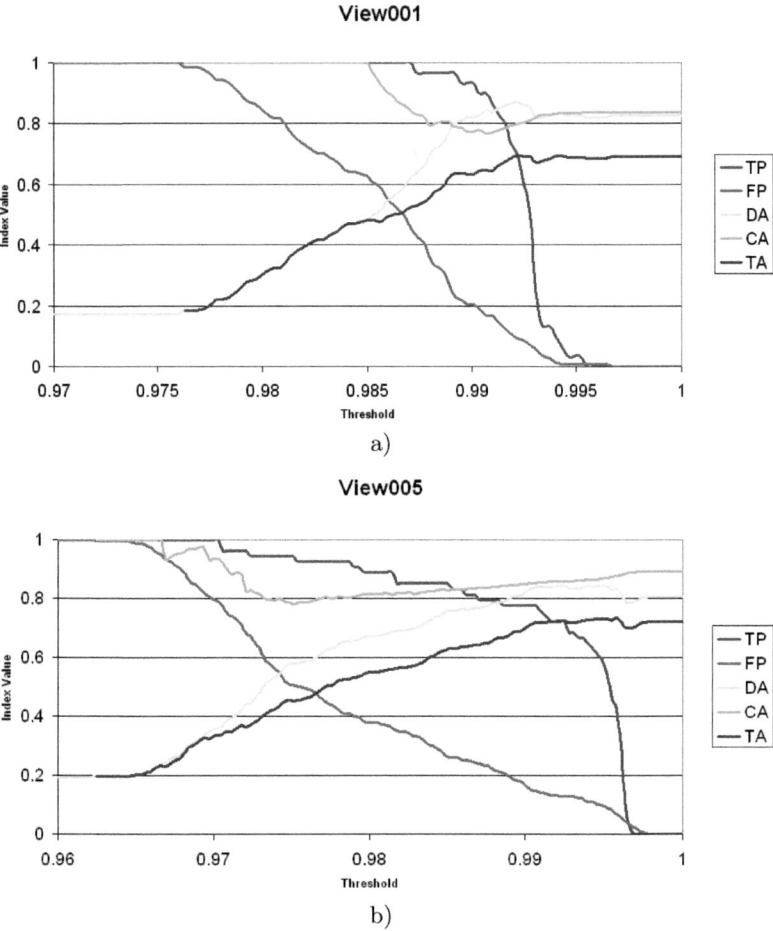

Fig. 3. Performances result on the view001 (a) and view005 (b) of the PETS2009 dataset

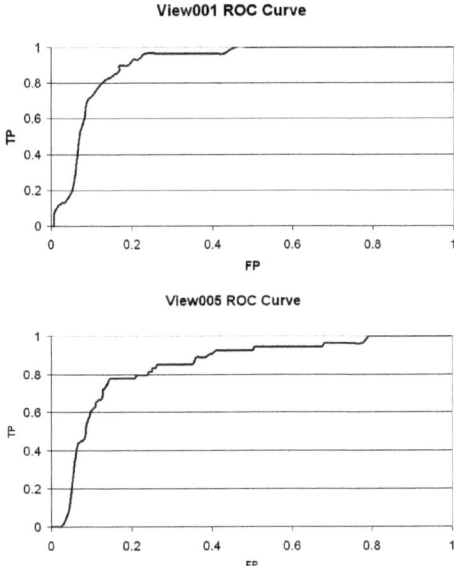

Fig. 4. ROC curves for the two sequences from the PETS2009 dataset

5 Conclusions

This paper presents a novel method for people re-identification based on a graph-based representation and a graph kernel. It combines our graph kernel with a novelty detection method based on Principal Component Analysis in order to detect if an incoming graph corresponds to a new person and, if not, to correctly assign the identity of a previously seen person.

Our future works will also extend the present method to people re-identification within groups. In such cases, a whole group is encoded by a single graph. Thus, the used kernel should be able to match subgraphs within larger graphs. We plan to study the ability of graphlet kernels to perform this task.

References

1. Database: Pets (2009), http://www.cvg.rdg.ac.uk/PETS2009/
2. Bak, S., Corvee, E., Brémond, F., Thonnat, M.: Person re-identification using haar-based and dcd-based signature. In: 2010 Seventh IEEE International Conference on Advanced Video and Signal Based Surveillance (2010)
3. Bazzani, L., Cristani, M., Perina, A., Farenzena, M., Murino, V.: Multiple-shot person re-identification by hpe signature. In: Proceedings of 20th International Conference on Pattern Recognition, ICPR 2010 (2010)
4. Bird, N., Masoud, O., Papanikolopoulos, N., Isaacs, A.: Detection of loitering individuals in public transportation areas. IEEE Transactions on Intelligent Transportation Systems 6(2), 167–177 (2005)

5. Brun, L., Conte, D., Foggia, P., Vento, M., Villemin, D.: Symbolic learning vs. graph kernels: An experimental comparison in a chemical application. In: 14th Conf. on Advances in Databases and Information Systems, ADBIS (2010)
6. Conte, D., Foggia, P., Percannella, G., Vento, M.: Performance evaluation of a people tracking system on pets2009 database. In: Seventh IEEE International Conference on Advanced Video and Signal Based Surveillance (2010)
7. Desobry, F., Davy, M., Doncarli, C.: An online kernel change detection algorithm. IEEE Transaction on Signal Processing 53(8), 2961–2974 (2005)
8. Dupé, F.X., Brun, L.: Tree covering within a graph kernel framework for shape classification. In: XV ICIAP (2009)
9. Gandhi, T., Trivedi, M.M.: Panoramic appearance map (pam) for multi-camera based person re-identification. In: IEEE International Conference on Video and Signal Based Surveillance, AVSS 2006 (2006)
10. Gauzere, B., Brun, L., Villemin, D.: Graph edit distance and treelet kernels for chemoinformatic. In: Graph Based Representation 2011. IAPR-TC15, Munster, Germany (May 2011) (submitted)
11. Haussler, D.: Convolution kernels on discrete structures. Tech. rep., Department of Computer Science, University of California at Santa Cruz (1999)
12. Hoffmann, H.: Kernel pca for novelty detection. Pattern Recognition 40(3), 863–874 (2007)
13. Kashima, H., Tsuda, K., Inokuchi, A.: Marginalized kernel between labeled graphs. In: Proc. of the Twentieth International Conference on Machine Learning (2003)
14. Mahé, P., Vert, J.P.: Graph kernels based on tree patterns for molecules. Machine Learning 75(1), 3–35 (2008)
15. Neuhaus, M., Bunke, H.: Bridging the Gap Between Graph Edit Distance and Kernel Machines. World Scientific Publishing Co., Inc., River Edge (2007)
16. Nock, R., Nielsen, F.: Statistical region merging. IEEE Transaction on Pattern Analysis and Machine Intelligence 26(11), 1452–1458 (2004)
17. de Oliveira, I.O., de Souza Pio, J.L.: People reidentification in a camera network. In: IEEE Int. Conf. on Dependable, Autonomic and Secure Computing (2009)
18. Riesen, K., Bunke, H.: Approximate graph edit distance computation by means of bipartite graph matching. Image Vision Computing 27(7), 950–959 (2009)
19. Riesen, K., Neuhaus, M., Bunke, H.: Bipartite graph matching for computing the edit distance of graphs. In: Escolano, F., Vento, M. (eds.) GbRPR 2007. LNCS, vol. 4538, Springer, Heidelberg (2007)
20. Scholkopf, B., Platt, J., Shawe-Taylor, J., Smola, A.J., Williamson, R.C.: Estimating the support of a high-dimensional distribution. Neural Computation 13, 1443–1471 (2001)
21. Shervashidze, N., Vishwanathan, S.V., Petri, T.H., Mehlhorn, K., Borgwardt, K.M.: Efficient graphlet kernels for large graph comparison. In: Twelfth International Conference on Artificial Intelligence and Statistics (2009)
22. Shervashidze, N., Borgwardt, K.: Fast subtree kernels on graphs. In: Advances in Neural Information Processing Systems, vol. 22. Curran Associates Inc. (2009)
23. Steinke, F., Schlkopf, B.: Kernels, regularization and differential equations. Pattern Recognition 41(11), 3271–3286 (2008)
24. Tax, D., Duin, R.: Support vector domain description. Pattern Recognition Letters 20, 1191–1199 (1999)
25. TruongCong, D.N., Khoudour, L., Achard, C., Meurie, C., Lezoray, O.: People re-identification by spectral classification of silhouettes. Signal Processing 90, 2362–2374 (2010)

Automatic Labeling of Handwritten Mathematical Symbols via Expression Matching

Nina S.T. Hirata* and Willian Y. Honda

Department of Computer Science,
Institute of Mathematics and Statistics,
University of São Paulo
nina@ime.usp.br, willianhonda@gmail.com

Abstract. Mathematical expression recognition is one of the challenging problems in the field of handwritten recognition. Public datasets are often used to evaluate and compare different computer solutions for recognition problems in several domains of applications. However, existing public datasets for handwritten mathematical expressions and symbols are still scarce both in number and in variety. Such scarcity makes large scale assessment of the existing techniques a difficult task. This paper proposes a novel approach, based on expression matching, for generating ground-truthed exemplars of expressions (and, therefore, of symbols). Matching is formulated as a graph matching problem in which symbols of input instances of a manually labeled model expression are matched to the symbols in the model. Pairwise matching cost considers both local and global features of the expression. Experimental results show achievement of high accuracy for several types of expressions, written by different users.

1   Introduction

The recent renewed advent of devices such as tablets, hand-held PDAs, and electronic whiteboards continues to spark interest in online handwriting recognition. These devices may work as a more suitable mechanism for inputting non-usual entries such as diagrams and equations into computer systems. However, in order to such devices serve fully as input mechanisms, handwriting recognition is crucial.

Recognition of mathematical expressions figures as one of the current challenging problems in the field of handwritten recognition. Many technical documents include some mathematical formula and their input is usually performed with a special typesetting command such as LaTeX or by using mechanisms such as symbol selection tools. Availability of mathematical expression recognition systems would allow users to enter mathematical expressions into computer systems naturally, in a similar way they are used to hand write them on a sheet of paper.

Common difficulties in mathematical expression recognition include the complexity of structural analysis (the meaning of each symbol in the expression is

---

* This work is supported by CNPq, Brazil (Grants 555418/2009-0 and 308217/2009-8).

X. Jiang, M. Ferrer, and A. Torsello (Eds.): GbRPR 2011, LNCS 6658, pp. 295–304, 2011.

determined by its relative position within a 2D arrangement of symbols), variations in the writing style from person to person, the large number of symbol types, and ambiguities in the notation (both structural and symbolic) [1,2,3].

The recognition process of mathematical expressions can be roughly divided in three steps: (i) symbol segmentation, (ii) symbol classification, and (iii) structural analysis. Most existing algorithms and systems impose some types of constraint (the way strokes should be written, a subset of allowed characters and expression types, etc). In the online recognition problem, step (i) is relatively simple because one can take advantage of the temporal information of the strokes. However the last two steps are not simple even in the online case. Attempts to solve these problems date back to early 1970s and continue to nowadays [2,3,4,5,6].

The design and performance assessment of recognition algorithms require ground-truthed dataset. Reported results are usually constrained to a reduced number of writers, expressions, and/or symbols. The available datasets, even considering offline data, are still quite scarce [7,6,8]. In order to allow large scale assessment of existing recognition techniques, it is important to develop procedures for generating ground-truthed data. Some existing approaches include writing symbols individually [5], grammar-based parsing [8], or building expressions from individually written symbols [6].

This work proposes a novel and simple approach for the generation of ground-truthed data for the online handwritten mathematical symbol recognition problem. The proposed technique assumes that writing a handful of expressions can be performed in a more natural way than writing several symbols individually. The main idea consists in matching a user written expression to the respective model expression, in order to find a one-to-one correspondence between unlabeled symbols in the input expression and labeled symbols in the model expression. By doing so, not only a large set of correctly labeled symbol samples but also expression samples whose symbols and structure are correctly identified can be easily generated.

A noteworthy aspect of this approach is the fact that since data is obtained from a handwritten expression, individual symbols are much likely to better resemble the way they are written within an expression rather than when they are written individually. Moreover, since no constraint is imposed to the model expressions, the proposed method has no restrictions with respect to, for instance, expression types, grammar rules, types of symbols, or notation conventions.

Model expressions are represented as graphs and input-model matching is formulated as a graph matching problem. Given an input instance of the model expression, a pairwise matching cost between symbols in the input and model expressions is established. To label the symbols in the input expression, a matching that minimizes the overall matching cost is selected. Experimental results obtained using graph deformation cost [9], that allows inclusion of local and global features of the expression, show that the proposed approach presents a good matching rate.

The rest of the paper is organized as follows. Section 2 describes how handwritten expressions are captured and how they are represented as graphs. Section 3 describes the proposed technique, which is composed of three basic steps: prepossessing applied to the input expression, matching cost computation, and optimal matching computation. Section 4 presents and discusses the results obtained for different experimental scenarios, including dataset variation, use of different connectivity in the model graph, and different configurations of the matching cost function. Section 5 summarizes the main contributions and lists some issues to be further investigated in the context of this work.

2   Graph Representation of an Expression

When using tablet-like devices, users can be induced to write traces of one symbol in such a way that the temporal gap between the end of a trace and the beginning of the next trace within the same symbol is no longer than the temporal gap between the end of a symbol and the beginning of the first trace of the next symbol. Experimentally, we have observed that when using a tablet like device together with a signal acquisition software with an 'undo' button, users can easily adapt themselves to write according to the above rule. Thus, in this work, we assume that symbols are correctly segmented.

Graph building. A graph is used to represent an expression. Symbols correspond to vertices and edges to relationships between symbols. Since expressions are 2D arrangements of symbols, the center of the bounding box of each symbol is taken as the corresponding vertex coordinate. Edges can be added to the graph using different criteria. Considering Delaunay triangulation, the resulting graph is planar. Another option is to consider complete graphs.

Graph features. Features can be added to vertices and edges in order to keep information of the expression. In the vertices, besides symbol spatial coordinates, one may consider local features such as (i) shape context [10] of the symbol, or of the neighboring symbols, or (ii) any feature set derived from the symbols (number of traces, curvature, relative size, etc). For the edges, a vector with the origin in one of the symbols and with the end in the other symbol can keep information such as relative orientation and distance between the symbols linked by the edge.

3   Expression Matching

Given a model expression whose symbols are individually labeled, the goal is to label symbols of an input instance of the same expression. The proposed technique consists of four steps: (i) build $G_M = (V_M, E_M)$, the graph of the model expression, where V_M is the set of vertices and E_M is the set of edges; (ii) normalize the input expression with respect to its location and scale, taking as reference the corresponding model expression; (iii) for each symbol v in the set of symbols V_I of the input expression, compute the matching cost $c(v, u)$ for all $u \in V_M$; and (iv) compute the minimum cost one-to-one matching between vertices in V_I and V_M.

Input expression normalization. Before further processing, input expression is spatially and scale aligned with the model expression. Specifically, the input expression is translated so that its bounding box center coincides with the model expression bounding box center. Next, both height and width of the input expression are scaled to match the model expression height and width, respectively. In this scale adjustment, the location of the center point of the bounding box of each symbol is also proportionally adjusted. Figure 1 shows an example of input normalization.

$$C_{ij} = \sum_{k=1}^{n} a_{ik} b_{kj} \qquad C_{ij} = \sum_{k=1}^{n} a_{ik} b_{kj} \qquad C_{ij} = \sum_{k=1}^{n} a_{ik} b_{kj}$$

(a) Input   (b) Model   (c) Normalized input

Fig. 1. Normalization of input expressions with respect to its corresponding model. Only symbol spatial coordinates are changed; symbol sizes are not modified.

Cost computation. We assume that the two expressions to be matched are similar (have the same number and type of symbols and symbols are roughly arranged in a similar way). Thus, the corresponding graph should also be similar.

If a vertex of the model expression is replaced by the corresponding vertex of the input expression, the deformation in the model graph should be small. Based on this reasoning, the best matching candidates for a model symbol would be those in the input expression that result in small deformation. Figure 2 illustrates the idea of graph deformation.

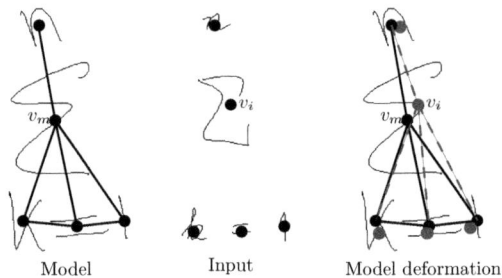

Model   Input   Model deformation

Fig. 2. Deformation induced to the model graph (dashed line edges) when one of its vertices (v_m) is replaced by a vertex (v_i) in the input graph

Among matchings that result in similar structural deformation, the preferred one should be the one that corresponds to the most similar symbol. Thus, a matching cost should take into consideration both symbol (local) and structural (global) features. The idea of deformation cost has been proposed in [9], and here we have adapted it to our application. Following [9], a general cost function for pairs of vertices (v_i, v_m), $v_i \in V_I$ and $v_m \in V_M$, may be expressed as the sum of two terms, as:

$$c(v_i, v_m) = \alpha \, c_1(v_i, v_m) + (1 - \alpha) \, c_2(v_i, v_m, G_M) \qquad (1)$$

where c_1 is relative to vertices, c_2 is relative to edges incident to v_m, and α is a real number between 0 and 1 used to weight the two terms.

In our context, **vertex cost** $c_1(v_i, v_m)$ should be small if the respective symbols have similar features. As for the second term, following [9], given two vectors $\mathbf{e}_1$ and $\mathbf{e}_2$ incident to a common vertex, we first define the cost:

$$c_{vec}(\mathbf{e}_1, \mathbf{e}_2) = \beta \frac{|cos\theta - 1|}{2} + (1 - \beta) \frac{||\mathbf{e}_1| - |\mathbf{e}_2||}{C} \qquad (2)$$

where θ is the angle between the two vectors, $|\cdot|$ is the length of the vector, β is a real number between 0 and 1, and C is a constant that represents the maximum distance between two symbols in the model expression. The first term, weighted by β, considers angular difference between the vectors and the second one, weighted by $1 - \beta$, considers length difference.

Then, the **edge cost** corresponding to the pair (v_i, v_m) is specified as:

$$c_2(v_i, v_m, G_m) = \frac{1}{|E(v_m)|} \sum_{e \in E(v_m)} c_{vec}(\tilde{\mathbf{e}}, \mathbf{e}) \qquad (3)$$

where $E(v_m)$ is the set of edges with one extremity in v_m and $|E(v_m)|$ is its size, $\mathbf{e}$ is the vector corresponding to edge e, and $\tilde{\mathbf{e}}$ is the vector obtained by replacing the extremity of $\mathbf{e}$ from v_m to v_i. The edge cost computes a kind of angular and length mean deformation of all edges incident to v_m.

Minimum cost matching computation. Given the pairwise matching costs between vertices of two graphs, an optimum matching can be computed using, for instance, the Hungarian algorithm [11]. The computation time is not critical in our application because the number of symbols in expressions are relatively small. In our case, the second graph (input) consists of only nodes (no edges) and the matching cost is the deformation cost described above.

4   Experimental Results

Dataset: Expressions were typeset in LATEX and volunteers were instructed to write them, without specific training. The only instruction given to the volunteers was to write the expressions in such a way as to leave a longer time gap between traces of two consecutive symbols than between two traces within a same symbol. Whenever traces of different symbols were joined, the writer could use an 'undo' option.

Dataset I consists of eight expressions (see Fig. 3), written on a tablet PC, by eight writers. Each of the eight expressions will be referred as expression

$$ax^2 + bx + c = 0$$

(a) Exp. 1

$$\sqrt{\alpha\beta\sigma} \geq \frac{1}{\varepsilon\delta}$$

(b) Exp. 2

$$C_{ij} = \sum_{k=1}^{n} a_{ik} b_{kj}$$

(c) Exp. 3

$$M = \begin{pmatrix} a & b & e \\ d & e & f \\ g & h & i \end{pmatrix}$$

(d) Exp. 5

$$P_a = \frac{1}{\sqrt{2\pi}} \int_a^{\infty} e^{-\frac{t^2}{2}} dt$$

(e) Exp. 6

$$\phi[n-k] = \begin{cases} 1, & n = k \\ 0, & n \neq k \end{cases}$$

(f) Exp. 7

$$p(w_i | x) = \frac{p(x/w_i) P(w_i)}{\sum_{j=1}^{c} p(x/w_j) P(w_j)}$$

(g) Exp. 4

$$\sum_{i=1}^{n} i^p = \frac{1}{p+1}\left[(n+1)^{p+1} - 1 - \sum_{i=1}^{p-1}\sum_{j=0}^{p-1}\binom{p+1}{j} i^j\right]$$

(h) Exp. 8

Fig. 3. The eight expressions of Dataset I, written by one of the eight writers

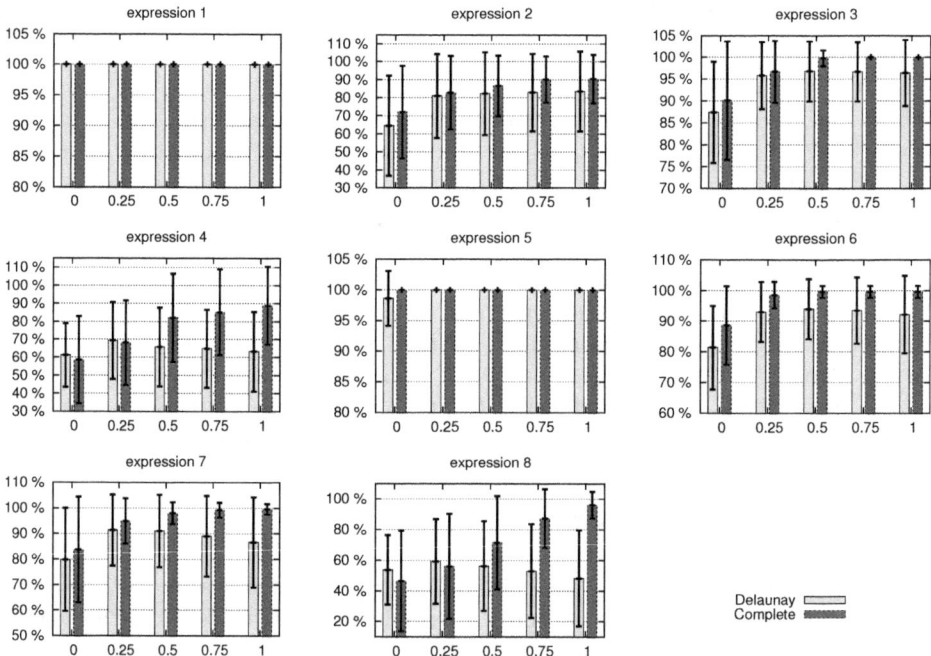

Fig. 4. Mean expression matching rates and respective standard deviations for $\alpha = 0$ and $\beta \in \{0, 0.25, 0.5, 0.75, 1\}$ for Dataset I

class. Therefore we have eight classes of expressions, containing eight expressions each. To assess the validity of the proposed technique, cross-matching tests were performed for each class: each expression in the class has been used once as the model expression to which the remaining expressions in the same class have been matched.

Let m be the number of expression classes, w the number of expressions in the classes, and E_{ij} the expression of class i written by writer j. In Dataset I, $m = 8$ and $w = 8$. Given an expression class i, we define

- **expression matching**, c_{ijk}, as the number of correct symbol matchings with respect to a specific model-input expression pair, that is, the number of correct symbol matchings when model is E_{ij} and input is E_{ik},
- **model matching**, c_{ij}, as the number of correct symbol matchings with respect to the model expression E_{ij}, that is, $c_{ij} = \sum_{k=1,k\neq j}^{w} c_{ijk}$,
- **mean expression matching** as $\sum_{j=1}^{w} c_{ij}/(w*(w-1))$, and **mean model matching** as $\sum_{j=1}^{w} c_{ij}/w$.

Figure 4 shows the mean expression matching rate and respective standard deviations for $\alpha = 0$ (no vertex cost) and different values of β (edge cost, Eq. 2). The x-axis of the graphs indicate the values of parameter β. Similar results were also obtained for a second dataset with seven expressions (due to space constraints, the dataset as well as the corresponding results are not described here).

Based on these results, the main conclusions are: (i) For complete graphs, as the value of β increases so does the correct matching rate. (ii) Best expression matching rates are obtained with $\beta = 1$ for complete graphs, indicating that only angular deformation can account for most of the relevant structural information of the expression. More than that, since no local feature is used, that means that only structural information is sufficient for a good matching rate. (iii) Results with complete graphs are in general superior than with Delaunay graphs. Thus, complete graphs can capture structural information in a more robust way. This is specially evident for more complex expressions (for instance, 4 and 8).

Figure 5 shows some examples of worst matching errors. In each case, the expression at the top is the model and the one at the bottom is the normalized input. Line segments indicate wrong symbol matchings. Matching errors can be explained based on some writing characteristics: in the first expression, the spacing between symbols in the right half of the input expression are smaller than the one in the left half, resulting in a strong misalignment of some parts of the expression between model and input after normalization (compare, for instance, the x-position of the symbol $=$); in the second case, symbol $\sqrt{}$ in the model expression is unusually long and normalization changes the relative position between $\sqrt{}$ and symbols inside it in the input expression; in the third case, model expression is small and thus normalized input symbols tend to become cluttered. To improve these results, one may consider local features such as symbol features. As a preliminary investigation, we considered shape context with four sections only, defined by the two diagonals passing through the vertex coordinates. The number of remaining vertices falling inside each section,

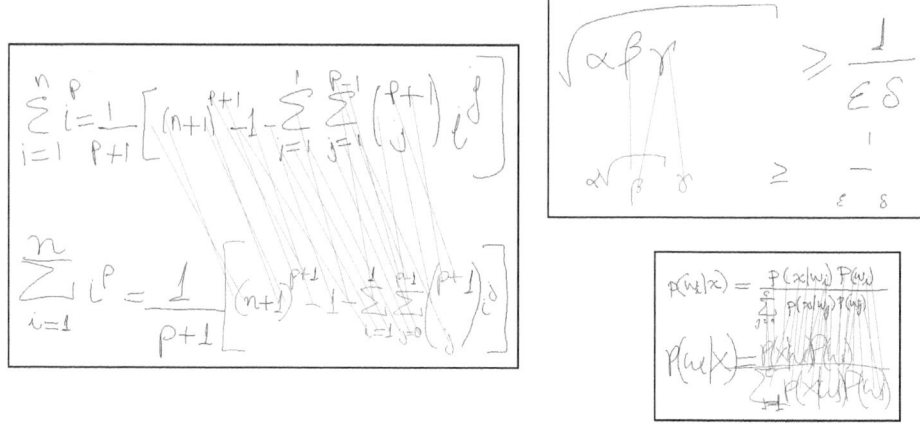

Fig. 5. Examples of worst matching errors. For each case, model expression is at the top and input expression is below it.

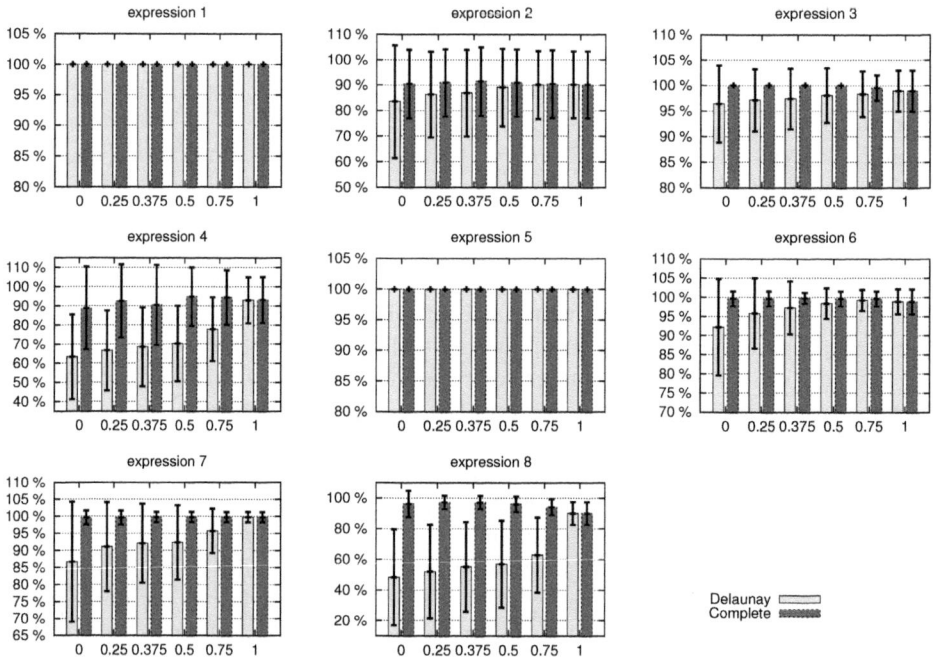

Fig. 6. Mean expression matching rates and respective standard deviations for Dataset I. Cost function is a weighted sum of vertex cost and edge cost. For edge cost, only angular cost was considered. x-axis of the graphs indicate the weight of vertex cost.

Mean model matching			
Exp. E_i	Total	Vertex cost only	Vertex + edge cost
1	70	70 ± 0	70 ± 0
2	63	57 ± 5.7	58 ± 5.3
3	105	105 ± 0	105 ± 0
4	266	236.2 ± 22	249 ± 18.6
5	91	91 ± 0	91 ± 0
6	133	132.5 ± 0.9	132.7 ± 0.7
7	126	125.5 ± 0.9	125.7 ± 0.7
8	322	309.6 ± 13.3	313.2 ± 6.5

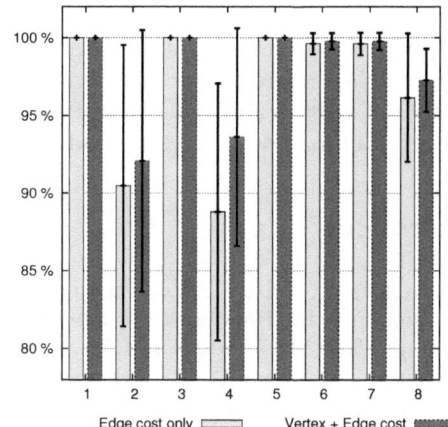

Fig. 7. Mean model matching with standard deviation, for matching cost with and without vertex cost. Column "Total" corresponds to the number of symbols in the expression times $w - 1 = 7$.

regardless their distance, were considered as the vertex feature (so that each vertex could capture information of symbols present respectively to its right, up, left, and down sides). Considering that when only edge costs were used the best results were obtained for $\beta = 1$, we tested different values for α with β fixed in 1. Results for Dataset I are shown in Figure 6. Better results are achieved for α between 0.25 and 0.5. One interesting fact is that with $\alpha = 1$ (therefore with no edge cost), the results are also quite good.

Table and graph in Fig. 7 show, respectively, the mean model matching and mean model matching rate with respective standard deviations, when deformation cost does not ($\alpha = 0$, $\beta = 1$) and does ($\alpha = 0.375$, $\beta = 1$) include vertex cost. As can be seen, conjunction of vertex and edge features not only improves mean slightly, but also diminishes standard deviation.

5 Concluding Remarks

A graph matching based approach to automatically label symbols in handwritten mathematical expressions have been presented. Vertices correspond to symbols and edges indicate spatial relationship between symbols. Given a model expression and an input expression, pairwise matching cost between vertices of both expressions are computed. In this work we have considered a simple deformation cost that takes into consideration both local (symbol) and global (structural) features.

Experimental results show that only structural information, obtained from edges or in the form of shape context, yields very encouraging matching rates. Thus, the proposed approach is a promising technique to help generation of ground-truthed handwritten mathematical symbols. We expect that the use of local information will generate better matching rates, with less variance. Besides finding simple and robust symbol features to be used in the cost function, using matching techniques such as the ones proposed in [12,13] are possible ways to proceed this research. We also plan experiments with a larger validation set (number of writers and types of expressions), and development of an interactive software tool for both matching and matching verification.

References

1. Blostein, D., Grbavec, A.: Recognition of mathematical notation. In: Bunke, H., Wang, P. (eds.) Handbook of Character Recognition and Document Image Analysis, pp. 557–582. World Scientific, Singapore (1997)
2. Chan, K.F., Yeung, D.Y.: Mathematical expression recognition: A survey. International Journal on Document Analysis and Recognition 3, 3–15 (2000)
3. Garain, U., Chaudhuri, B.B.: Recognition of online handwritten mathematical expressions. IEEE Trans Syst., Man, and Cybernetics Part B: Cybernetics 34(6), 2366–2376 (2004)
4. Tapia, E., Rojas, R.: Recognition of on-line handwritten mathematical expressions using a minimal spanning tree construction and symbol dominance. In: Lladós, J., Kwon, Y.-B. (eds.) GREC 2003. LNCS, vol. 3088, pp. 329–340. Springer, Heidelberg (2004)
5. LaViola Jr., J.J., Zeleznik, R.C.: A practical approach for writer-dependent symbol recognition using a writer-independent symbol recognizer. IEEE Trans. Pattern Anal. Mach. Intell. 29, 1917–1926 (2007)
6. Awal, A.M., Mouchère, H., Viard-Gaudin, C.: Towards handwritten mathematical expression recognition. In: Proceedings of 10th International Conference on Document Analysis and Recognition, pp. 1046–1050 (2009)
7. Suzuki, M., Uchida, S., Nomura, A.: A ground-truthed mathematical character and symbol image database. In: Proceedings of the Eighth International Conference on Document Analysis and Recognition, pp. 675–679. IEEE Computer Society, Los Alamitos (2005)
8. MacLean, S., Labahn, G., Lank, E., Marzouk, M., Tausky, D.: Grammar-based techniques for creating ground-truthed sketch corpora. International Journal on Document Analysis and Recognition, 1–10 (2010)
9. Noma, A., Pardo, A., Cesar Jr, R.M.: Structural matching of 2D electrophoresis gels using deformed graphs. Pattern Recognition Letters 32(1), 3–11 (2011)
10. Belongie, S., Malik, J., Puzicha, J.: Shape matching and object recognition using shape contexts. IEEE Trans. Pattern Anal. Mach. Intell. 24(4), 509–522 (2002)
11. Kuhn, H.W.: The hungarian method for the assignment problem. Naval Res. Logist. Quart. 2, 83–97 (1955)
12. Jouili, S., Mili, I., Tabbone, S.: Attributed graph matching using local descriptions. In: Blanc-Talon, J., Philips, W., Popescu, D., Scheunders, P. (eds.) ACIVS 2009. LNCS, vol. 5807, pp. 89–99. Springer, Heidelberg (2009)
13. Riesen, K., Bunke, H.: Approximate graph edit distance computation by means of bipartite graph matching. Image Vision Comput. 27, 950–959 (2009)

Structure-Based Evaluation Methodology for Curvilinear Structure Detection Algorithms

Xiaoyi Jiang[1], Martin Lambers[2], and Horst Bunke[3]

[1] Department of Computer Science, University of Münster, Germany
[2] Computer Graphics Group, University of Siegen, Germany
[3] Institute of Computer Science and Applied Mathematics,
University of Bern, Switzerland

Abstract. Curvilinear structures are useful features, particularly in medical image analysis. Typically, a pixel-wise comparison with manually specified ground truth is used for performance evaluation. In this paper we propose a novel structure-based methodology for evaluating the performance of curvilinear structure detection algorithms. We consider the two aspects of performance, namely detection rate and detection accuracy, separately. This is in contrast to their mixed handling in earlier approaches that typically produces biased impression of detection quality. The proposed performance measures provide a more informative and precise performance characterization. A series of experiments in the context of retinal vessel detection are presented to demonstrate the advantages of our approach.

1 Introduction

The term *curvilinear structure* denotes a line or a curve with some *width*. Particularly in medical imaging, curvilinear structures belong to the most widely observed and important features; examples are blood vessels, bones, airway trees, and other thin structures. There is only very few work on evaluation methodology for curvilinear structure detection. It is the purpose of this work to discuss the weaknesses of an approach that is widely used in medical image analysis literature and to propose an improved evaluation methodology.

Throughout this paper our discussion will be exemplified by the task of detecting blood vessels in retinal images. Reliable segmentation of the vasculature in retinal images is a nontrivial task for image analysis and has immense clinical relevance. It is important to remark that our approach is in no way bounded to retinal images only, but instead applicable in the general context of evaluating of 2D/3D curvilinear structure detection algorithms.

We motivate our work with a detailed description of the weaknesses of the early non-structural approach (Section 2). Then, we describe an improved, structure-based evaluation methodology in Section 3. Experimental work demonstrating the advantages of our approach follows in Section 4. Finally, some discussions conclude the paper.

X. Jiang, M. Ferrer, and A. Torsello (Eds.): GbRPR 2011, LNCS 6658, pp. 305–314, 2011.

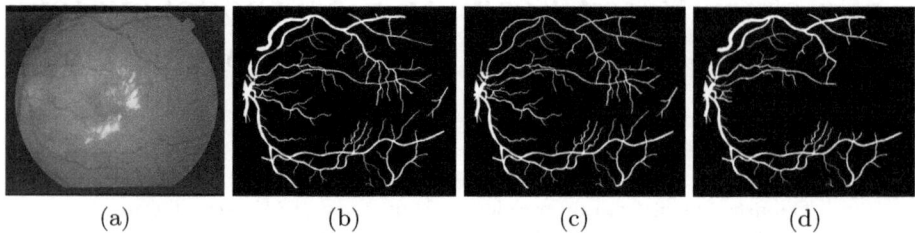

Fig. 1. (a) retinal image; (b) GT; (c) MS_{thin}; (d) MS_{del}

2   Drawbacks of Non-structural Performance Evaluation

Many algorithms have been proposed for vessel segmentation in retinal images; see [6,7] and the references therein. A standard practice is to report on experimental results based on data sets with manually specified ground truth. Popular databases are STARE [4] and DRIVE [10].

Typically, a straightforward method is used for performance evaluation. Given a machine-segmented result image (MS) and its corresponding ground truth image (GT), any pixel marked as vessel in both MS and GT is counted as a true positive. Any pixel marked as vessel in MS but not in GT is counted as a false positive. The true positive rate (TPR) is established by dividing the number of true positives by the total number of vessel pixels in GT. The false positive rate (FPR) is computed by dividing the number of false positives by the total number of non-vessel pixels in GT. As an alternative, the FPR can also be based on the total number of non-vessels pixels within the circular field of view (FOV) only. This latter version is more reasonable and thus will be consistently used in this work. If different pairs of sensitivity and specificity can be achieved, for instance by thresholding a soft classification or various parameter sets, the performance of a vessel detection algorithm can be investigated by receiver operating curves (ROC). The closer a ROC approaches the top left corner (TPR=100%, FPR=0%), the better the performance.

A fundamental weakness of this approach is illustrated in Figure 1 with two modified versions of the GT. MS_{thin} results from thinning the GT at some places while in MS_{del} some vessel sections are deleted and others remain unchanged. For both MS_{thin} and MS_{del} we obtain TPR=85.1% and FPR=0.0%, indicating an equal rate of 85.1% correct detection and no spurious vessels. But in reality there are substantial differences between the two MS images. In MS_{thin} the entire vessel network is correctly detected, but some vessels have a smaller width than GT. In contrast MS_{del} perfectly equals GT except the deleted parts. A more objective performance measure would be TPR(MS_{thin})=1.00 and TPR(MS_{del})<1.00, indicating the percentage of the correctly detected part of the vessel network. These correctly detected parts can be further evaluated with respect to the detection accuracy, i.e. the width error. Then, we would expect a non-zero width error for MS_{thin} and zero width error for MS_{del}, respectively.

Due to the nature of curvilinear structures being thin and elongated regions, a pixel-wise comparison is obviously not the most meaningful way of performance assessment. In fact, the overall performance measures TPR and FPR are both a mixture of two different aspects of performance, namely detection rate (how much of the vessel network structure is detected) and detection accuracy (what is the accuracy of the detected network structure). As shown in the examples above, such a mixture may result in a biased impression of the detection quality.

The key of a more meaningful way of performance assessment is to separate the detection rate and detection accuracy. In our previous work [5] we represent the structure of a vessel network by its thinned version of midline points of one pixel width. The structures of the GT and the MS vessel network are compared by a point matching process. While this approach is exactly what we need to alleviate the problems addressed above, the point matching process is an ad-hoc one and results in many non-optimal matchings. In this work we further develop the approach of [5] by introducing an optimal point matching process.

3 Structure-Based Evaluation Methodology

Given a binary data set V, for example a blood vessel image, we define its structure as the set of midline points along with the width information. Such a representation fully characterizes the curvilinear network by two information sources, allowing us to investigate the detection rate and the detection accuracy separately. We extract this structure in the following way:

- Find the midline points by computing the skeleton V_s of V. We use the method from [1] because it guarantees that (a) the skeleton is connected and thin (single-pixel wide) and (b) the skeleton can be used to reconstruct the original image with a tolerance of one pixel. Furthermore, it can be generalized to higher dimensions.
- Compute a distance map of V. Each structure point is assigned its Euclidean distance d to the background. Then, each structure point $p \in V_s$ receives a width value $w_p = 2d_p$. We apply the linear-time method from [8] for computing an exact Euclidean distance transform.

Given a MS and GT, we propose to measure the detection rate by comparing MS_s and GT_s only, i.e. how much of the GT curvilinear network structure is detected in MS. In a second step the width of matched MS_s and GT_s structure points is compared to give a measure of detection accuracy. The most crucial part of our approach is how to match GT_s and MS_s. We formulate this problem as one of optimal graph matching.

3.1 Graph Matching

The disjoint structure point sets GT_s and MS_s form a bipartite graph G_{gm} if every edge connects a structure point $p \in GT_s$ with a structure point $q \in MS_s$. Each such edge represents a *candidate* for a match between the structure points p and q, and is associated with costs that depend on the distance of p and q and on the difference of their width information.

A match (called a *structural matching* in the following) between the two disjoint vertex sets of a bipartite graph is a set of edges so that each vertex is endpoint of at most one edge. In such a structural matching each structure point in GT_s is matched to at most one structure point in MS_s and vice versa. The task can now be expressed as one of *maximum-cardinality minimum-cost matching*, i.e. finding an optimal structural matching with minimum costs among all structural matchings with the maximum number of edges in G_{gm}.

To build the graph G_{gm} we need to determine the set of match candidates and to specify their costs. Given G_{gm}, we have to develop a procedure for finding its optimal matching. Based on the optimal matching we finally define a new set of performance measures. The details of these steps are given in the following.

Selecting match candidates. Not every pair (p, q) should be a match candidate. We reduce the number of pairs significantly by considering only match candidates that make sense. For a match candidate (p, q) the Euclidean distance $d(p, q)$ should not be too high and p, q should not represent structures of very different width. A pair (p, q) is a match candidate if and only if $d(p, q) \leq d_{\max} \wedge |w_p - w_q| \leq w_{\max}$. The two thresholds are not independent of each other. In the case of thick structures, the allowed difference in position may be higher than in the case of thin structures, where it is more important to match the position exactly. To reflect this, $w_{\max}$ is determined from GT_s: $w_{\max} = c_w \cdot \max\{w_p \mid p \in GT_s\}$. Then, $d_{\max}$ is determined from $w_{\max}$: $d_{\max} = c_d \cdot w_{\max}$. Details of choosing parameters c_w and c_d will be discussed in Section 3.3.

Costs of match candidates. For each match candidate $(p, q) \in G_{gm}$, $p \in GT_s$, $q \in MS_s$, its cost $c(p, q)$ should be proportional both to the Euclidean distance $d(p, q)$ and to the difference $|w_p - w_q|$ of the structure widths. Additionally, the costs should be normalized to $[0, 1]$ to ease the task of defining the quality measures. Because $d(p, q)$ is bounded by $d_{\max}$ and $|w_p - w_q|$ is bounded by $w_{\max}$, the following definition fulfills these requirements:

$$c(p, q) = 1 - \left(1 - \frac{d(p, q)}{d_{\max}}\right) \cdot \left(1 - \frac{|w_p - w_q|}{w_{\max}}\right)$$

Then, the cost of a structural matching M between GT_s and MS_s is defined as:

$$C(M) = \sum_{(p,q) \in M} c(p, q)$$

Computing optimal structural matching. The problem of determining a maximum-cardinality minimum-cost matching on G_{gm} can be reduced to the computation of a minimum-cost perfect match in an auxiliary graph G'_{gm}; see [2]. (A perfect match in a bipartite graph $G = A \cup B$ is a match so that each vertex of A is matched to exactly one vertex of B and vice versa.)

From the bipartite graph G_{gm}, we form another graph G'_{gm} by putting G_{gm} and a copy of G_{gm} together. Then, we connect each vertex in G_{gm} with its copy and each such new edge is assigned the costs $N \cdot c_{max}$, where N is the number of vertices in G_{gm} and c_{max} is the maximum cost assigned to an edge in G_{gm} (in our case $c_{max} = 1$). G'_{gm} is again bipartite with the two disjoint vertex sets $\mathrm{GT}^1_s \bigcup \mathrm{MS}^2_s$ and $\mathrm{MS}^1_s \bigcup \mathrm{GT}^2_s$, where GT^1_s and MS^1_s are the vertices of GT and MS part of G_{gm}, respectively, and GT^2_s and MS^2_s are the corresponding vertices from the copy of G_{gm}. It can be shown that G'_{gm} contains a minimum-cost perfect match, which corresponds to a maximum-cardinality minimum-cost match in G_{gm} when all edges that end in a vertex of the copy of G_{gm} are eliminated (Interested readers are referred to [2] for the proof). The problem of finding a minimum-cost perfect match in G'_{gm} can be solved for example using the CSA algorithm from [3].

3.2   Quality Measures

The optimal matching $\mathcal{M}$ enables us to define the following quality measures.

True positives. The successfully detected structure points of GT_s are those that have a corresponding structure point in MS_s according to $\mathcal{M}$. The *true positives rate* (TPR) can thus be defined as:

$$\mathrm{TPR} = \frac{|\mathcal{M}|}{\#\ \text{stucture points in GT}_s}$$

TPR tells us how much of the GT curvilinear network structure is successfully detected in the machine segmentation. A measure of the matching quality of the true positives is the *detection error* (DE):

$$\mathrm{DE} = \frac{C(\mathcal{M})}{|\mathcal{M}|} \in [0,1]$$

The detection error can be split into two values to separately measure the *position error* (PE) and *width error* (WE):

$$\mathrm{PE} = \frac{1}{|\mathcal{M}|} \sum_{(p,q)\in\mathcal{M}} d(p,q); \quad \mathrm{WE} = \frac{1}{|\mathcal{M}|} \sum_{(p,q)\in\mathcal{M}} |w_p - w_q|$$

False positives. Those structure points of MS_s that have no match in GT_s according to $\mathcal{M}$ are false positives. The *false positives rate* (FPR) is defined as:

$$\mathrm{FPR} = \frac{\#\ \text{structure points in MS}_s - |\mathcal{M}|}{\#\ \text{non-structure points in FOV of GT}_s}$$

In addition to computing FPR it is also interesting to ask about the characteristic, for instance the width, of these spurious structures. It is probably more problematic to erroneously detect thick structures than thin ones. To obtain this information we can establish a width histogram of the false positives.

False negatives. Those structure points of GT_s that have no match in MS_s according to $\mathcal{M}$ are false negatives. The *false negatives rate* (FNR) is defined as:

$$\mathrm{FNR} = \frac{\#\ \text{structure points in GT}_s - |\mathcal{M}|}{\#\ \text{structure points in GT}_s}$$

Table 1. Evaluation results for MS_{thin} and MS_{del}

	MS_{thin}	MS_{del}
TPR	99.6%	77.6%
Detection error DE	0.061	0.001
Position error PE	0.202	0.054
Width error WE	0.452	0.002
False positives	0	0
False negatives	32	1729
FN histogram	–	1–2:50.3%, 2–3:11.9%, 3–4:25.0%, 4–5:9.4%

This measure indicates how much of the GT structure is missing in the machine segmentation. Likewise we can investigate the width characteristics of these false negatives by a width histogram.

3.3   Choosing Parameter Values

Two parameters c_d and c_w are used during the selection of match candidates. They affect the number of match candidates as well as their costs, and therefore also the optimal match $\mathcal{M}$ and the induced quality measures. Fortunately, it turns out that the quality measures are fairly robust to parameter changes.

Since there is no stringent reason to treat distance in position differently from difference in width, $c_d = 1$ and therefore $d_{max} = w_{max}$ is a suitable choice. This leaves only c_w to be determined.

The influence on the number of match candidates is more critical than the influence on their costs. Since the optimal matching $\mathcal{M}$ is a maximum-cardinality match, too many match candidates inevitably lead to nonsense matches in $\mathcal{M}$. Therefore, c_w must not be chosen too large. On the other hand, c_w must not be chosen too small either, to avoid the exclusion of reasonable match candidates. Suppose the segmentation algorithm tends to mark structures wider than they really are. Then a small value of c_w quickly leads to the exclusion of reasonable match candidates. It turns out that all reasonable match candidates are already included for $c_w = 0.5$: The true positives rate conforms to the expectations. The higher TPR observed for increasing values of c_w comes at the cost of match quality: The position error PE and the width error WE increase rapidly already for $c_w \approx 1$. For these reasons, a good parameter choice is $c_w = 0.5$, $c_d = 1$.

4   Experimental Results

A series of experiments using both synthetic and real data have been conducted to demonstrate the effectiveness of our approach.

4.1   Synthetic Data

First we show how our method evaluates the two images in Figure 1(c)–(d), see Table 1. As wanted, MS_{thin} has TPR near 100%, implying a full detection

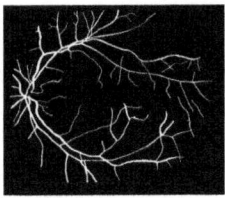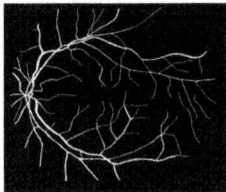

Fig. 2. Two hand-labelings of a retinal image from STARE database

of the vessel network structure. The fact that the detected vessels are thinner than GT is expressed by the width error 0.452 (pixel). The width error indirectly results in a position error 0.202. In contrast MS_{del} leads to TPR=77.6% only and accordingly 22.4% of the vessel network structure undetected. Since no error has been added to the correctly detected 77.6% of the vessel network in synthesizing MS_{del}, the error measures are all negligible in this case. The missing vessels in MS_{del} are expressed by the high number of false negatives 1729, meaning that 1729 of the structure points of GT_s cannot be matched to the segmentation result. In comparison MS_{thin} only has 32 missing structure points. Based on the histogram of false negatives we see further that the missing vessels are relatively thin; 87.2% of the missing parts have a width up to 4. The interpretation of these evaluation measures is exactly what we postulated for more informative and precise performance evaluation in contrast to the pixel-wise evaluation method.

4.2   STARE Database

The STARE database [4] contains 20 images[1] (700 × 605 pixels, 8 bits per color channel). There are two hand-labelings made by two different persons, see Figure 2 for an example. The first hand-labeling, which is usually used as ground truth in performance evaluation [4,6,10], took a more conservative view of the vessel boundaries and in the identification of small vessels than the second hand-labeling.

We compare our evaluation approach with the early pixel-wise method in three different situations with the first hand-labeling being used as ground truth in all three of them. The results averaged over 20 images are summarized in Figure 3.

Verification-based adaptive local thresholding [6]. This vessel detection method has been evaluated on the STARE database. Based on eight parameter sets the ROC is plotted in Figure 3 ("multi-threshold probing"). Note that the 20 retinal images are divided into a subset of normal and a subset of abnormal cases. The performance study thus can be done for three test instances (all, normals, abnormals). In this case both evaluation methods have similar TPR values. The reason lies in the fact that the results from [6] tend to be thicker than the ground truth. Therefore, as soon as some part of the vessel network

---

[1] Available at http://www.parl.clemson.edu/stare/probing/

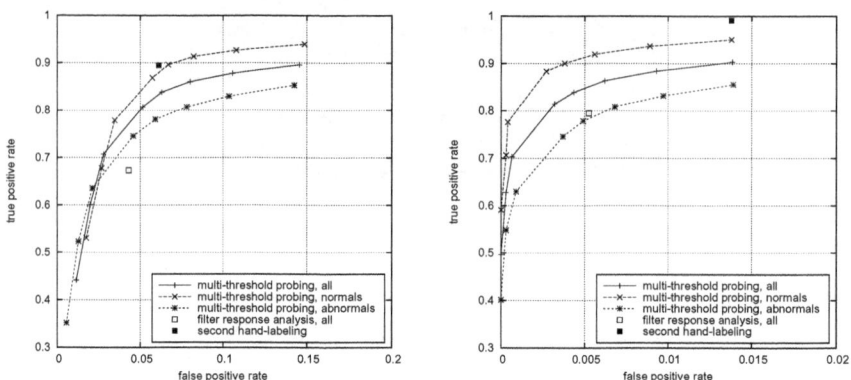

Fig. 3. Evaluation of multi-threshold probing, filter response analysis, and second hand-labeling on STARE database: pixel-wise evaluation (left) and our approach (right)

is detected, most of the vessel pixels of that part will be marked, leading to a local TPR value near 100% comparable to the local TPR from our evaluation approach. On the other hand, the FPR has much smaller values due to the use of spurious midline pixels only in our approach instead of all spurious vessel pixels.

Piecewise threshold probing of matched filter response [4]. For this method only one result per image for a particular parameter set is available. Looking at Figure 3 ("filter response analysis"), we see that our evaluation method rates the TPR considerably more positive (from lower than 70% to almost 80%). The low TPR value of pixel-wise evaluation is caused by the algorithm's tendency of not fully marking all local vessel pixels even if the middle part, thus the local network structure, is correctly found. Our structure-based evaluation approach considers the aspects of structure detection and local detection accuracy separately and is therefore able to characterize the behavior of an algorithm more precisely.

Second hand-labeling. In [4,6] the second hand-labeling has been used as "machine-segmented result images" and compared to the first hand-labeling. The detection performance measures are then regarded as a target performance level. In Figure 3 this level is indicated by an isolated mark in each plot ("second hand-labeling"). Although the second observer masked the vessels more completely, the pixel-wise TPR only amounts to about 90% because the second labeling is partly thinner than the first one. This assessment is obviously against our intuition and expectation. Using our approach the TPR increases to almost 100%.

Table 2 gives the details of this comparison for a single retinal image (shown in Figure 2). The pixel-wise evaluation results in a TPR value of only 66.0% for this image. On the other hand, our approach indicates that a much higher rate of 92.2% of the vessel network structure has been correctly segmented by the second observer. The large divergence is caused by the differences in position and width of the marked vessels by the two observers, which is signified by quite

Table 2. Evaluation results for comparing the second labeling in Figure 2 against the first labeling

TPR	92.2%
Detection error DE	0.375
Position error PE	1.619
Width error WE	0.634
False positives	954
FP histogram	1–2:90.0%, ≥ 2:10.0%
False negatives	557
FN histogram	1–2:70.2%, 2–3:5.9%, 3–4:17.2%, ≥ 3:6.7%

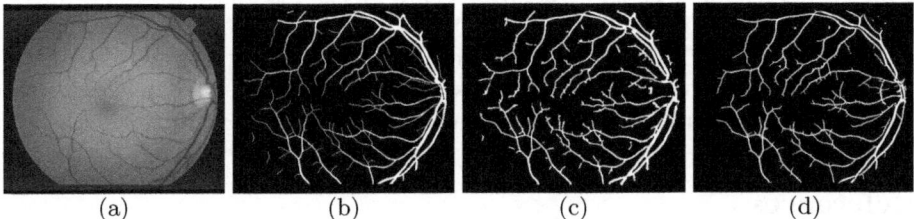

(a)   (b)   (c)   (d)

Fig. 4. (a) retinal image; (b) GT; (c) first detection results MS_1; (d) second detection result MS_2

large position and width errors in our case. The second observer labels more small vessels. This is documented by the number of false positives, namely 954. Among them 90.0% are midline pixels of thin vessels of one pixel width. This example makes once more our way of assessing the detection quality clear.

4.3 Further Validations

As another example, Figure 4 shows a retinal image (from the STARE database), the corresponding ground truth, and vessel detection MS_1 and MS_2 from two different algorithms. Using the pixel-wise evaluation we obtain: MS_1: TPR=91.9%, MS_2: TPR=80.3%. There is a large difference (11.6%) in TPR. The evaluation measures based on our approach are: MS_1: TPR=89.3%, MS_2: TPR=87.2%. Actually, MS_1 only detects 2.1% more of the vessel network structure than MS_2. The much larger difference of 11.6% above is explained by the fact that MS_1 tends to be thicker than GT. Thus, it produces a better pixel-wise matching results. Measured by our method, this is expressed by a larger width error for MS_1 (1.129) than MS_2 (0.693). Here our performance measures provide again a more precise description of the differences between algorithmic results and ground truth.

The DRIVE database [10] consists of 40 images[2] (768×584 pixels, 8 bits per color channel). The pixel classification approach to vessel detection from [9] has been evaluated using both methods and the results support the same conclusions

---

[2] Available at http://www.isi.uu.nl/Research/Databases/DRIVE/

as in case of STARE database. Due to the space limitation we do not show the performance measures here.

5 Conclusion

In this paper we have proposed a novel structure-based methodology for performance evaluation of algorithms for curvilinear structure detection. Our evaluation framework is a reasonable alternative to the pixel-wise comparison. We plan to conduct a large-scale comparison study, which should include a large number of recent detection algorithms.

The description of the evaluation methodology and the experimental work have been embedded in the context of blood vessel detection in retinal images. It is important to point out that our approach is applicable in the general context of evaluating curvilinear structure detection algorithms. In particular, extraction of airway tree and other thin structures in volumetric data is a challenging task and our evaluation technique will help assess the algorithm performance as well.

References

1. Cardoner, R., Thomas, F.: Residuals + directional gaps = skeletons. Pattern Recognition Letters 18(4), 343–353 (1997)
2. Gabow, H., Tarjan, R.: Faster scaling algorithms for network problems. SIAM Journal on Computing 18(5), 1013–1036 (1989)
3. Goldberg, A., Kennedy, R.: An efficient cost scaling Algorithm for the assignment problem. Math. Prog. 71, 153–178 (1995)
4. Hoover, A., et al.: Locating blood vessels in retinal images by piece-wise threshold probing of a matched filter response. IEEE Trans. on Medical Imaging 19(3), 203–210 (2000)
5. Jiang, X., Mojon, D.: Supervised evaluation methodology for curvilinear structure detection algorithms. In: Proc. of 16th Int. Conf. on Pattern Recognition, vol. I, pp. 103–106 (2002)
6. Jiang, X., Mojon, D.: Adaptive local thresholding by verification-based multi-threshold probing with application to vessel detection in retinal images. IEEE Trans. on PAMI 25(1), 131–137 (2003)
7. Lam, B.S.Y.: General retinal vessel segmentation using regularization-based multiconcavity modeling. IEEE Trans. Med. Imaging 29(7), 1369–1381 (2010)
8. Maurer, C.R., et al.: Exact Euclidean distance transforms of binary images in arbitrary dimensions. IEEE Trans. on PAMI 25(2), 265–270 (2003)
9. Niemeijer, M., et al.: Comparative study of retinal vessel segmentation methods on a new publicly available database. In: Fitzpatrick, J., Sonka, M. (eds.) SPIE Medical Imaging, vol. 5370, pp. 648–656 (2004)
10. Staal, J., et al.: Ridge-based vessel segmentation in color images of the retina. IEEE Trans. on Medical Imaging 23(4), 501–509 (2004)

Keygraphs for Sign Detection in Indoor Environments by Mobile Phones

Henrique Morimitsu, Marcelo Hashimoto, Rodrigo B. Pimentel,
Roberto M. Cesar-Jr., and Roberto Hirata-Jr.

Instituto de Matemática e Estatística - IME, Universidade de São Paulo - USP, Brazil
{henriquem87,mh,rbp,roberto.cesar,hirata}@vision.ime.usp.br

Abstract. We present an application for mobile phones to detect indoor
signs and help in localization. Because it depends only on device capabilities, it is flexible and unconstrained. Detection is accomplished online
by keygraph matching between sign images collected offline and the image from a mobile camera phone. After detection we apply a simple localization method based on a comparison between the detected sign and a
dataset, consisting of images of the whole environment taken at different
positions. We show the results obtained using the application in a local
indoor environment.

1 Introduction

Object detection is a subject that, despite extensive study and different approaches already developed, it still poses some challenges in computer vision.
More specifically, given the constant technological development of processing
capacity and shrinking of physical components, it is becoming very important
to adapt and develop methods for mobile devices. These equipments can be specially interesting to work with, because they are very easy to carry around and
offer a great flexibility for physical handling.

The choice of mobile devices, mainly cell phones, to perform object detection
is an area that is receiving increasing attention. Some authors showed successful
approaches using an external computer connected via wireless network to perform heavy processing in real time [4,13]. However, this method is somewhat
restrictive, as it is limited to regions where network is available.

The detection of artificial markers has proven to be an interesting choice for
mobile devices [8,11,14]. This approach is advantageous because markers are
usually easier to detect and results are more robust. On the other hand, it
naturally suffers from the problem that such markers need to be pre-attached
wherever the user want to go, and that may not be viable in practice. The
natural choice to overcome this problem is to detect objects that are part of
the environment. Even though this task can be more complex and demand more
computer resources, results obtained in previous works by other authors [1,12,16]
show it is already a feasible option for mobile devices.

Regardless of the choice of which object to detect, or how to process the data,
we can see that one of the most common choices to perform actual detection

X. Jiang, M. Ferrer, and A. Torsello (Eds.): GbRPR 2011, LNCS 6658, pp. 315–324, 2011.

relies on keypoints methods [1,4,12,16], such as SIFT [7] or SURF [2]. The main drawback of using such rich and robust keypoint detectors is that they are usually complex or computationally expensive, what is considerably sensible in small low power devices such as cell phones.

In this paper, we propose to perform detection by keygraph matching [6]. Keygraphs are built over a set of keypoints and allow us, by taking advantage of their natural structural properties, to create richer and more discriminating descriptors. Such richness does not necessarily come with an increase in complexity, in fact, if well tuned keygraph descriptors can be much simpler than those used by complex keypoint detectors. Also, as shall be explained in Sect. 2.3, their properties can be specially interesting to treat the problem at hand.

Our work will focus on detection of pre-existing signs in indoor environments by a mobile phone. The whole processing is performed in the device, without relying on an external computer for additional processing. Besides detection, we also show a localization method based on a comparison between the image being captured by our camera phone and a pre-built image dataset of the environment we are working in. It is a simple localization method implemented to show a direct application of object detection to estimate user location without any additional heavy computer processing, but with no intention of obtaining results comparable to top-notch stereo vision and 3D techniques [5] or state of the art localization works.

This paper is organized as follows. In Sect. 2 we explain the methodology and concepts needed to implement the application. Section 3 shows some experimental results obtained in a practical real-time sign detection. Finally, in Sect. 4 we present our conclusions.

2   Mobile Keygraph System

The system consists of one set of manually acquired sign images which are our *models* and a dataset of images of a corridor in a local building. Figure 1 shows a picture of the corridor and some of the signs we want to detect. More details about the dataset acquisition and how it is used in our application are presented in Sect. 2.1 and Sect. 2.2. Image detection and localization estimation is done online, while a user holds the phone and walks along the corridor.

2.1   Building Image Dataset

The images used in this research is part of a dataset collected at the Instituto de Matemática e Estatística, Universidade de São Paulo (IME - USP)[1].

The data acquisition rig consisted of a laptop computer placed approximately 1.60m above the floor, atop a wheeled structure. The camera faced the direction of the motion of the rig. At each acquisition step, the rig was moved 60cm forward and brought to a complete stop. The computer then stored one picture from the camera. The rig was then moved another 60cm forward, and the process was

---

[1] `http://www.vision.ime.usp.br/VisionDataset/AlbumDetail?albumId=36`

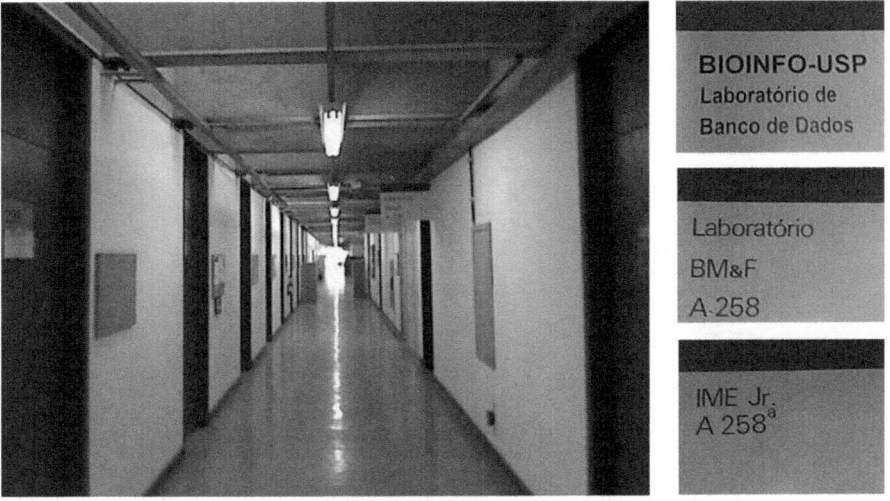

Fig. 1. Corridor and signs used in tests

repeated. Thus, all images were obtained while the rig was at rest. Also, since the rig's motion followed a straight line along a corridor, most of the elements on the pictures are at a fixed distance from the camera. This procedure led to the acquisition of 134 images of the corridor.

2.2 Localization

To estimate the user's position, assuming we can identify a sign in the image presented by the user, we use the relative proportion of the sign area in the whole image, i.e. we compute sign area (in pixels) in each picture and then divide it by the image area (also in pixels). Figure 2 shows some images from dataset, their relative sign proportion (r) in each one and the distance (d) between camera and sign in centimeters.

Using these results as samples and plotting them, we obtain the points shown in Fig. 3. Then, by applying a least square minimization, we get the distance

(a) $r = 0.024$; $d = 330cm$ (b) $r = 0.035$; $d = 270cm$ (c) $r = 0.058$; $d = 210cm$

Fig. 2. Examples of images and signs ratios

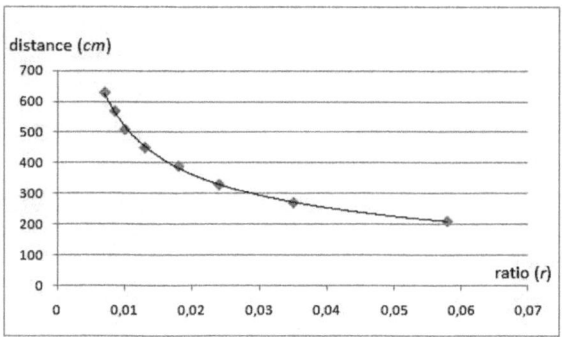

Fig. 3. Distance function d

function that, given r, we can compute distance between user and the sign, obtaining location.

Even though simple, this result is good enough for our localization problem and can be computed very quickly. The ratio r can be easily computed in real time because the sign area can be immediately obtained as a result from our detection algorithm.

2.3 Keygraphs

A keygraph is a digraph, built by taking a set of keypoints as its vertices and adding directed arcs between them. More formally we can define a keygraph as a pair (V, A), where V is a subset of a set of keypoints K and A is a subset of V^2. The set of keypoints K can be obtained by using any traditional keypoint detector, such as [9,15].

Unlike individual points, there are many ways to find correspondences between two graphs. In this work a correspondence is an isomorphism and therefore all keygraphs are isomorphic. Specifically, they are circuits of length 3, whose graphs can be easily obtained from a triangulation. Isomorphisms between circuits are trivial to find without the need to enumerate all possible bijections.

Matching keygraphs, like keypoints, is usually divided in two phases: indexing and searching. The indexing phase is done offline and consists of building keygraphs from model images and obtaining their respective descriptors. Searching is done online and consists of applying these same procedures, but on a scene image, and then selecting matches between model and scene keygraphs and performing pose estimation. Figure 4 shows an example of a sign being detected.

The main motivation for using keygraphs over keypoints lies on the fact that, by taking into consideration a set of points, we can obtain structural information that is not available when analyzing only one point at a time. Even though it may seem that this additional information will overload even more the matching process and preclude real time detection, it does not necessarily mean so. Of course, if we use a very complex keypoint detector, such as SIFT, to generate

Fig. 4. Example of detection using keygraphs. The white lines between the image on the left and right shows the correspondences found. Green triangles on both images shows the keygraphs matched, and red ones on the image on the right shows keygraphs not matched. Blue bounding box indicates estimated pose.

our keygraphs vertices, we cannot hope to speed up things very much since we would be bounded by its detection time. However, if we apply a simpler, but faster detector and rely on keygraph properties to build a robust descriptor efficiently, we could expect better results. One thing that should be noticed is that whereas the number of keypoints in an image is linear on its size, the amount of keygraphs is exponential. Thus, to achieve real time performance, we must impose some restrictions on arc generation. These, however, can be easily obtained when taking into consideration the geometrical properties of keygraphs, as shall be explained ahead.

There are many ways of creating descriptors from keygraphs. One possibility would be to use the keypoints descriptors and add additional information obtained from the graphs. However, each keypoint descriptor has its singularities and it would be very difficult to create a model that would fit to all of them. In that sense, we chose to discard completely keypoints descriptors and build new ones based on arcs.

Arc descriptors can be computed by taking into consideration the intensity profile [17,18], i.e., the grayscale values of the pixels along the arc of the keygraph. This choice of descriptor has several advantages. First, intensity profiles are naturally robust to rotation. Second, even though they are not originally invariant to scale, the arcs have a well-defined concept of length that can be used to scale the profiles accordingly and obtain robustness. Finally, as they are unidimensional, effective descriptors do not need to be very large and can be computed very fast.

Intensity profiles also have some problems, as they are not robust to perspective transforms, and also impose the restriction of planarity over the object. However, these difficulties can be circumvented by applying some restrictions. We can use the geometrical properties of keygraphs structure to discard arcs and obtain more robust and significant descriptors. Not only that, these restrictions are also important to reduce the amount of keygraphs and make real time processing feasible.

One simple criterion that can be used is the distance between vertices. If the distance is too short, the resulting arc descriptor will be very poor, and hence such arcs are not considered. On the other hand, very long arcs should also not be included. The main reason for that is due to the intensity profiles used to create descriptors. By removing long arcs, we obtain more locally restricted descriptors that are more likely to be according to the planarity restriction. Besides, the effects of perspective transforms are softened on short profiles. The optimum choice of values may depend on the problem being treated, as well as on the size of the images. In our application, we found out that choosing arcs whose length was between 10 and 100 pixels yielded good results.

Another possible restriction is to consider the relative positioning of the keygraphs. As we chose to work with keygraphs that are circuits with 3 vertices, we can impose over them the restriction that all arcs in a keygraph are clockwise oriented. This can be used to avoid some unnecessary comparisons, unless we need to take mirroring into account, which is not the case. This constraint can also be efficiently verified by simply computing a cross product of the vertices.

Even though the above criteria may somewhat constrain the amount of keygraphs, it does not really decrease the asymptotic complexity of the problem. Hence, while this restriction alone may be sufficient to be applied on model images during offline indexing phase, it most probably will not be enough to achieve good results during online searching. In this sense, we propose the insertion of an additional constraint for scene keygraphs, restricting arcs via Delaunay triangulation. The number of triangles in such triangulation is linear in the number of points [3], which makes the problem treatable.

It should be noted, however, that, by applying a hard constraint on keygraphs generation, we also impoverish image description, which can make matching difficult. For that reason, we only applied triangulation restriction to scene keygraphs and let the model have a richer description. In doing so, we allowed the model to be found under more diverse changes in scene image. In order to guarantee that the number of keygraphs on the model does not greatly affect matching performance, we used an indexing library [10] to store model descriptors.

3   Experimental Results

The implementation was done in C++ with OpenCV^2 library on a Nokia ®N900 cell phone. It is equipped with an ARM ®Cortex ^{TM}A8 600Mhz processor, 256MB of RAM and runs Maemo 5 OS. The application captures images via built-in camera phone in 640x480 resolution and executes keygraph matching and localization estimation on the fly.

Figure 5 shows the graphic user interface (GUI) shown by the application on the device screen. In the left side, a map of the corridor is presented, showing user's current position in the corridor (white dot) and existing signs (yellow

---

^{2} http://opencv.willowgarage.com/wiki/

Fig. 5. User interface shown by device. The white dot on the left represents the user position in the corridor.

rectangles) and doors (in red). The rest of the screen shows the image being captured by the camera phone as well as its keygraphs (red and green lines, green keygraphs indicates a match was found) and the detected sign (blue bounding box).

For keypoint detection we chose the MSER detector [9]. Empirical results showed that this detector worked better than the others for our approach. As explained in Sect. 2.3, keygraphs have the important properties of considering structural information and also are robust to scale and rotation changes. These are specially important for the problem at hand because, as the user will be walking around with the device, our application must be able to identify the signs even under diverse distance and point of view changes. Besides, as signs are very well structured objects, keygraphs are ideal to describe them and perform detection more easily.

Figure 6 shows detection results obtained by our application. It can be seen that detection works well even under partial occlusion and with rotation changes. Robustness to scale is also obtained, but the sign is not detected until the user comes closer. The blue bounding box on screen indicates when a sign was detected.

Localization, even though simple, was also satisfactory, as shown in the map on the left part of each picture in Fig. 7, correctly displaying user location when a sign is detected.

Besides accuracy, we also verified that memory allocation of the application was low and did not cause any overflow in the device memory. The observed running time of our application was about 651 milliseconds per frame or nearly 2 frames per second. This result was obtained by computing the average time spent in each frame while processing the videos used to obtain the results shown above.

Fig. 6. Results showing robustness to scale, rotation and partial occlusion

Fig. 7. Detection and respective location

4   Conclusion

We showed that keygraphs are a viable option to perform sign detection by cell phones. Our application detects them correctly and also allows the user to know its location in the environment online. For further work, we are interested in doing quantitative analysis of the experimental results we obtained. For that we will: (i) use our application to estimate user position directly on the images from the environment dataset and (ii) compare the position obtained by our application with a position estimated by a completely different method based on an analysis of wi-fi networks signals. This way we can use those results as a groundtruth to check the precision of our method. Besides, we would like to improve detection and allow the application to detect signs even farther away.

This could be accomplished by relaxing keygraphs thresholds and allowing richer description of images, but doing so without additional bounds would compromise online detection. One possible option to overcome this problem is to integrate our localization result to restrict detection. Our localization method is still very simple, and can only tell the distance between the user and a sign when the sign is seen straight. But it can be further improved by taking into consideration the homography obtained after performing pose estimation. In this way it is possible not only to know the distance, but also the angle of view between the user and the detected sign, obtaining greater precision in location estimation. User position could also be tracked using a Kalman filter, allowing more robust estimation, even when detection fails. Having an accurate estimation of user position would allow us to decide what signs are likely to be detected and thus decrease matching complexity.

Acknowledgments. We would like to thank FAPESP, CAPES, FINEP and CNPq for the support.

References

1. Arth, C., Wagner, D., Klopschitz, M., Irschara, A., Schmalstieg, D.: Wide area localization on mobile phones. In: Proceedings of IEEE International Symposium on Mixed and Augmented Reality (ISMAR), pp. 73–82 (2009)
2. Bay, H., Ess, A., Tuytelaars, T., Gool, L.V.: Surf: Speeded up robust features. In: Computer Vision and Image Understanding (CVIU), vol. 110(3), pp. 346–359 (2008)
3. de Berg, M., Cheong, O., van Kreveld, M., Overmars, M.: Computational Geometry: Algorithms and Applications, 3rd edn. Springer, Heidelberg (2008)
4. Gammeter, S., Gassmann, A., Bossard, L., Quack, T., Gool, L.V.: Server-side object recognition and client-side object tracking for mobile augmented reality. In: IEEE International Workshop on Mobile Vision (2010)
5. Hartley, R., Zisserman, A.: Multiple View Geometry in Computer Vision, 2nd edn. Cambridge University Press, Cambridge (2004)
6. Hashimoto, M., Cesar Jr, R.M.: Object detection by keygraph classification. Graph-Based Representations in Pattern Recognition 5534, 223–232 (2009)
7. Lowe, D.G.: Distinctive image features from scale-invariant keypoints. International Journal of Computer Vision 60, 91–110 (2004)
8. Manduchi, R., Coughlan, J., Ivanchenko, V.: Search strategies of visually impaired persons using a camera phone wayfinding system. In: 11th International Conference on Computers Helping People with Special Needs, ICCHP (2008)
9. Matas, J., Chum, O., Urba, M., Padja, T.: Robust wide baseline stereo from maximally stable extremal regions. In: Proceedings of British Machine Vision Conference, pp. 384–396 (2002)
10. Muja, M., Lowe, D.G.: Fast approximate nearest neighbors with automatic algorithm configuration. In: International Conference on Computer Vision Theory and Application (VISSAPP), pp. 331–340. INSTICC Press (2009)
11. Mulloni, A., Wagner, D., Barakonyi, I., Schmalstieg, D.: Indoor positioning and navigation with camera phones. IEEE Pervasive Computing 8, 22–31 (2009)

12. Paucher, R., Turk, M.: Location-based augmented reality on mobile phones. In: IEEE International Workshop on Mobile Vision (2010)
13. Ravi, N., Shankar, P., Frankel, A., Elgammal, A., Iftode, L.: Indoor localization using camera phones. In: Proceedings of the Seventh IEEE Workshop on Mobile Computing Systems and Applications, pp. 1–7 (2006)
14. Rohs, M., Zweifel, P.: A conceptual framework for camera phone-based interaction techniques. In: Gellersen, H.-W., Want, R., Schmidt, A. (eds.) PERVASIVE 2005. LNCS, vol. 3468, pp. 171–189. Springer, Heidelberg (2005)
15. Shi, J., Tomasi, C.: Good features to track. In: IEEE Conference on Computer Vision and Pattern Recognition, CVPR (1994)
16. Takacs, G., Chandrasekhar, V., Gelfand, N., Xiong, Y., Chen, W.C., Bismpigiannis, T., Grzeszczuk, R., Pulli, K., Girod, B.: Outdoors augmented reality on mobile phones using loxel-based visual feature organization. In: Proceedings of the 1st ACM Internation Conference on Multimedia Information Retrieval (MIR), pp. 427–434 (2008)
17. Tell, D., Carlsson, S.: Wide baseline point matching using affine invariants computed from intensity profiles. In: Vernon, D. (ed.) ECCV 2000. LNCS, vol. 1842, pp. 814–828. Springer, Heidelberg (2000)
18. Tell, D., Carlsson, S.: Combining appearance and topology for wide baseline matching. In: Heyden, A., Sparr, G., Nielsen, M., Johansen, P. (eds.) ECCV 2002. LNCS, vol. 2350, pp. 68–81. Springer, Heidelberg (2002)

Classification of Graph Sequences Utilizing the Eigenvalues of the Distance Matrices and Hidden Markov Models

Miriam Schmidt and Friedhelm Schwenker

Institute of Neural Information Processing,
University of Ulm, 89069 Ulm, Germany
{miriam.k.schmidt,friedhelm.schwenker}@uni-ulm.de
http://www.uni-ulm.de/in/neuroinformatik.html

Abstract. In this paper, the classification of human activities based on sequences of camera images utilizing hidden Markov models is investigated. In the first step of the proposed data processing procedure, the locations of the person's body parts (hand, head, etc.) and objects (table, cup, etc.) which are relevant for the classification of the person's activity have to be estimated for each camera image. In the next processing step, the distances between all pairs of detected objects are computed and the eigenvalues of this Euclidean distance matrix are calculated. This set of eigenvalues built the input for a single camera image and serve as the inputs to Gaussian mixture models, which are utilized to estimate the emission probabilities of hidden Markov models. It could be demonstrated, that the eigenvalues are powerful features, which are invariant with respect to the labeling of the nodes (if they are utilized sorted by size) and can also deal with graphs, which differ in the number of their nodes.

Keywords: eigenvalues, weighted adjacency matrix, graph classification, hidden Markov models.

1   Introduction

Graph structures naturally appear in many situations of our everyday life, e.g. street maps with junctions (nodes) and streets (edges), communication networks, such as the world wide web or the telephone network, and therefore graphs with their special characteristics play also an important role in various research fields. Cvetković et al. presented numerous applications in chemistry, physics, mathematics and computer science in their book *Application of Graph Spectra* [1]. In computer vision and pattern recognition, the classification of graphs is a fundamental issue with many applications such as scene analyses or optical character recognition. For instance, Bunke et al. [2] proposed graphical representation of the letters for hand-writing recognition. In bioinformatics, graph classification is used to cluster proteins concerning their structure because the individual function is closely related to the shape of the protein.

X. Jiang, M. Ferrer, and A. Torsello (Eds.): GbRPR 2011, LNCS 6658, pp. 325–334, 2011.

Two main approaches to classify graphs have been suggested in the literature. In the first approach, a distance measure between graphs is defined, e.g. the graph edit distance. Utilizing this distance, graphs can be clustered or classified in this metric space [3]. In the second one, special discriminative features are extracted from the graphs, embedded into a finite dimensional pattern space or vector space and then classified [4]. One possible feature of graphs is the set of the eigenvalues of the graph's adjacency matrix or the Laplacian matrix. These features build the so-called spectrum of the graph [5] and serve as input vectors for subsequent classifiers.

In this study, we investigated the approach of extracting the eigenvalues of the weighted adjacency matrix (on the one hand with zeros on the diagonal, on the other hand with the sum of the distances) for graph sequence classification (for more information about the eigenvalues of Euclidean distance matrices see [6]). The underlying data provides sequences of fully connected graphs of labeled nodes. Here, a node is representing a particular object and the label of the node is the object's position given as relative coordinates in the camera images. The Euclidean distances between the nodes build the matrices for the spectral computation. The spectra was utilized as feature for the sequence classification by hidden Markov models (HMMs). HMMs have been applied for the classification of sequential data, for instance in speech recognition [7] or recognition of facial expressions [8], because they are able to model temporal dependencies. It could be revealed that the eigenvalues are powerful features: for small graphs, they are easy to compute and contain information about the topology and the shape of the graphs. The values are invariant with respect to the labeling of the nodes, if they are used in increasing or descending order, because the exchange of two columns (and the corresponding rows) don't affect the values themselves. Because each eigenvalue includes information about all nodes, it is possible to use just a subset of the eigenvalues (e.g. the largest n) for the classification. Therefore, in the case of different number of nodes in the data set, only a subset of the eigenvalues can lead to satisfying results.

The rest of the paper is organized as follows: Section 2 provides a brief overview of the hidden Markov models and the Gaussian mixture models. In Sect. 3 the video data and the extraction of the feature vectors are described. The application of the classifier, the results and particularly the advantages of the features are introduced in Sect. 4, followed by a brief summary of the paper in Sect. 5.

2 Stochastic and Functional Principles

2.1 Hidden Markov Models

A hidden Markov model $\lambda = (Z, V, \pi, A, E)$ is a statistical model, which is composed of two random processes [9][10]. The first process is a Markov chain consisting of a fixed number of states $Z = (z_1, \ldots, z_n)$ and corresponding state transition probabilities compiled to the transition matrix $A = (a_{ij})$, where a_{ij} designates the probability of a transition from state z_i to state z_j (see Fig. 1). The initial probability vector π, which defines the probabilities of the states to

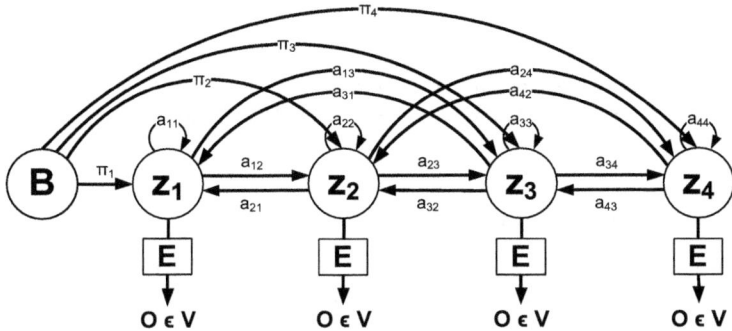

Fig. 1. A graphical representation of a HMM with four states $z_1, ..., z_4$ and an additional initial state B to include the initial probabilities $\pi_1, ..., \pi_4$ to the scheme. Label a_{ij} of the edge from node z_i to z_j refers to the corresponding state transition probability. The emission probabilities in matrix E influence the output $O \in V$.

be the initial state, forms the third component of this Markov chain process. The second random process determines the output: it consists of possible observations $V = (v_1, ..., v_m)$ and the observation matrix $E = \{e_j(k)\}$, where $e_j(k)$ is defined as the probability of observation v_k in state z_j. The sequence of observations provides information about the sequence of the hidden states. The topology of the transition matrix defines the structure of the model. A connection between two states z_i and z_j is given, if the corresponding entry a_{ij} is greater than 0.

There are three basic problems associated with HMMs [10]. Each of it can be solved with a specific dynamic programming algorithm:

Decoding Problem: Given the parameters of the model $\lambda = (Z, V, \pi, A, E)$ and an observed output sequence $O = O_1, ..., O_L$. Evaluate the most likely state sequence which could have generated the output sequence.
Solution: *Viterbi algorithm.*

Evaluation Problem: Given the parameters of the model $\lambda = (Z, V, \pi, A, E)$. Compute the probability of an observed output sequence $O = O_1, ..., O_L$.
Solution: *forward algorithm.*

Learning Problem: Given a set of output sequences $O_1, ..., O_L$ and the structure of the model $\lambda = (Z, V)$. Determine the transition matrix A, the observation matrix E and the initial probability vector π so that the probability for this HMM, producing $O_1, ..., O_L$, is the maximum value.
Solution: *Baum-Welch algorithm.*[1]

The HMMs were utilized for classification of sequential data, e.g. in speech recognition, recognition of gestures and bioinformatics. In this type of application, usually one single HMM λ_i for every class i $(i = 1, ..., n)$ is trained

---

[1] The Baum-Welch algorithm is an instance of the expectation-maximization algorithm.

using only data of this class. The probabilities $P(O \mid \lambda_i)$ for an unclassified observation sequence $O = O_1, \ldots, O_L$ and the i-th HMM were estimated with the forward algorithm. For numerical reasons, because the probabilities could become very small, the logarithm was used in the computation. The maximum of the n achieved values $P(O \mid \lambda_i)$, one for each HMM, leads to the most likely class.

2.2   Gaussian Mixture Models

A multivariate Gaussian mixture model (GMM) $g(f_1, \ldots, f_m)$ is a probabilistic model for the estimation of probability density functions [11]. It combines m Gaussian density functions $f_1, \ldots, f_m$, where $f_i = (\mu_i, \Sigma_i, \alpha_i)$ is given by a mean vector $\mu_i \in \mathbb{R}^d$ and the covariance matrix $\Sigma_i \in \mathbb{R}^d \times \mathbb{R}^d$ of the i-th component. Σ_i could be a full (symmetric) covariance matrix or a diagonal matrix $\Sigma_i = diag(\sigma_1^2, \ldots, \sigma_d^2)$, where $(\sigma_1^2, \ldots, \sigma_d^2) \in \mathbb{R}^d$ is the vector of variances of the d-th input dimensions. The third value α_i stands for the weight of the i-th Gaussian in the linear combination with $\sum_{i=1}^{m} \alpha_i = 1$.

The individual Gaussians f_i are assumed to be stochastically independent and defined by

$$f_i(x \mid \mu_i, \Sigma_i) = \frac{1}{\sqrt{(2\pi)^{N/2} \mid \Sigma_i \mid}} exp\left(-\frac{1}{2}(x - \mu_i)^T \Sigma_i^{-1}(x - \mu_i) \right). \tag{1}$$

This defines the total probability density function $P(X \mid f_1, \ldots, f_m)$ with:

$$P(X \mid f_1, \ldots, f_m) = \sum_{i=1}^{m} \alpha_i f_i(X, \mu_i, \Sigma_i). \tag{2}$$

In this study, one GMM in each state of the HMM was utilized to define the observation probabilities E, which were mentioned in Sect. 2.1. It should be noted that GMMs cannot take temporal dependencies into account as HMMs do. The parameters of the GMM are trained with an expectation maximization algorithm.

3   Application Data and Selected Features

3.1   Data Collection

The data was developed at the Institute of Neural Information Processing at the University of Ulm. The data set consists of 289 video clips with a length between one and eight seconds. Each video was recorded with a resolution of 356×474 pixels and a frame rate of 24 fps. Figure 2 shows a freeze image of one video. A male person is recorded, sitting at a table, playing one of three different action classes. Every sequence belongs to one of the classes:

Drink: The person drinks out of the cup (89 samples).
Move: The person moves a cup on the table (100 samples).
Scratch: The person scratches his head (100 samples).

3.2   Graph and Feature Extraction

The person always wore a red cap, a blue glove and used a yellow cup. Therefore, the possibly complex object recognition could be simplified by calculating the center of the colored areas. Each center refers to one node of a graph, labeled with the coordinates of the center (see Fig. 3). All nodes are connected to each other with undirected edges and every edge is labeled with the Euclidean distance of the two connected nodes. These pairwise distances, extracted from the graph, build the weighted adjacency matrix A for the calculation of the spectrum. We also investigated an extension A^+ of this matrix with the sum of the distances on the diagonal (instead of zeros). Therefore, the features, which were used by the HMMs for the classification, are sequences of spectra of the weighted adjacency matrix A (or its extension A^+), one spectrum for each single image. It should be noted, that the eigenvalues were sorted in descending order within the feature vector.

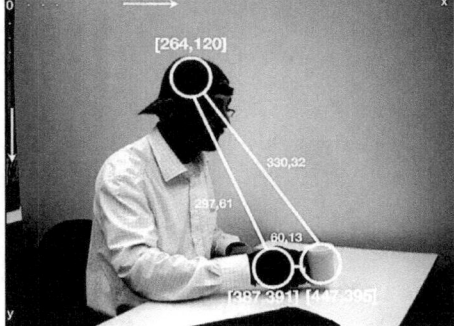

Fig. 2. Each action is played by a male person, who is sitting at a table with the camera on his right side. He is wearing a red cap, a blue glove and is using a yellow cup.

Fig. 3. The head, the hand and the cup build the three nodes of a fully connected graph. The nodes are labeled with the co-ordinates, the edges with the distances between the two connected nodes.

4   Experimental Settings

In this study, the goal was to investigate the potential and flexibility of the eigenvalues as input features for probability density estimation. Therefore, the chosen dataset (see Sect. 3.1) is simple but easily expandable to more complex variations to demonstrate the power of the spectrum for graph classification. In the first part of this section, the settings of the HMMs and GMMs are determined before in Sect. 4.2 the advantages of the spectrum are verified.

4.1   Settings of the HMMs and GMMs

For the implementation of the HMMs and GMMs the *HMM Toolbox for Matlab*, written by Murphy [12], was utilized and adapted. After testing different settings for the HMMs and GMMs, a number of 15 states for the HMMs and four Gaussian distributions for the GMMs provided the highest recognition rates. For the HMMs, we tested both *fully connected* models and *leftright* models (which means only connections to the state itself and to the following state are allowed: $a_{ij} = 0$ for $i > j$ and $j > i + 1$ (see Sect. 2.1). Also different types for the covariance matrices (*full* and *diagonal*) were investigated (see Sect. 2.2). The classification results for the different settings are shown in Table 1.

To verify the test results, a 10-fold cross validation was utilized. In each step, one HMM for each of the three classes, with the settings as mentioned before, was trained using just data of one class. For each sequence O in the remaining test set, the probability $P(O \mid \lambda_i), i = 1..3$, that the HMM λ_i generated this sequence, was estimated with the *forward algorithm* (for further information see Sect. 2.1 or [10]). The maximum of the three achieved values $P(O \mid \lambda_i)$ leads to the most likely class.

Table 1. Classification results for the different model types of the HMMs and the types of the covariance matrix of the GMMs. The first part of the table refers to the results using the spectrum of the weighted adjacency matrix A, the second part to the extension A^+.

Underlying matrix	Model type	Covariance matrix	Classification rates		
			drink	move	scratch
	fully connected	full	1.00	0.94	1.00
A	fully connected	diagonal	1.00	0.93	1.00
	leftright	full	1.00	0.87	1.00
	leftright	diagonal	1.00	0.92	1.00
	fully connected	full	1.00	0.95	1.00
A^+	fully connected	diagonal	1.00	0.95	1.00
	leftright	full	1.00	0.91	1.00
	leftright	diagonal	1.00	0.90	1.00

4.2 Adequacy and Performance of the Features

To investigate, if the eigenvalues of the Euclidean distance matrix are an adequate feature for the classification in this case, first their temporal development was determined (see Figure 4). The eigenvalues change over time (but never their order), whereas the development of the eigenvalues of the class *drink* should be emphasized.

The classification results accentuate the good performance of the spectrum as feature. In the case with a fully connected model and a full covariance matrix, the class *drink* and the class *scratch* have a recognition rate of 1.00. The class *move* achieves 0.94 (three sequences classified as *drink*, three as *scratch*). The spectrum of the matrix A^+ used as feature, achieved 0.95 for the class *move*.

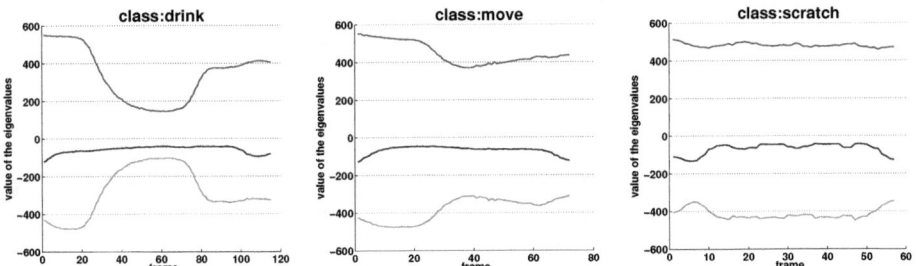

Fig. 4. Temporal development of the three eigenvalues of three example sequences

It is not necessary to use all eigenvalues: Figure 4 shows, that even only the highest eigenvalue has a different development in the three classes. By using just the first eigenvalue in the spectrum (the eigenvalues are sorted my size), or the two first eigenvalues, the following classification results can be achieved(see Table 2). For the results, fully connected models and full covariance matrices were utilized.

Table 2. Classification rates for the different number of eigenvalues used as features for the HMMs (results for the matrices A and A^+

Underlying matrix	No.ev	Classification rates		
		drink	move	scratch
A	1	1.00	0.77	0.84
	2	1.00	0.88	0.98
	3	1.00	0.94	1.00
A^+	1	1.00	0.82	0.85
	2	0.99	0.87	0.99
	3	1.00	0.95	1.00

Different number of nodes has no influence: To investigate the power of the spectrum as feature concerning insertion of nodes and different number of nodes in the graph, extra nodes were artificially inserted in the graph (see Fig. 5). On the one hand, a second cup was placed on the table by inserting a node in the graph with the corresponding coordinates of [300,400]. On the other hand, a second person was sat at the table in the same manner (coordinates are [350,150]). Thus, the following settings were generated:

Cup: One person sitting at the table with two cups → 4 nodes.
Person: Two persons sitting at the table with one cup → 4 nodes.
Cup/Person: Either a second cup or a second person is added → 4 nodes.
Cup+person: Two persons sitting at the table with two cups → 5 nodes.

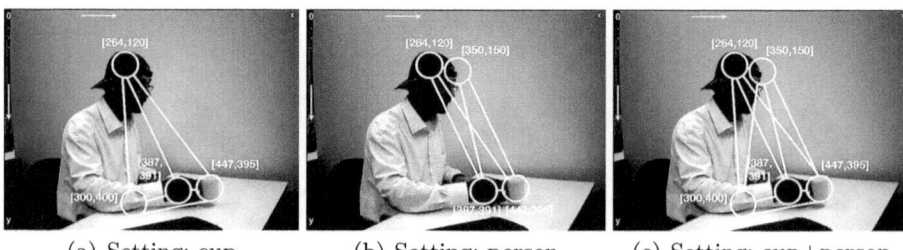

| (a) Setting: cup | (b) Setting: person | (c) Setting: cup+person |

Fig. 5. Freeze images with inserted artificial nodes. The coordinates for the second cup are [300,400], for the second person [350,150]. These coordinates don't change over time.

Table 3. Summary of the experiments and their classification rates. The experiments differ in the number of nodes (No.nodes) and their position (see setting). Also both distance matrices (A, A^+) were used for the calculation of the spectrum.

Underlying matrix	Setting	No.nodes	Classification rates		
			drink	move	scratch
	Cup	3-4	1.00	0.89	1.00
A	Person	3-4	1.00	0.96	0.99
	Cup/Person	3-4	0.99	0.94	0.99
	Cup+Person	3-5	0.99	0.91	0.98
	Cup	3-4	0.98	0.91	0.99
A^+	Person	3-4	1.00	0.98	0.99
	Cup/Person	3-4	0.98	0.91	0.98
	Cup+Person	3-5	1.00	0.91	0.98

In the case *Cup* and the case *Person*, the data set includes the normal data set with three nodes and the artificial new data set with four nodes. In the setting *Cup/Person*, the normal data set and the two data sets *Cup* and *Person* were utilized. In the last setting *Cup+Person*, the sequences of the setting *Cup/Person* were used plus sequences with two cups and two persons. Because all sequences contain at least three nodes, the first three eigenvalues are utilized for the classification even though the number of eigenvalues can be greater.

Table 3 shows the recognition rates of all experiments, which were mentioned before. The table is divided again in two parts: the first half refers to the weighted adjacency matrix A as feature, the second half to the matrix A^+. The second column names the setting of the experiment. The third column shows the number of nodes (No.nodes), which appear in the data set. The last three columns contain the classification rates for the three classes. The results show, that even though the settings are getting more complex, the classification still works satisfactorily (with an average classification rate for the setting *cup+person* of nearly 0.96).

The order of the nodes has no influence: One important advantage of the spectrum is, that the order of the nodes in the graph has no influence on the sorted spectrum itself, because by changing rows and the corresponding columns in the distance matrix, the eigenvalues remain the same.

5   Summary

In this study the applicability and potential of the spectrum of the distance matrix for sequential graph classification was investigated. Therefore, a simple but expandable data set, with video clips, was utilized (see Sect. 3.1). The videos show three different actions played by a male person sitting at a table. The first action is *drink*, whereas the person drinks out of a cup. During the second action, the cup is moved on the table (*move*) and during the third action the person is scratching his head (*scratch*).

After extracting the three objects *head*, *hand* and *cup* with the corresponding coordinates from the single images, one fully connected graph for each image was developed with three nodes referring to the three objects (see Sect. 3.2). The undirected edges were labeled with the Euclidean distance between the two connected edges and the distance matrix A was generated. We also investigated the power of an extension A^+ to the matrix A with the sum of the distances on the diagonal. In the next step, the eigenvalues of the distance matrices were calculated, which build the spectrum of the matrix. Hence, the features for the classification are sequences of spectra (one spectrum for each single image).

Because the dataset has a temporal development, HMMs in combination with GMMs were used to classify the sequences (for more specific information about HMMs and GMMs see Sect. 2).

The HMMs were tested as *fully connected* models and as *leftright* models (only connections to the state itself and to the following state are allowed) with 15 states, the GMMs with four distributions for the linear combination and both *full* and *diagonal* covariance matrices. To verify the rest results, a 10-fold cross validation was applied.

To investigate the power and flexibility of the spectrum, different experiments were accomplished. It was revealed, that it is not necessary to use all present eigenvalues, but just the first one leads to adequate results. Therefore, graphs with different number of nodes can be classified by using just the greatest possible number of eigenvalues. Another advantage of the spectrum is, that it is invariant with respect to the order of the nodes.

Future work within this ongoing research project concerns the investigation of the adequacy and classification performance of the eigenvalues as features with more complex scenes and user activities.

Acknowledgments. This paper is based on work done within the *Information Fusion* subproject of the Transregional Collaborative Research Center *SFB/ TRR 62 Companion-Technology for Cognitive Technical Systems*, funded by the German Research Foundation (DFG). The work of Miriam Schmidt is supported by a scholarship of the Graduate School *Mathematical Analysis of Evolution, Information and Complexity* of the University of Ulm.

The authors want to thank Lutz Bigalke for recording, preprocessing and providing the data.

References

1. Cvetković, D.M., Doob, M., Horst, S.: Spectra of Graphs. Theory and Applications. Vch Verlagsgesellschaft Mbh (1998)
2. Bunke, H., Roth, M., Schukat-Talamazzini, E.G.: Off-line Cursive Handwriting Recognition Using Hidden Markov Models. Pattern Recognition 28, 1399–1413 (1995)
3. Sanfeliu, A., Fu, K.: A Distance Measure between Attributed Relational Graphs for Pattern Recognition. IEEE Transactions on Systems, Man, and Cybernetics 13, 353–362 (1983)
4. Luo, B., Wilson, R.C., Hancock, E.R.: Spectral Embedding of Graphs. Pattern Recognition 36, 2213–2230 (2003)
5. Chung, F.R.K.: Spectral Graph Theory. CBMS Regional Conference Series in Mathematics, vol. 92. American Mathematical Society, Providence (1997)
6. Alfakih, A.Y.: On the Eigenvalues of Euclidean Distance Matrices. Computational & Applied Mathematics 27, 237 250 (2008)
7. Rabiner, L.R., Juang, B.H.: Fundamentals of Speech Recognition. Prentice Hall PTR, Englewood Cliffs (1993)
8. Schmidt, M., Schels, M., Schwenker, F.: A Hidden Markov Model Based Approach for Facial Expression Recognition in Image Sequences. In: Schwenker, F., El Gayar, N. (eds.) ANNPR 2010. LNCS, vol. 5998, pp. 149–160. Springer, Heidelberg (2010)
9. Durbin, R., Eddy, S., Krogh, A., Mitchison, G.: Biological Sequence Analysis: Probabilistic Models of Proteins and Nucleic Acids. Cambridge University Press, Cambridge (1998)
10. Rabiner, L.R.: A Tutorial on Hidden Markov Models and Selected Applications in Speech Recognition. Proceedings of the IEEE 77, 257–286 (1989)
11. Bishop, C.M.: Pattern Recognition and Machine Learning. Springer, New York (2007)
12. Murphy, K.: Hidden Markov Model Toolbox for Matlab (1998), http://www.cs.ubc.ca/murphyk/Software/HMM/hmm.html

Using Kernels on Hierarchical Graphs in Automatic Classification of Designs

Barbara Strug

Department of Physics, Astronomy and Applied Computer Science,
Jagiellonian University, Reymonta 4, Krakow, Poland
barbara.strug@uj.edu.pl

Abstract. In this paper the use of kernel methods in automatic classification of hierarchical graphs is presented. The classification is used as
a basis for evaluation of designs in a computer aided design system. A
kernel for hierarchical graphs based on a combination of tree and graph
kernels is proposed. Hierarchical graphs-based representation used in design problems is briefly presented. The proposed approach is tested in
experiment on a flat layout design task and preliminary result are also
presented.

1 Introduction

Designing is an important process. Whether it is computer aided or traditional,
it usually has an iterative approach consisting of a number of steps. Starting
from a preliminary or conceptual design, which is then analyzed or tested in
order to find out what has to be redesigned or refined, the process of evaluation
and optimization is repeated until an acceptable solution is found. The longer
the process the more expensive it is. Thus there is a need for a method able to
speed up this process, and to lower the costs.

In the computer aided design there is a need for an adequate design representation. Graphs have been shown to be a very useful way of representing complex
objects in different domains [29]. Their ability to represent the structure of an
object as well as the relations of different types between its components makes
them particularly useful in the domain of computer aided design. Graphs can
represent any artifact being designed and they can take into account the inter-
related structure of many design objects i.e. the fact that parts of an object
can be related to other parts in different ways. As designing new artifacts requires a method of generating representing them graphs many methods for graph
generation were researched.

Many of these methods were based on the approach using the theory of formal languages to the computer aided design [28,12,13,15], in particular the graph
based representation jointly with graph grammars [3,12,13,14,16,26], and grammar systems [5,8,9,30,24,21]. Other methods used to generate graphs representing
designs include evolutionary computations that were used in different domains of
design [3,16,26].

X. Jiang, M. Ferrer, and A. Torsello (Eds.): GbRPR 2011, LNCS 6658, pp. 335–344, 2011.

All these generation methods result in producing a number of graphs representing designs. Using graph grammars and evolutionary methods, either separately or combined together, we can build large database of graphs and thus a database of designs.

The main problem lies in the complexity and size (understood as the number of graphs) of such database. It is difficult to automatically evaluate and analyze the quality of these graphs, where the quality of graph is defined as the quality of the design a given graph represents in respect to a design problem being solved. Thus the process of evaluation usually requires the presence of a human designer who is responsible for choosing the best solution or give some numerical values to each design. Unfortunately a human "evaluator" needs all graphs to be rendered (visualized) to designs in order to evaluate them. Until the design problem is very simple the rendering of designs is usually a complex and costly process resulting in long execution times. This problem is especially important in evolutionary design systems where large numbers of graphs are generated in each population and a speed of fitness evaluation is a key factor in getting a usable design system. In such situations it would be useful to be able to evaluate graphs without the need of visualizing them.

One of the methods for approaching the problem is using a number of graphs representing designs for which a human "evaluator" has defined a quality value as a basis for evaluating other designs in the same design problem (a training set). As it can be noticed that the designs getting higher quality values usually have some common elements, finding frequent substructures in graphs is a useful approach. The results obtained with the use of the FFSM algorithm were presented in [31], and the results obtained with the use of the gSpan algorithms and a comparison of both results was presented in [32].

However, there is a problems that can be noticed in this method: the number of frequent subgraphs even for high support parameters is very large. Taking into account the fact that the evaluation of a graph representing a new design consists in checking how many of the frequent subgraphs are also subgraphs of the new graph the evaluation requires a huge number of subgraph isomorphism checking operations. Even taking into account the fact that the graphs used are labelled, what lowers considerably the computational cost of these operations, it still is a costly and time consuming process.

Yet in majority of the situations what is needed is not a numerical quality value, as the final decision would anyway be taken by a human, but a more general notion of classifying each graph into good, acceptable, poor or unacceptable group. Then the poor and unacceptable designs could be discarded and the good ones presented to the designer for final evaluation/decision. To be able to classify graphs a similarity measure is needed. One of the possibilities is the use of kernels. There has been a lot of research on different kernels for structured data, including tree and graph kernels [2,4,10,11,20,25,27]. Yet, to the author's knowledge, there has been no attempt to propose a kernel for hierarchical graphs.

This paper is organized in the following way:in Section 2 a hierarchical graphs used in design representation are presented, Section 3 contains a short review of existing graph kernels and then describes a proposed kernel for hierarchical graphs, some experiments and their results and in Section 4 some conclusions and possible future research is briefly described.

2   Hierarchical Graphs in Design Representation

There is a number of representations used in CAD problems like boundary representations, sweep-volume representation, surface representations or CSG (constructive solid geometry) [22,18,23] but they allow only for the expression of geometry of an object being designed and do not take into account the structure of design objects i.e. the fact that parts of an object can be related to other parts in some ways. In this paper a representations based on graphs is used.

Different types of graphs have been researched and used in this domain, in addition to simple graphs, also hierarchical graphs, hypergraphs and hierarchical hypergraphs were used [15,14,31,32].

In this paper an extension of graphs called hierarchical graphs is used. Such graphs can represent an artifact being designed at different levels of detail at different stages of the design process. In floor layout example used in this paper using hierarchical graphs as the representation model makes it parallel to a way a designer views a floor layout i.e. it starts by dividing the space into functional areas (sleeping. eating, living etc.) which in turn can be divided further. Such an approach makes it also possible to add more detailed descriptions of particular spaces and even include, at lower level of hierarchy, elements of interior design (like furniture or some appliances).

Hierarchical graphs (HGs) are an extension of traditional graphs. They consist of nodes and edges. What makes them different from standard graphs is that nodes in HGs can contain internal nodes. These nodes, called children, can in turn contain other internal nodes and can be connected to any other nodes with only exception being their ancestors.

A node in a hierarchical graph may represent a geometrical object or a groups of objects. It may also be used to hide certain details of a designed object that are not needed at a given stage of design or to group object having some common features (geometrical or functional). A node that represents a single object is called an object node. Nodes that do not represent actual geometric entities but are used to represent hierarchical structure or other relations are called group nodes. It is important to note that the edges between nodes are by no means limited to edges between descendants of the same node. In other words there may exist edges between nodes having different ancestors.

Nodes and edges in hierarchical graphs can be labelled and attributed. Labels are assigned to nodes and edges by means of node and edge labelling functions respectively, and attributes - by node and edge attributing functions. Attributes represent properties (for example size, position, colour, length, orientation or material) of a component represented by a given node.

As each node may either belong to the top level or be nested in other nodes a context based labelling is defined. Three types of labelling are proposed. A node may be referred to by its own label, by a *weakly-context label*, which contains a

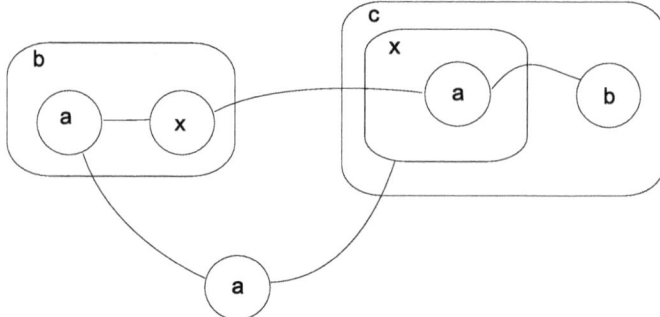

Fig. 1. A hierarchical graph

label of its direct ancestor and its own label separated by a dot, or by a so called *full label* which contains labels of its all ancestors in the hierarchical structures separated by a dot. For example a node labelled a in fig. 1 being a direct child of a node labelled x, and a descendant of a node labelled c has a simple label a, a weakly-context label $x.a$ and a full label $c.x.a$.

In computer aided design a labelled attributed hierarchical graph may represent a potentially infinite number of designs. A given hierarchical graph G can represent a, potentially, infinite subset of designs having the same structure. To represent an actual design we must define an instance of a graph. An instance of a hierarchical graph is a hierarchical labelled attributed graph in which to each attribute a value has been assigned from the set of possible values of a given attribute.

As such a hierarchical graph defines only a structure of a design to create a visualisation of an object an interpretation is necessary. Such interpretation determines the assignments of geometrical objects to nodes and correspondence between edges and sets of relations between objects (components of a design). The objects assigned to nodes are usually called primitives. So the process of visualisation requires a lot of computation and any method that would limit the number of graphs to be visualised would also significantly lower cost and time of developing new designs.

3   A Kernel on Hierarchical Graphs

A hierarchical graph contains more information then a simple graph because it adds to it information about ancestor/child relation. But such a graph can be viewed as a combination of a tree and an underlying graph. Taking into account that in design applications only the lowest level nodes represent actual geometric objects and the remaining nodes serve as grouping containers the underlying graph would include only the lowest level nodes. As the hierarchical graph does not have to include a "supernode" containing all nodes a special node is added to the tree and labelled *root*. A process of separating the hierarchy and underlaying structure results in a flattening of a hierarchical graph. An example of flattening of a graph from Fig. 1 is depicted in Fig. 2.

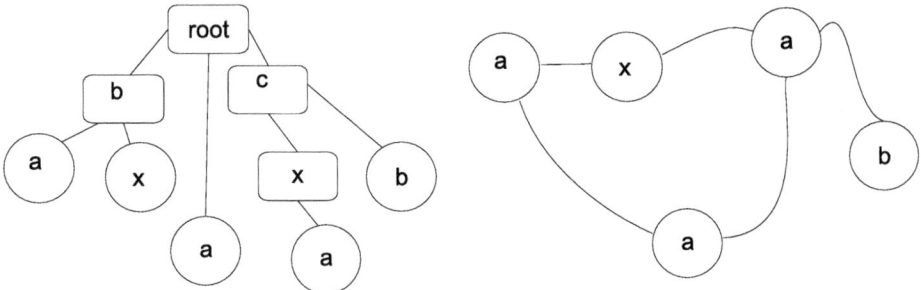

Fig. 2. A flattening of a graph from fig.1, consisting of a tree and an underlying graph

Let H_i be a hierarchical graph, T_i - a tree representing a hierarchical structure of this graph and G_i - an underlying graph. Thus a combination of a tree and a graph kernel can be used as a kernel for hierarchical graphs. Let $K_T(T_1, T_2)$ be a tree kernel and $K_G(G_1, G_2)$ be a graph kernel. Thus both $K(H_1, H_2) = K_T(T_1, T_2)K_G(G_1, G_2)$ and $K(H_1, H_2) = K_T(T_1, T_2) + K_G(G_1, G_2)$ are also kernels [11].

There has been a lot of research on kernels for structured data like trees or graphs. Tree kernels were proposed by Collins and Duffy [4] and applied to natural language processing. The basic idea is to consider all subtrees of the tree, where a subtree is defined as a connected subgraph of a tree containing either all children of a vertex or none. This kernel is computable in $O(|V_1||V_2|)$, where $|V_i|$ is the number of nodes in the $i-th$ tree [4,11].

In case of graph kernels there is a choice of several different ones proposed so far. One of them is based on enumerating all subgraphs of graphs G_i and calculating the number of isomorphic ones. An all subgraph kernel was shown to be NP-hard by Gartner et al [10]. Although taking into account that in case of labelled graphs the computational time is significantly lower such a kernel is feasible in design applications.

Another interesting group of graph kernels is based on computing random walks on both graphs. It includes the product graph kernel [10] and the marginalized kernels [20]. In product graph kernel a number of common walks in two graphs is counted. The marginalized kernel on the other hand is defined as the expectation of a kernel over all pairs of label sequences from two graphs. These kernels are computable in polynomial time, $(O(n^6)$ [11]), although for small graphs it may be worse then 2^n, when the neglected constant factors contribute stronger.

Taking into account the computational cost of different graph and tree kernels a kernel for hierarchical graph would be computable in the same time as the selected graph kernel, as the cost for tree kernel is smaller.

In this paper two kernels for hierarchical graphs are tested:

$$K_{HG}^i(H_1, H_2) = K_T(T_1, T_2) + K_G^i(G_1, G_2) \tag{1}$$

where i denotes the type of graph kernel used (i =1, for a random walk kernel and i=2 for an all subgraph kernel).

3.1   Experiments and Results

Two kernels K_{HG} were tested on a design task from the domain of flat layout design. A training set consists of 10 hierarchical graphs for which a classification was defined by a human designer. These set contains five graphs representing good designs, three - not so good but acceptable and two graphs representing bad designs. Graphs representing good and acceptable designs were generated from actual floor layouts. Graphs representing bad designs had to be generated artificially either by "spoiling" good designs or from scratch as it is difficult to come across really bad designs.

In Fig. 3 four hierarchical graphs from the training set are presented. Graphs depicted in Figs. 3a,b and c represent good designs. Graph depicted in Fig. 3d represents a bad designs; it can be noticed that there is a number of problems with this design. Firstly one of the bedrooms has no doors (there is no edge labelled *acc* linked to the top left node labelled B), nodes labelled H and ER, representing a hall and an eating room, are linked with an edge labelled *jnt* what represent an open connection - while having no doors from hall to eating room is not a desired property. Moreover, there is an accessibility relation between a

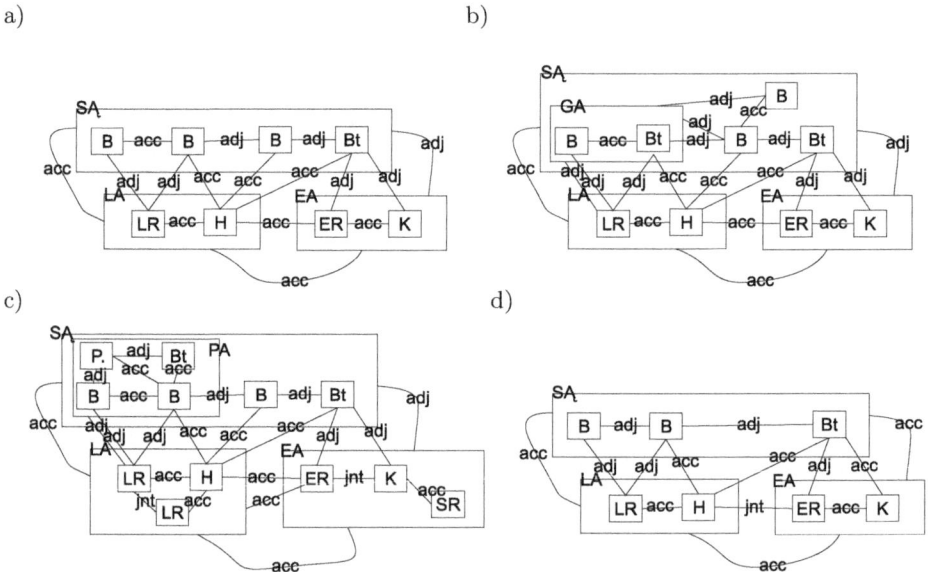

Fig. 3. Examples of hierarchical graphs representing floor layouts (nodes labelled B represent bedrooms, Bt - bathrooms, K - kitchens, ER - eating room, LR - living room, H- hall, St - storage, P -play room, SA - sleeping area, GA -guest area, EA -eating area and LA - living area, edges labelled by *adj* represent adjacency between spaces, *acc* - accessibility (wall with doors) and *jnt* - open access (no walls))

kitchen and a bathroom, while normally there should not exist doors between spaces where eating is prepared and a bathroom.

Then a set of 50 hierarchical graphs generated by a hierarchical graph grammar, being a part of a larger design application [14], was analyzed.

The results are presented in Tables 1, 2 and 3, for three different node kernels: a simple node label, a weakly-context label and a full label kernel, respectively. The percentage values are for the kernel using random-walk kernel and all-subgraph kernel as a graph kernel part of hierarchical graph kernel.

Table 1. The classification of hierarchical graphs with simple node kernel

algorithm	Random walk	All subgraph
correctly classified as good	89%	91%
wrongly classified as poor	7%	3%

Table 2. The classification of hierarchical graphs with weakly-context node kernel

algorithm	Random walk	All subgraph
correctly classified as good	91%	93%
wrongly classified as poor	5%	3%

Table 3. The classification of hierarchical graphs with full label node kernel

algorithm	Random walk	All subgraph
correctly classified as good	92%	95%
wrongly classified as poor	4%	2%

The most interesting from the application point of view are two values and only they are presented in the tables. The first is how many of the graphs classified as good are actually good. It is important because these graphs are visualized and presented to the human designer for final decision. Thus large proportion of poor graphs classified as good results in large cost of unnecessary visualization and, moreover, a designer is forced to analyze poor designs. So the higher the percentage of correct classification the less computational time is wasted on rendering and, equally importantly, less human time is wasted. Second important value is the number of graphs wrongly classified as poor, as these graphs are not rendered and hence the possibly good, and maybe also interesting, design is lost.

It can be noticed that the use of all-subgraphs graph kernel gives consistently slightly better results. The all subgraph kernel was also slightly faster then a random walk kernel, what is most likely the result of the fact that graphs used in the experiment were smaller then 40 nodes. Thus the neglected constants in $O(n^6)$ were actually contributing strongly to the computational time. Moreover the labelling of nodes contributing to the time of all subgraphs kernel being significantly smaller then the worst possible 2^n.

It can also be observed that the use of node kernel based on full labels results in consistently better classifications. It results from the fact that nodes having the same full label are more similar then those having only simple labels identical. In context of design tasks it can be explained by the fact that two elements are more similar if they are not only geometrically similar but also are placed in similar part of a design. For example a bedroom, labelled by B, is on the level of simple label identical with any other bedroom, but using a weakly-context label allows for differentiating between just a bedroom and a bedroom within the guest area (and a flat with guest area may be evaluated higher).

It had to be noted here that attributes of nodes were not taken into account for the kernel computation. So only structural similarity of graphs is actually analysed. Such an approach may not eliminated problems with size of shape of spaces but it ensures the elimination of graphs representing structurally invalid layouts (like rooms with no doors for example). Moreover, some aspects of geometrical properties turned out to be implicitly taken into account by these kernels. One of the training graphs had a very large room which, as a result, was adjacent to nearly all other rooms within a flat. As this graph was evaluated as a bad one, existence of node with a large number of edges within a test graph also lowered its classification.

4   Conclusions and Future Research

In this paper a use of a kernel for hierarchical graphs was proposed. It uses well known graph and tree kernels combined together with three different node kernels. Two of the node kernels also take into account the hierarchical structure of the representation. The proposed kernel was tested on several examples from the domain of computer aided design in which a hierarchical graphs are a very useful representation. The results presented in this paper seem promising and the application of kernel based classification methods to design patterns analysis seems encouraging.

The kernel proposed in this paper uses two kernels, one for a tree and one for a graph and takes a value being a sum of the two kernels. So the influence of each parts of hierarchical graph is equal. It seems interesting to replace a simple sum by a weighted one and experiment with different values of the weights. It would be possible to make either the hierarchy or the underlying structure more important.

The future plans include also experimenting with other graph kernels, especially with the shortest paths based graph kernel [2] as it has good computational properties, and with the frequent pattern based kernel. Another research direction currently being investigated is based on using different methods of flattening of a hierarchical graph which could result in a smaller number of nodes.

As in some design tasks hypergraphs are better suited as a representation model a kernel based approach to hypergraphs classification is also currently being researched. An combination of such a kernel with kernels presented in this papers is then planed to be applied to the representation using hierarchical hypergraphs.

References

1. Agrawal, R., Imielinski, T., Swami, A.: Mining association rules between sets of items in large databases. In: Proc. 1993 ACM-SIGMOD Int. Conf. Management of Data (SIGMOD 1993), Washington, DC, pp. 207–216 (1993)
2. Borgwardt, K.M., Kriegel, H.P.: Shortest-path kernels on graphs. In: ICDM 2005, pp. 74–81 (2005)
3. Borkowski, A., Grabska, E., Nikodem, P., Strug, B.: Searching for Innovative Structural Layouts by Means of Graph Grammars and Esvolutionary Optimization. In: Proc. 2nd Int. Structural Eng. And Constr. Conf., Rome (2003)
4. Collins, M., Duffy, N.: New Ranking Algorithms for Parsing and Tagging: Kernels over Discrete Structures, and the Voted Perceptron. In: Proceedings of ACL 2002 (2002)
5. Csuhaj-Varj, E., Dassow, J., Kelemen, J., Paun, G.: Grammar systems. A grammatical approach to distribution and cooperation. In: Topics in Computer Mathematics, vol. 8. Gordon and Breach Science Publishers, Yverdon (1994)
6. Csuhaj-Varj, E., Dassow, J., Paun, G.: Dynamically controlled cooperating/distributed grammar systems. Information Sciences 69(1-2), 1–25 (1993)
7. Csuhaj-Varj, E., Vaszil, G.: On context-free parallel communicating grammar systems: Synchronization, communication, and normal forms. Theoretical Computer Science 255(1-2), 511–538 (2001)
8. Csuhaj-Varj, E.: Grammar systems: A short survey. In: Proceedings of Grammar Systems Week 2004, Budapest, Hungary, July 5-9, pp. 141–157 (2004)
9. Dassow, J., Paun, G., Rozenberg, G.: Grammar systems. In: Salomaa, A., Rozenberg, G. (eds.) Handbook of Formal Languages, vol. 2, ch. 4, pp. 155–213. Springer, Heidelberg (1997)
10. Gartner, T.: A survey of kernels for structured data. SIGKDD Explorations 5(1), 49–58 (2003)
11. Gartner, T.: Kernels for structured data. Series in Machine Perception and Artificial Intelligence. World Scientific, Singapore (2009)
12. Grabska, E.: Theoretical Concepts of Graphical Modelling. Part one: Realization of CP-graphs. Machine GRAPHICS and VISION 2(1), 3–38 (1993)
13. Grabska, E.: Theoretical Concepts of Graphical Modelling. Part two: CP-graph Grammars and Languages. Machine GRAPHICS and VISION 2(2), 149–178 (1993)
14. Grabska, E., Palacz, W.: Hierarchical graphs in creative design. MG&V 9(1/2), 115–123 (2000)
15. Grabska, E.: Graphs and designing. In: Ehrig, H., Schneider, H.-J. (eds.) Dagstuhl Seminar 1993. LNCS, vol. 776, Springer, Heidelberg (1994)
16. Grabska, E., Nikodem, P., Strug, B.: Evolutionary Methods and Graph Grammars in Design and Optimization of Skeletal Structures Weimar. In: 11th International Workshop on Intelligent Computing in Engineering, Weimar (2004)
17. Han, J., Pei, J., Yin, Y., Mao, R.: Mining Frequent Patterns without Candidate Generation: A Frequent-pattern Tree Approach. Data Mining and Knowledge Discovery: An International Journal 8(1), 53–87 (2004)
18. Hoffman, C.M.: Geometric and Solid Modeling: An Introduction. Morgan Kaufmann, San Francisco (1989)
19. Inokuchi, A., Washio, T., Motoda, H.: An Apriori-Based Algorithm for Mining Frequent Substructures from Graph Data. In: Zighed, D.A., Komorowski, J., Żytkow, J.M. (eds.) PKDD 2000. LNCS (LNAI), vol. 1910, pp. 13–23. Springer, Heidelberg (2000)

20. Kashima, H., Tsuda, K., Inokuchi, A.: Marginalized Kernels Between Labeled Graphs. In: ICML 2003, pp. 321–328 (2003)
21. Kelemen, J.: Syntactical models of cooperating/distributed problem solving. Journal of Experimental and Theoretical AI 3(1), 1–10 (1991)
22. Mantyla, M.: An Introduction To Solid Modeling, vol. 87. Computer Science Press, Rockville (1988)
23. Martin, R.R., Stephenson, P.C.: Sweeping of Three-dimensional Objects. Computer Aided Design 22(4), 223–234 (1990)
24. Martn-Vide, C., Mitrana, V.: Cooperation in contextual grammars. In: Kelemenov, A. (ed.) Proceedings of the MFCS 1998 Satellite Workshop on Grammar Systems, pp. 289–302. Silesian University, Opava (1998)
25. Moschitti, A., Pighin, D., Basili, R.: Tree kernels for Semantic Role Labeling, Special Issue on Semantic Role Labeling, Computational Linguistics Journal (2008)
26. Nikodem, P., Strug, B.: Graph Transformations in Evolutionary Design, Lecture Notes in Computer Science,vol 3070. In: Rutkowski, L., Siekmann, J.H., Tadeusiewicz, R., Zadeh, L.A. (eds.) ICAISC 2004. LNCS (LNAI), vol. 3070, pp. 456–461. Springer, Heidelberg (2004)
27. Pighin, D., Moschitti, A.: On Reverse Feature Engineering of Syntactic Tree Kernels. In: Proceedings of the 2010 Conference on Natural Language Learning, Upsala, Sweden, July 2010. Association for Computational Linguistics (2010)
28. Rozenberg, G.: Handbook of Graph Grammars and Computing by Graph. In: Transformations. Fundations, vol. 1. World Scientific, London (1997)
29. Rozenberg, G.: Handbook of Graph Grammars and Computing by Graph. In: Transformations. Applications, Languages and Tools, vol. 2. World Scientific, London (1999)
30. Simeoni, M., Staniszkis, M.: Cooperating graph grammar systems. In: Paun, G., Salomaa, A. (eds.) Grammatical models of multi-agent systems, pp. 193–217. Gordon and Breach, Amsterdam (1999)
31. Strug, B., Slusarczyk, G.: Reasoning about designs through frequent patterns mining. Advanced Engineering Informatics 23, 361–369 (2009)
32. Strug, B., Slusarczyk, G.: Frequent Pattern Mining in a Design Supporting System. Key Engineering Materials 450, 1–4 (2011)
33. Vishwanathan, S.V.N., Borgwardt, K.M., Schraudolph, N.N.: Fast Computation of Graph Kernels. In: NIPS 2006, pp. 1449–1456 (2006)
34. Yan, X., Yu, P.S., Han, J.: Substructure Similarity Search in Graph Databases. In: SIGMOD 2005 (Proc. of 2005 Int. Conf. on Management of Data) (2005)
35. Yan, X., Yu, P.S., Han, J.: Graph Indexing: A Frequent Structure-based Approach. In: SIGMOD 2004 (Proc. of 2004 Int. Conf. on Management of Data) (2004)

Author Index